A CULTURAL HISTORY OF PLANTS

VOLUME 1

A Cultural History of Plants
General Editors: Annette Giesecke and David J. Mabberley

Volume 1
A Cultural History of Plants in Antiquity
Edited by Annette Giesecke

Volume 2
A Cultural History of Plants in the Post-Classical Era
Edited by Alain Touwaide

Volume 3
A Cultural History of Plants in the Early Modern Era
Edited by Andrew Dalby and Annette Giesecke

Volume 4
A Cultural History of Plants in the Seventeenth and Eighteenth Centuries
Edited by Jennifer Milam

Volume 5
A Cultural History of Plants in the Nineteenth Century
Edited by David J. Mabberley

Volume 6
A Cultural History of Plants in the Modern Era
Edited by Stephen Forbes

A CULTURAL HISTORY OF PLANTS

IN ANTIQUITY

VOLUME 1

Edited by
Annette Giesecke

BLOOMSBURY ACADEMIC
LONDON • NEW YORK • OXFORD • NEW DELHI • SYDNEY

BLOOMSBURY ACADEMIC
Bloomsbury Publishing Plc
50 Bedford Square, London, WC1B 3DP, UK
1385 Broadway, New York, NY 10018, USA
29 Earlsfort Terrace, Dublin 2, Ireland

BLOOMSBURY, BLOOMSBURY ACADEMIC and the Diana logo
are trademarks of Bloomsbury Publishing Plc

First published in Great Britain 2022
Paperback edition published in 2025

Series design: Raven Design
Cover image © DEA/A. DAGLI ORTI/De Agostini/Getty Images

A catalogue record for this book is available from the British Library.

Library of Congress Control Number: 2021932844

ISBN: PB: 978-1-3505-5053-7
Set: 978-1-3505-5064-3
ePUB: 978-1-3502-5927-0
ePDF: 978-1-3502-5926-3

Series: The Cultural History Series

Typeset by Integra Software Services Pvt. Ltd.
Printed and bound in Great Britain

CONTENTS

ILLUSTRATIONS

SERIES PREFACE

The connectedness of humans to plants is the most fundamental of human relationships. Plants are, and historically have been, sources of food, shelter, bedding, tools, medicine, and, most importantly, the very air we breathe. Plants have inspired awe, a sense of well-being, religious fervor, and acquisitiveness alike. They have been collected, propagated, and mutated, as well as endangered or driven into extinction by human impacts such as global warming, deforestation, fire suppression, and over-grazing. *A Cultural History of Plants* traces the global dependence of human life and civilization on plants from antiquity to the twenty-first century and comprises contributions by experts and scholars in a wide range of fields, including anthropology, archaeology, art history, botany, classics, garden history, history, literature, and environmental studies more broadly. The series consists of six illustrated volumes, each devoted to an examination of plants as grounded in, and shaping, the cultural experiences of a particular historical period. Each of the six volumes, in turn, is structured in the same way, beginning with an introductory chapter that offers a sweeping view of the cultural history of plants in the period in question, followed by chapters on plants as staple foods, plants as luxury foods, trade and exploration, plant technology and science, plants and medicine, plants in (popular) culture, plants as natural ornaments, and the representation of plants. This cohesive structure offers readers the opportunity both to explore a meaningful cross-section of humans' uses of plants in a given period and to trace a particular use—as in medicine, for example—through time from volume to volume. The six volumes comprising *A Cultural History of Plants* are as follows:

Volume 1: *A Cultural History of Plants in Antiquity* (*c.* 10,000 BCE–500 CE)
Volume 2: *A Cultural History of Plants in the Post-Classical Era* (500–1400)
Volume 3: *A Cultural History of Plants in the Early Modern Era* (1400–1650)
Volume 4: *A Cultural History of Plants in the Seventeenth and Eighteenth Centuries* (1650–1800)
Volume 5: *A Cultural History of Plants in the Nineteenth Century* (1800–1920)
Volume 6: *A Cultural History of Plants in the Modern Era* (1920–present).

By way of guidance to our readers, it should be noted that the plant names used in these volumes accord with those in the fourth edition of *Mabberley's Plant-book* (Cambridge University Press, 2017). When they are discussed, individual plants are identified using their common names and, at their first mention in each chapter, with their scientific names: e.g. bay laurel (*Laurus nobilis*). As is recommended for general works such as this, the authorities to whom the scientific names are attributed (e.g. *Laurus nobilis* L., where L. identifies Linnaeus as the identifying authority) have been omitted.

Annette Giesecke and David J. Mabberley,
General Editors

ABBREVIATIONS

CLASSICAL AUTHORS AND THEIR WORKS

Ael.	Aelianus
Ep.	*Epistulae*
NA	*De natura animalium*
VH	*Varia historia*
Aen.	*Aeneid*
Aesch.	Aeschylus
Ambros.	Ambrosius
Lc.	*Expositio in Lucam*
Anacr.	Anacreon
Anth. Pal.	*Anthologia Palatina*
Apollod.	Apollodorus mythographus
Bibl.	*Bibliotheca*
Epit.	*Epitome*
App.	Appian
Apul.	Apuleius
Met.	*Metamorphoses*
Ap. Rhod.	Apollonius Rhodius
Argon.	*Argonautica*
Ar.	Aristophanes
Lys.	*Lysistrata*
Nub.	*Nubes*
Pl.	*Plutus*
Ran.	*Ranae*
Thesm.	*Thesmophoriazusae*
Arist.	Aristotle
De an.	*De anima*
De motu an.	*De motu animalium*
Div. somn.	*De divinatione per somnia*
Eth. Eud.	*Ethica Eudemia*
Eth. Nic.	*Ethica Nicomachea*
Gen. an.	*De generatione animalium*
Gen. corr.	*De generatione et corruptione*
Hist. an.	*Historia animalium*
IA	*De incessu animalium*
Int.	*De interpretatione*
Metaph.	*Metaphysica*
Mete.	*Meteorologica*
Mir. ausc.	*De mirabilibus auscultationibus*

Part. an.	*De partibus animalium*
Ph.	*Physica*
Pol.	*Politica*
[*Pr.*]	*Problemata*
Rh.	*Rhetorica*
Sens.	*De sensu*
Arr.	Arrian
Anab.	*Anabasis*
Cyn.	*Cynegeticus*
Peripl. M. Eux.	*Periplus Maris Euxini*
Artem.	Artemidorus Daldianus
Ath.	Athenaeus
Deip.	*Deipnosophistae*
Aug.	Augustus
RG	*Res gestae*
Bion	
Id.	*Idylls*
Caes. Arelat.	Caesarius Arelatensis
Serm.	*Sermones*
Callim.	Callimachus
Cant.	*Cantus* (Canticle, *Song of Solomon*)
Cass.	Dio Cassius Dio
Cassiod.	Cassiodorus
in Ps.	*Expositio psalmorum*
Inst.	*Institutiones*
Var.	*Variae*
Cato	
Agr.	*De agricultura* or *De re rustica*
Orig.	*Origines*
Catull.	Catullus
CCG	Corpus christianorum, series graeca
CCL	Corpus christianorum, series latina
Chromat. Aquil.	Chromatius Aquileiensis
Cic.	Cicero (Marcus Tullius)
Ad Brut.	*Epistulae ad Brutum*
Amic.	*De amicitia*
Att.	*Epistulae ad Atticum*
Div.	*De divinatione*
Fam.	*Epistulae ad familiares*
Fat.	*De fato*
Fin.	*De finibus*
Leg.	*De legibus*
Nat. D.	*De natura deorum*
Off.	*De officiis*
Orat.	*Orator ad M. Brutum*
QFr.	*Epistulae ad Quintum fratrem*
Tusc.	*Tusculanae disputationes*
Verr.	*In Verrem*

Clem. Al.	Clemens Alexandrinus
Protr.	*Protrepticus*
Columella	
Rust.	*De re rustica*
Cornut.	Cornutus
Nat. deor.	*Natura deorum*
CSEL	Corpus scriptorum ecclesiasticorum latinorum
Cyr.	Cyrillus Hierosolymitanus
Catech. ad illuminandos	*Catecheses ad illuminandos*
Dio Cass.	Dio Cassius
Diod. Sic.	Diodorus Siculus
Diog. Laert.	Diogenes Laertius
Dion.	Hal. Dionysius Halicarnassensis
Ant. Rom.	*Antiquitates Romanae*
Dion.	Thrax Dionysius Thrax
Dsc.	Dioscorides
De mat. med.	*De materia medica*
Donat.	Aelius Donatus
Vit. Verg.	*Vita Vergilii*
Dtn.	Deuteronomy
Ep.	*Epistulae*
Eur.	Euripides
Bacch.	*Bacchae*
IA	*Iphigeneia Aulidensis*
Med.	*Medea*
Euseb.	Eusebius Caesariensis
in Cant.	*In Canticum Canticorum expositiones*
Chron.	*Chronica*
Hist. eccl.	*Historia ecclesiastica*
FGrH	*Die Fragmente der griechischen Historiker*
Fest.	Festus
Fortunat. Aquil.	Fortunatinus Aquileiensis
Comm. in Ev.	*Commentaria in Evangelia*
Gal.	Galen
De antid.	*De antidoto*
De indol.	*De indolentia*
Gell.	Aulus Gellius
NA	*Noctes Atticae*
Gen.	Genesis
Geopon.	*Geoponica*
Greg. M.	Gregorius Magnus
in Hes. Hom.	*Homiliae in Hiezchielem prophetam*
Greg. Nyss.	Gregorius Nyssenus
Comm. in Cant.	*Commentarius in Canticum Canticorum*
Ep.	*Epistulae*
Or.	*Orationes*
Hdt.	Herodotus
Hist.	*Historiae*

Hes.	Hesiod
Op.	*Opera et Dies*
Theog.	*Theogonia*
Hieron.	Hieronymus, see Jerome
Hippoc.	Hippocrates
Acut.	*De diaeta in morbis acutis*
Aer.	*De aera, aquis, locis*
Art.	*De articulis*
[*Ep.*]	*Epistulae*
Epid.	*Epidemiae*
Genit.	*De genitura*
Morb.	*De morbis*
Morb. sacr.	*De morbo sacro*
Mul.	*De mulierum affectibus*
Nat. mul.	*De natura muliebri*
Nat. puer.	*De natura pueri*
Off.	*De officina medici*
Virg.	*De virginibus morbis*
VM	*De vetere medicina*
Hippol.	Hippolytus
frg. arm. in Cant.	*Fragmentum Armenium in Canticum Canticorum*
Hist. Aug.	*Historia Augusta* (see SHA)
Hom.	Homer
Il.	*Iliad*
Od.	*Odyssey*
Hom. Hymn Dem.	*Homeric Hymn to Demeter*
Hom. Hymn Apoll.	*Homeric Hymn to Apollo*
Hor.	Horace
Carm.	*Carmina* or *Odes*
Epist.	*Epistulae*
Epod.	*Epodi*
Sat.	*Satirae* or *Sermones*
Hyg.	Hyginus
Fab.	*Fabulae*
Poet. astr.	*Poetica astronomica*
IG	*Inscriptiones Graecae*
Isid.	Isidorus
De vir. ill.	*De viris illustribus*
Etym.	*Etymologiae*
Jer.	Jerome
in Eccl.	*Commentarius in Ecclesiasten*
Comm. in Ioel.	*Commentarius in Ioel*
Comm. in Zech.	*Commentarius in Zechariah*
Jdc.	Liber Iudicum (Judges)
Joh. Lyd.	Ioannes Laurentius Lydus
Mens.	*De mensibus*
Joseph.	Josephus
Ant.	*Antiquitates Judaicae*

Ap.	*Contra Apionem*
Bell. Jud.	*Bellum Judaicum*
Juv.	Juvenal
LCL	Loeb Classical Library
Liv.	Livy
Lk.	Luke
Luc.	Lucan
Lucian	
De luct.	*De luctu*
Dial. meret.	*Dialogi meretricii*
Lucr.	Lucretius
DRN	*De rerum natura*
Lycoph.	Lycophron
Alex.	*Alexandra*
Macrob.	Macrobius
Sat.	*Saturnalia*
Mart.	Martial
Epigr.	*Epigrams*
Men.	Menander
Pk.	*Perikeiromenē*
Mosch.	Moschus
Eur.	*Europa*
Mt.	Matthew
Nic.	Nicander
Alex.	*Alexipharmaca*
Nonnus	
Dion.	*Dionysiaca*
Num.	Numbers
Olympiod.	Olympiodorus
Comm. in eccl.	*Commentarius in ecclesiasten*
Orig.	Origenes
in Cant. Hom.	*In Canticum Canticorum homiliae*
Schol. in Cant.	*Scholia in Canticum Canticorum.*
Ov.	Ovid
Am.	*Amores*
Ars am.	*Ars amatoria*
Fast.	*Fasti*
Her.	*Heroides*
Medic.	*Medicamina faciei*
Met.	*Metamorphoses*
Rem. am.	*Remedia amoris*
Tr.	*Tristia*
Paus.	Pausanias
Petron.	Petronius
Sat.	*Satyrica*
PG	Patrologia Graecia
PGM K.	Preisendanz and others (eds.), *Papyri Graecae Magicae: Die griechischen Zauberpapyri*

Philostr.	Philostratus
Imag.	*Imagines*
VA	*Vita Apollonii*
Philox.	Philoxenus
Pind.	Pindar
Isthm.	*Isthmian Odes*
Nem.	*Nemean*
Ol.	*Olympian*
Pyth.	*Pythian*
Pl.	Plato
Ap.	*Apologia*
Leg.	*Leges*
Phdr.	*Phaedrus*
Tim.	*Timaeus*
Plin.	Pliny (the Younger)
Ep.	*Epistulae*
Plin.	Pliny (the Elder)
HN	*Historia naturalis*
Plut.	Plutarch
Aem.	*Aemilius Paulus*
Alc.	*Alcibiades*
Alex.	*Alexander*
Amat.	*Amatorius*
Ant.	*Antonius*
Caes.	*Caesar*
Cat. Mai.	*Cato Maior*
Cim.	*Cimon*
Conv. sept. sap.	*Convivium septem sapientium*
De Alex. fort.	*De fortuna Alexandri*
De Is. et Os.	*De Iside et Osiride*
De soll. an.	*De sollertia animalium*
Marc.	*Marcellus*
Mor.	*Moralia*
Ner.	*Nero*
Num.	*Numa*
Per.	*Pericles*
Pomp.	*Pompeius*
Quaest. Rom.	*Quaestiones Romanae*
Sol.	*Solon*
Sull.	*Sulla*
Tim.	*Timoleon*
Vit.	*Vitae parallelae*
Poll.	Pollux
Onom.	*Onomasticon*
Pompon.	Pomponius
POxy.	*Oxyrhynchus Papyri*
PSI	*Papiri della Società Italiana*
PsJoh. Chrys.	Pseudo-Johannes Chrysostomos

in Eccl.	*Commentarius in Ecclesiasten*
PsPaulin. Nol.	Pseudo-Paulinus Nolae
Carm. in App.	*Carmina 4 Appendicis*
Reg.	Liber regum (Kings)
Rev.	Revelation
SC	Sources Chrétiennes
Sen.	Seneca
Clem.	*de Clementia*
Ep.	*Epistulae*
Serv.	Servius
SHA	Scriptores Historiae Augustae
Ael.	*Aelius*
Alex. Sev.	*Alexander Severus*
Ant.	*Pius Antoninus Pius*
Aurel.	*Aurelian*
Hadr.	*Hadrian*
Heliogab.	*Heliogabalus*
M. Ant.	*Marcus Aurelius Antoninus* (Caracalla)
Soph.	Sophocles
OC	*Oedipus Coloneus*
OT	*Oedipus Tyrannus*
Stat.	Statius
Silv.	*Silvae*
Strab.	Strabo
Geograph.	*Geographica*
Suda	Greek Lexicon formerly known as *Suidas*
Suet.	Suetonius
Aug.	*Divus Augustus*
Tac.	Tacitus
Agr.	*Agricola*
Ann.	*Annales*
Germ.	*Germania*
Hist.	*Historiae*
Tert.	Tertullian
Ad nat.	*Ad nationes*
Them. *Or.*	Themistius, *Orationes*
Theoc.	Theocritus
Id.	*Idylls*
Theophr.	*Theophrastus*
Caus. pl.	*De causis plantarum*
De odor.	*De odoribus*
Hist. pl.	*Historia plantarum*
Val. Flac.	Valerius Flaccus
Varro	
Ling.	*De lingua Latina*
Rust.	*De re rustica*
Verg.	Virgil
Aen.	*Aeneid*

Catal.	*Catalepton*
Ecl.	*Eclogues*
G.	*Georgics*
Vitr.	Vitruvius
De arch.	*De architectura*
Xen.	Xenophon
An.	*Anabasis*
Cyr.	*Cyropaedia*
Oec.	*Oeconomicus*

Introduction

Plants and Culture in Antiquity

ANNETTE GIESECKE

This is the first in a series of six volumes dedicated to the dependence of human life and civilization on plants, from antiquity to the twenty-first century. Each volume examines plants as grounded in, and shaping, the cultural experiences of one of six historical periods (Antiquity, the Post-Classical Era, the Early Modern Era, the Seventeenth and Eighteenth Centuries, the Nineteenth Century, and the Modern Era). In terms of temporal coverage, this volume is the most extensive, ranging from 10,000 BCE to 500 CE—broadly speaking, the period that witnessed transitions from hunter-gatherer subsistence to the practice of agriculture in the Fertile Crescent and elsewhere, and culminated in the fall of the Roman Empire, the end of the Han Dynasty in China, the rise of Byzantium, the first flowering of Mayan civilization, and, not long after, the birth of Islam. Geographically, this volume aims to span the globe, but most chapters inevitably focus on the civilizations of the ancient Near East, Egypt, Greece, and Rome (together with their empires), as these have yielded the greatest number and diversity of cultural artifacts—written documents, paintings, sculpture, architecture, pottery, jewelry, textiles, remains of planted gardens, and so forth—that inform our understanding of the human–plant relationship in antiquity. This density of artifacts was a reflex of the density of advanced civilizations that flourished in the lands surrounding the Mediterranean in this period.

Importantly, it was the human relationship with plants that drove cultural evolution. In this context, it is worth reflecting on the derivation and meaning of the word "culture," which we now understand to signify what distinguishes one social group from another: particular customs, arts, beliefs, social institutions, and ways of life more generally. "Culture" and "cultivation" share an etymology: they derive from the Latin *colo*, *colere* ("to care for, honor, till, make grow"), which in turn is derived from the Indo-European root *k^{w}el-*, meaning "to revolve," "circle," "turn over"—here one thinks of the tilling of the soil—and also "to inhabit" and "to dwell" (Finley 2015: 170). To dwell on earth, then, is to garden; embedded in these words is the notion that the success of our lives depends on our success in maintaining the planet and its produce.

The domestication of plants, together with that of animals, allowed for a quantum leap in population growth and for the rise of permanent settlements and cities, as evidenced in Mesopotamia—part of the "Fertile Crescent" and the so-called "Cradle of Civilization"—which has yielded some of the earliest evidence of the domestication of cereals. These domesticates became staple foods, readily available and consumed on a regular basis. As growing conditions dictated what could and could not grow, staples in Mesopotamia necessarily differed from those of sub-Saharan Africa and North America in this period, for example. Staple food plants everywhere in the ancient world were distinguished from those "luxuries" that were more highly prized, less readily available, and often deriving from distant lands. Desire for the latter, as well as for plants most useful in applications ranging from ship-, house-, and temple-building to making tools and weapons; weaving textiles to serve as clothing, nets, or bedding; and the extraction of fragrant essences and resins for unguents, incense, and perfume drove trade and exploration on a local and "global" scale. The usefulness of plants prompted the naming and cataloguing of them, especially those discovered through trial and error to be efficacious in curing human ailments. Their medicinal potency reflected their inherent link with the divine: the gods to whom they were sacred or who inhabited them. As indicators of learnedness, wealth, prestige, and the favor of the gods, exotic and native plants alike were gathered in botanic gardens and decorated structures or objects ranging from tombstones and temples to cups and other tableware. As their life cycles ran parallel to those of humans—for they were observed to sprout, grow, mature, and wither in old age—they engendered narratives explaining and memorializing the human proximity to plants.

As already will have become apparent, a given plant's manifold cultural associations and applications were intertwined, its sacred aspects being linked with usage as a decorative motif or as poetic symbol, for example. Understanding any aspect of a plant's cultural history thus entails a well-equipped scholarly toolkit: the knowledge of a botanist, an art historian, a philologist, an anthropologist, an archaeologist, architectural historian, historian, and more. Just how culturally rich plant-related objects in antiquity were can be illustrated by examining any number of artifacts from different parts of the world. Exemplary among these, however, is an expansive garden fresco, analysis of which can serve as an apposite introduction to unlocking the cultural code(s) of plants more generally. The fresco is Roman, and thus a product of a multi-ethnic, global cultural group, for at its height the Roman Empire included all the lands bordering on the Mediterranean Sea and extended into northern Europe and western Asia so as to include diverse peoples dwelling in Britain, the Iberian Peninsula, Gaul, Italy, Greece, the Balkans, northern Africa, Egypt, Asia Minor, and Syria, for example. As a consequence, Roman Italy impacted a sizeable part of the world and was, in turn, impacted by it: goods and information traversed the Empire, notably including plants and knowledge of their uses, medical and otherwise—knowledge gathered over millennia by other groups, whose arts and wisdom were absorbed into the fabric of Roman life. This body of botanical knowledge, collected and enhanced by the Romans, is an important but often overlooked aspect of Rome's legacy to modernity.

FLUTTERING LEAVES, FICTIVE BREEZES

Among the myriad ancient marvels housed in Rome is a series of garden frescoes that once fully covered all four walls of a spacious, semi-subterranean dining room (11.7 × 5.9 m, 38.4 × 19.3 ft) in the so-called House of Livia at Prima Porta, an imperial villa located about 15 km (9 mi) north of the city (see Figure 0.1). Almost invariably, the frescoes elicit a gasp—or hushed "oh"—when visitors to Rome's Museo Nazionale Romano di Palazzo Massimo come upon them, especially if they do so with no sense of the paintings'

FIGURE 0.1 Garden fresco from the Villa of Livia at Prima Porta, detail, *c.* 20 BCE. Palazzo Massimo alle Terme, Museo Nazionale Romano. Photo by Adam Eastland. Courtesy of Alamy Stock Photo.

ancient provenance or awareness that such images existed, much less in proliferation, in Roman homes of the first century BCE and later. The nearly overwhelming abundance and diversity of plant life represented here has led to the characterization of this fictive garden as paradisiacal: at least two dozen species of plant belonging to twenty families have been identified, including spruce, cypress, palm, pine, bay laurel, quince, box, arbutus, myrtle, oleander, viburnum, fern, ivy, pomegranate, oak, acanthus, iris, chamomile, aster, poppy, rose, violet, and daisy (Caneva and Bohuny 2003; Gabriel 1955). The perception of this garden as Edenic is intensified by the realization that the plants portrayed bear fruit and bloom outside the parameters of real-word seasonal constraints. The quince, pomegranate, and arbutus, heavy with fruit, produce their bounty in autumn, not in coincidence with the springtime blossoming of poppy, rose, violet, daisy, and viburnum or with the summer flowering of oleander. Not only this, but, while necessarily static, frozen by the painter's brush, the garden is curiously, or perhaps preternaturally, animate. Leaves flutter and branches bend to fictive breezes while an abundance of birds—sixty-nine are counted, with thrushes, orioles, larks, goldfinches, martins, blackbirds, jays, quails, and partridges in their number—flit or rest (one caged) among the foliage. Some plants are bathed in light and others are obscured by shadows to suggest the varied output of a waxing and waning sun.

The frescoes raise a host of questions. Is this painted garden in any sense a reliable reflection of "real" designed landscapes in ancient Rome? When, precisely, was it painted? What meaning or meanings did the painted garden hold for its owner and her guests? Is the lushness of the plants a reflection of her providence or the favor of deities that she worshipped? Could this assemblage of plants be a botanic garden? Were all these plants readily available in Italy in the first century of the Common Era when the fresco was created? Were they native or were they imported? By what names did the Romans call them? What techniques were deployed to propagate and maintain them? Were the species represented here "useful" or ornamental? What other facets of Roman culture might be encoded here? As will quickly become apparent, addressing these questions reveals the Romans' deep understanding of the interconnectedness between plants and every facet of human life.

The Prima Porta painting, dated variously between 30 and 20 BCE, is the earliest known, and also the most famous, exemplar of the Roman garden painting "genre." Long viewed as distinct from other styles of wall painting in Roman houses, garden paintings are now understood to be an offshoot of the so-called Architectural or Second Pompeian Style, which was characterized by bold illusionism whereby painters "pierced" built surfaces to produce vistas into landscapes (temple precincts, colonnaded courts, and sacred grottoes among them) framed by a range of architectural forms (Giesecke 2012; Kuttner 1999b). It was also long thought that garden paintings were pure fantasies, bearing little relation to planted gardens. This notion, too, has been recently disproven. At first glance the Prima Porta garden is a chaotic botanical mélange—a seemingly unreal proliferation of unruly, unchecked growth—but closer inspection reveals an organic degree of control, a harmonizing rhythm. Not only is the wild garden triply bounded by reed fence, low wall, and close-shorn verdant walk (punctuated by sparse clumps of ivy, fern, or violet), but one plant emerges as dominant within the apparent chaos of the copse: recurring both as tree and shrub, the bay laurel anchors the composition. There is also a graduated arrangement of plants, with smaller, flowering or fruit-bearing plants in the foreground and taller, predominantly evergreen plants in the background. All of this—the linear placement of plants, their gradation of heights, and their careful enclosure in clearly outlined, fenced beds—aligns with reconstructed planting schemes in the recently excavated Great Peristyle of the Villa Arianna, Stabiae.

PLANTS AND PAINTINGS

Stabiae offers the largest concentration of well-preserved Roman coastal sea-view villas in the Mediterranean, as such presenting a unique opportunity to view at least a small cross-section of the luxury villas of Rome's elite that once that lined the Bay of Naples, the so-called *Crater Delicatus* (Crater of Luxurious Delights). The Villa Arianna, the oldest of the Stabian villas, dates to the first century BCE and features both a subterranean passageway to the sea and an enormous peristyle garden (approximately 50 × 100 m, 164 × 328 ft). The garden was found to contain eight wide, linear, densely planted, fenced beds (two fully excavated being 25 × 5 m, 82 × 16.4 ft, at a minimum) between and around which visitors could stroll (Gleason 2016, 2019).

Gardens in Pompeian town houses were necessarily smaller and less densely planted than that of the Villa Arianna, offering no exact parallel to the Prima Porta fresco's dense, extensive plantings, but excavated gardens in Pompeii do align with garden frescoes from that site, allowing one to conclude that frescoes *are* good indicators of the true appearance of Roman garden spaces. The garden of the House of the Chaste Lovers, Pompeii, with its reed-fence-enclosed, symmetrical planting beds, aligns with the general landscape scheme depicted on the rear wall of the House of the Marine Venus' garden, for example, which

FIGURE 0.2 House of the Chaste Lovers, peristyle garden, first century CE. Pompeii. Courtesy of Annette Giesecke.

is one of the best-preserved garden paintings still in situ (see Figures 0.2 and 0.3). On the basis of carbonized plant remains and pollens, together with root cavities left in the pyroclastic matter that buried the House of the Chaste Lovers in 79 CE, it was determined that its garden contained juniper and rose as well as grasses, carnations, asters, and ferns—that is, a combination of flowering and evergreen shrubs as well as an assortment of smaller, decorative plants planted along the inside of the beds' enclosure (Ciarallo and Mariotti Lippi 1994). The Marine Venus garden fresco, for its part, depicts shrubs of oleander and myrtle in flower, rose bushes laden with red blooms, diminutive pines, a clump of southernwood (*Artemisia abrotanum*), as well as fruit-bearing strawberry trees and cherry plums (*Prunus cerasifera*) all enclosed by woven fencing. Smaller plants appear in the foreground and larger specimens to the rear. However, here, as in all Roman garden paintings, the artist has disregarded fidelity to the plants' actual life cycle. Painted gardens thus are both real and unreal reflections of their living counterparts.

HOMEGROWN OR EXOTIC?

Returning to the expansive Prima Porta fresco, what clues might it contain about the provenance of "typical" Roman garden plants? Fortunately, the artists painted the plants with sufficient detail to allow for reasonably close identification. The majority of the flora represented are, broadly speaking, Mediterranean, some indigenous to the more temperate east, others to the west, and others still to warmer regions to the south: *Acanthus mollis* (acanthus), native to the Mediterranean region from what are now Portugal and northwest Africa east to Croatia; *Anthemis* spp. (chamomile, perhaps *Athemis arvenis*,

FIGURE 0.3 Garden fresco, House of the Marine Venus, detail, first century CE. Pompeii. © Werner Forman. Courtesy of Getty Images.

corn chamomile), native to the Mediterranean region and Southwest Asia east to Iran; *Arbutus unedo* (strawberry tree), native to the Mediterranean region and western Europe; *Buxus sempervirens* (box), native to western and southern Europe, northwest Africa, Southwest Asia, southern England south to northern Morocco, and east through the northern Mediterranean region to Turkey; *Cornus mas* (cornelian cherry, cherry dogwood), native to southern Europe and southwestern Asia; *Cupressus sempervirens* (Italian or Mediterranean cypress), native to the eastern Mediterranean region, including northeast Libya, southern Albania, coastal Bulgaria, southern coastal Croatia, southern Montenegro, southern Bosnia and Herzegovina, southern Greece, southern Turkey, Cyprus, northern Egypt, western Syria, Lebanon, Malta, Italy, Israel, western Jordan, and Iran; *Cydonia oblonga* (quince), native to western Asia, Armenia, Turkey, Georgia, and northern Iran to Afghanistan; *Glebionis coronaria* (garland chrysanthemum or crown daisy), native to the Mediterranean region; *Hedera helix* (common or English ivy), native to Europe and western Asia; *Iris*, most species being found in temperate Northern Hemisphere zones, from Europe to Asia and across North America; *Laurus nobilis* (bay laurel), native to the Mediterranean region; *Myrtus communis* (myrtle), native to southern Europe, North Africa, western Asia, Macaronesia, and the Indian Subcontinent; *Nerium oleander* (oleander), native to the Mediterranean region (including Europe) extending to China and Japan; *Papaver somniferum* (opium poppy), likely native to the eastern Mediterranean; *Phoenix dactylifera* (date palm), believed to be native to the Persian Gulf region and abundant from Mesopotamia to Egypt; *Pinus pinea* (stone pine), native to the northern Mediterranean coastal region (southern Europe to Turkey and Lebanon); *Punica granatum* (pomegranate), native to middle and western Asia, extending from modern-day Iran to northern India; *Quercus robur* (common or European oak), native from the British Isles to the Caucasus; *Quercus ilex* (holly or holm oak), native to the Mediterranean region (southern Europe and northern Africa); *Rosa gallica* (French rose), native from southern and central Europe eastwards to Turkey and the Caucasus; *Viburnum tinus* (laurustine viburnum), native to the Mediterranean area of Europe and North Africa; and *Viola reichenbachiana* (early dog or pale wood violet), found in western, southern, and central Europe (Brickell and Zuk 1997; Caneva and Bohuny 2003; Mabberley 2017). However, from a geographic perspective, there is one distinct outlier if its botanical identification is correct: *Picea abies* (Norway spruce), native to northern, eastern, and central Europe. Nonetheless, all these plants may in fact be perceived as having a common geography in that they hail from parts of the world that had come under Roman control or influence during the Augustan period, the very "moment" in history that yielded the Prima Porta paintings. Put differently, these plants could be seen as a botanical map of the early Empire, with quince, pomegranate, and cypress, for example, reflecting Persia and the eastern Mediterranean; spruce reflecting northern and northeastern Europe; and date palm reflecting Egypt, North Africa, and the Levant (Tally-Schumacher and Niemeier 2016: 69). Importantly "Romanized" exotic plants retained their local cultural associations, making them all the more appealing to increasingly worldly consumers.

WHAT'S IN A NAME?

Before delving more deeply into provenance, with consequent discussion of sourcing, trade, and exploration, it is worth reflecting on the issue of plant identification and nomenclature. Can we be confident in our identification of plants in the Prima Porta fresco? And do the names we today assign botanical specimens correspond to names used in antiquity? Take the pomegranate, *Punica granatum*, for example. As noted earlier, this particular fresco is painted with sufficient realistic detail to allow for positive plant identification; the fictive pomegranate is unmistakably a pomegranate. Not only this, but the Prima Porta garden offers a rare glimpse into the skill of the Roman gardener, or *topiarius*, whose "shears" have left their mark on pruned branches, shaping and dwarfing specimen plants to best effect (Gleason 2019). As for nomenclature, the modern English "common" and Latin "scientific" names for the pomegranate do, in fact, derive from classical antiquity. In classical Latin, the fruit was known either as *malum punicum* or *malum granatum* (also *melogranatum*). *Malum* is the Latin word and *mēlon* the Greek for "apple" (which could be used for a range of tree fruit). *Granatum* derives from *granum*, "grain," and means "(multi-) grained," alluding to the fruit's numerous grain-like seeds

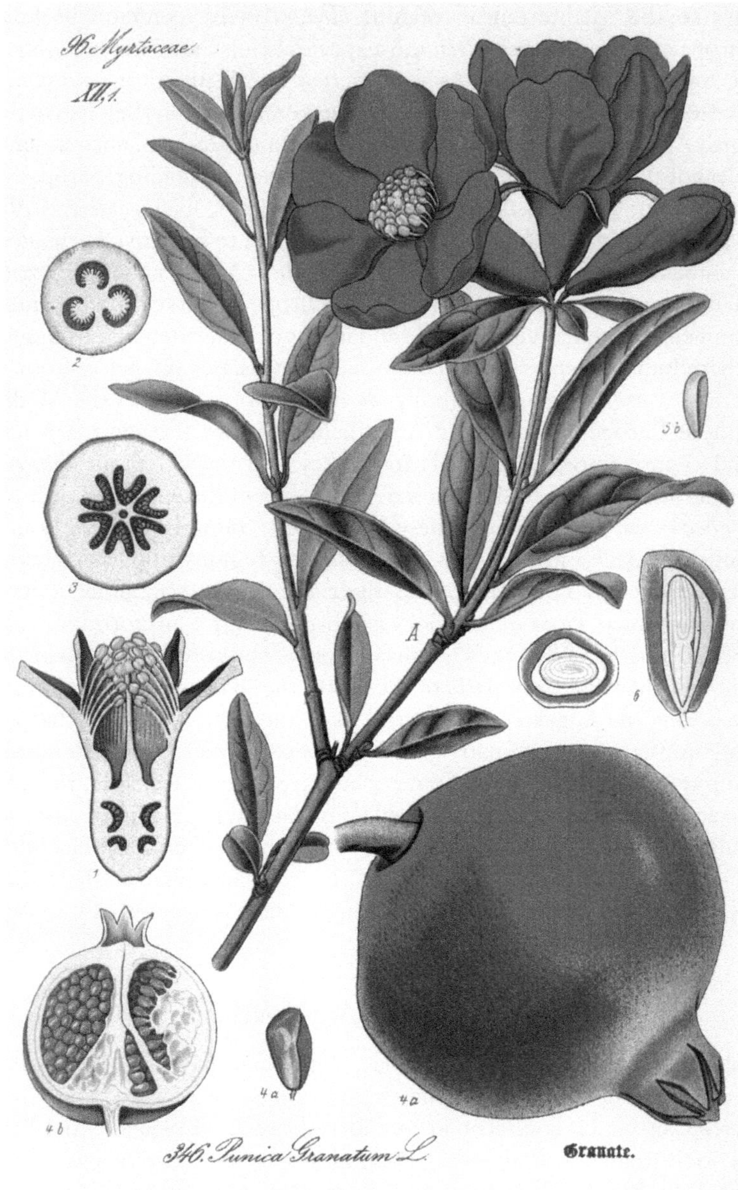

FIGURE 0.4 Pomegranate (*Punica granatum*). From Professor Dr. Otto Wilhelm Thomé, *Flora von Deutschland, Österreich und der Schweiz*, 1885, Gera, Germany. © Kurt Stueber's online library. Courtesy of www.biolib.de.

(see Figure 0.4). The adjective *punicus*, strictly speaking, refers to Phoenicia in Asia Minor but was more frequently used with respect to Carthage, the Phoenician colony in northern Africa, the birthplace of Hannibal and long Rome's mortal enemy; the pomegranate was believed to be of African origin. The Romans thus called the pomegranate by at least two names, "Punic apple" and "many-seeded apple." The Greeks, incidentally, called the pomegranate *rhóā* but also referred to it as *mēlon*.

While this plant's modern names are indeed derived from the classical, it is important to note that they do not reflect actual ancient usage. The common English name "pomegranate" is a combination of Latin *pomum*, "fruit," and *granatum*, "many-seeded." The modern scientific name *Punica granatum* (the feminine ending *-a* being appropriate for a fruit-bearing and thus "feminine" tree) borrows from both *malum punicum* and *malum granatum* but reflects modern plant taxonomy, the science that finds, identifies, describes, classifies, and names plants by designating, among other things, a plant's Family (group of genera that share a set of underlying features); Genus (group of one or more plants sharing a wide range of characteristics); Species (group of plants capable of breeding together and producing offspring similar to themselves); Subspecies (naturally occurring distinct variants of a species often in isolated populations and indicated by the abbreviation "subsp." followed by a descriptor); Variety (minor subdivisions of a species differing slightly in botanical structure and indicated by the abbreviation "var." followed by a descriptor); and Cultivar (distinct variant or hybrid known only in cultivation and indicated by a name in single quotes) (Brickell and Zuk 1997: 11). In particular, *Punica granatum* is a binomial, adhering to the formal two-part naming system indicating genus and species established by Carolus Linnaeus in the eighteenth century to classify all life. In antiquity, plant names were neither consistently binomial—indeed, they were often not—nor were they necessarily consistent even within a given culture or language. In fact, some plants were not named (if wild and not "useful"), and others were known by multiple names: by Latin names, Greek names, and names given them in their places of origin and in the language of that region, for instance (Hardy and Totelin 2016: 93–104). This is not to suggest that names assigned in antiquity were random or meaningless, far from it. Rather, it is the case that naming is a direct reflex of the desire to "own," and this desire is predicated on usefulness. It is also the case that plants' names, however flexible, appear largely to have reflected their physical characteristics, their place of origin, or their agency in medicinal or alimentary contexts. An example, discussed further in Chapter 5 of this volume, is *Artemisia* spp. ("spp." indicating a range of species), perhaps most likely the species *vulgaris* known commonly as "mugwort." Here again the Linnaean name is derived from its ancient Greek name, *artemisía*, which was employed to treat gynecological conditions and was named after the goddess Artemis, goddess of childbirth.

It is not the case that ancient names always found their way into modern nomenclature, and they can be misleading to a modern reader of an ancient text. The rhododendron is an example. What we know today as *Rhododendron* spp., a genus of some one thousand species of woody plants in the heath family (Ericaceae), is a very different plant from what Roman encyclopedist Pliny the Elder in the sixteenth book of his *Natural History* called *rhododendron* (from the Greek meaning "red tree"), namely the *Nerium oleander*, a shrub or small tree in the dogbane family Apocynaceae—though both, notably, are shrubs or small trees that can bear red or reddish flowers:

> Belonging to this last class, there are the following trees that do not lose their leaves: the olive, the bay laurel, the palm, the myrtle, the cypress, the pine, the ivy, the *rhododendron* .... The *rhododendron*, as its name indicates, comes from Greece. By some it is known as the *nerium*, and by others as the *rhododaphne*. It is an evergreen, bearing a strong resemblance to the rose-tree, and throwing out numerous branches from the stem; to beasts of burden, goats, and sheep it is poisonous, but for man it is

> an antidote against the venom of serpents.
>
> (*HN* 16.33.79)

Yet elsewhere, Pliny uses *rhododendron* to name what has been identified as *Rhododendron ponticum*:

> In the country of the Sanni, in the same part of Pontus, there is another kind of honey, which, from the madness it produces, has received the name of "*maenomenon*" [maddening]. This evil effect is generally attributed to the flowers of the *rhododendron*, with which the woods there abound; and that people, though it pays a tribute to the Romans in wax, derives no profit whatever from its honey, in consequence of these dangerous properties.
>
> (*HN* 21.45.77)

Do the names we assign botanical specimens correspond to names used in antiquity? The answer, then, is "sometimes in part, and sometimes not."

LUXURY AND ORNAMENT

Returning to the Prima Porta fresco and its mélange of natives and "nativized" exotics, is it meaningful to try to differentiate common plants from rare, humble from luxurious, useful from ornamental? Did the Romans make such distinctions? The luxury status of certain plants was indeed recognized, and they became important vehicles of self-display on the part of the elite and the socially mobile. Certainly there was caché in introducing new varieties of plants from distant lands; had this not been the case, victorious generals would not have paraded "captive" plant specimens in Triumph, the public celebration of a military commander's successful campaigns. The first Roman general recorded as having introduced an exotic plant to Rome consequent to foreign conquest was the profligate Lucius Licinius Lucullus, though in this instance there is no mention of a triumph:

> The cherry did not exist in Italy before the period of the victory gained over Mithridates by L. Lucullus, up to the year 680 [79 BCE]. He was the first to introduce this tree from Pontus [in northeast Asia Minor], and now, in the course of one hundred and twenty years, it has travelled across the ocean to arrive even in Britannia.
>
> (*HN* 15.30)

The cherry in question is the bitter cherry (*Prunus cerasus*), and its Latin name, *cerasus*, was derived from Cerasus, the Pontine city from which Lucullus introduced this plant to Italy. Regarding another import, balsam, the elder Pliny writes:

> This variety of shrub was exhibited to the capital by the emperors Vespasian and Titus; and it is a remarkable fact that ever since the time of Pompey the Great even trees have figured among the captives in our triumphal processions. The balsam-tree is now a subject of Rome, and pays tribute together with the race to which it belongs …
>
> (*HN* 12.54)

While it is uncertain whether balsam (*Commiphora gileadensis*) was paraded in triumph—presumably it *was*—we do know that in his great Asiatic triumph of 61 BCE, Pompey exhibited ebony trees. In any case, both trees species had clear links with specific geographic areas that had fallen under Roman control: the balsam with Judea and the ebony tree with Ethiopia. It is also true that these had considerable commercial value, yielding high-priced luxury goods. Resin from the balsam tree was highly sought after in the manufacture of medicines and perfumes. Ebony, on the other hand, was employed in the making of fine furnishings, and had been prized for this purpose by the kingdoms of the Near East, including Egypt, Mesopotamia, and Persia as well as by the Greeks (Meiggs 1982: 282–5; Östenberg 2009: 185–7). As these and other examples reveal, it is safe to assume that plants hailing from distant lands were considered luxuries not only because of their distant, exotic origins but also because the collection of "foreign" plants was inextricably linked with notions of dominion over nature and over the plants' originary lands; many or most of these were likely introduced by those who had served in military campaigns overseas (von Stackelberg 2009: 43). The pomegranate (Punic apple) and quince (known by the Romans as the Cydonian apple or *struthium*, depending on variety), which are both depicted on the Prima Porta fresco, fall into this category.

To be sure, not all prestige plants were exotics. People also vied to produce new, better strains of local produce that could be named after them or their estates. As so often in the case of plants and much else, the elder Pliny offers a wealth of information. Grafting, he states, was means to ensure celebrity (*HN* 15.15):

> Why should I hesitate to indicate by name the remaining varieties of fruit, seeing that they have prolonged the memory of those who established them for all time, as though on account of some outstanding achievement in life? Unless I am mistaken, the recital will reveal the ingenuity exercised in grafting, and will show that nothing is so trifling as to be incapable of producing celebrity. Well then, there are kinds of fruit that have their origin from Matius and Cestius, from Mallius, and likewise from Scaudius; and on the last a member of the Claudian family named Appius grafted the quince, producing the fruit called Appian; this has the smell of a quince, the size of a Scaudian apple, and a ruddy color. And in order that nobody may imagine that it has gained its position by influence due to distinction and family, there is also a Sceptian apple named from a freedman who discovered it, which is remarkable for its round shape.

There is also reason to believe that a good many of those plants and their products that were regularly imported by the time of Augustus remained luxury or prestige items, especially if they could not readily be grown in parts of the Empire beyond those regions to which they were native and/or if they had particular symbolic importance in the context of Roman religion, politics, and family life more generally.

Some exotic fruit and plant products became so integrated into Roman life that they might be classed as "staple" or "mass" luxuries. Among them were the pomegranate and date, which, incidentally, are among a particularly interesting group of archaeobotanical finds from a Roman military encampment in Vindonissa, modern Windich, Switzerland. Pomegranates do best in a semi-arid, mild-temperate to subtropical climate; wet climates and temperatures below minus 10 degrees Celsius (14° F), on the other hand, will cause serious damage. Growing wild in regions south of the Caspian Sea (Iran), northeastern Turkey, and the southern Balkans, the pomegranate was likely first cultivated in western Asia, spreading to Syria, Palestine, Mesopotamia, and Cyprus by the third millennium BCE. From Mesopotamia, it was brought to Egypt by 1500 BCE via military campaigns of the pharaoh Thoutmosis III, and likely earlier; from Cyprus it appears to have been conveyed to the western Mediterranean and northwestern Africa (Dalby 2003: 266; Jacomet et al. 2002; Zohary and Hopf 2012: 133–4). As for the date palm, hot, dry conditions, coupled with a steady supply of water, are ideal for its growth and flourishing. Domesticated in the Persian Gulf, Mesopotamia, and Palestine from about 5000 BCE, it could be grown in Greece and elsewhere north of this region but did not fruit (Dalby 2003: 113; Zohary and Hopf 2012: 131). It certainly could neither be grown nor bear fruit in Alpine climates, but unlike the pomegranate, it was an important staple food where it was domesticated (see Chapters 1 and 2 of this volume for further discussion). In Vindonissa, carbonized remains of date and pomegranate accompanied other exotics that included olive, walnut, pine (*Pinus pinea*), pistachio, peach, and cherry. All of these are edible plants and were found in considerable quantities, suggesting that, as in the case of other military encampments, Romans stationed "overseas" viewed them as essential (Jacomet et al. 2002). Roman literary sources indicate that the pomegranate—its seeds, both in fresh and dried form, and its juice—was a prized ingredient in cooking, but its uses outside the kitchen were manifold and diverse. Being many-seeded, as mentioned above, the pomegranate was an important fertility symbol sacred to the goddesses Aphrodite, Hera, and Persephone, all of whom were associated with fecundity and the cycle of life and death (Murr 1890: 50–5). Not surprisingly, it found application in the cure of a host of feminine ailments. It was useful, too, in the tanning process and as a source of dye. Dates, for their part, were an important source of sugar and could yield both "flour" and sweet wine. They pervade the cookbook of the first-century CE epicure Apicius, where, among other preparations, they are stuffed and crushed to make sauces, like that used to prepare flamingo and parrot, which the Romans considered delicacies:

> Scald the flamingo, wash and dress it, put it in a pot, add water, salt, dill, and a little vinegar, to be parboiled. Finish cooking with a bunch of leeks and coriander, and add some reduced must to give it color. In the mortar crush pepper, cumin, coriander, laser [silphium] root, mint, rue, moisten with vinegar, add dates, and the fond of the braised bird, thicken, strain, cover the bird with the sauce and serve. Parrot is prepared in the same manner.
>
> (6.231)

Yielding as much as 100–200 kg (200–400 lbs) of fruit per tree annually, the date palm was an important symbol of bounty and life, especially in Mesopotamia, where it was identified as the Tree of Life (see Chapters 6 and 8 of this volume for further discussion). As in the case of the pomegranate, every part of the date palm was considered useful in antiquity: its trunk could be used as timber, its foliage as roofing or for weaving baskets and mats, and its fibers to make rope (*HN* 13.6–9; Zohary and Hopf 2012: 131). Regarding medicinal applications, dates were found to be particularly helpful in treating digestive ailments but could also be used to cure afflictions ranging from hemorrhoids to dysentery, according to first-century CE physician and botanist Dioscorides in his *De materia medica* (1.148). For the Romans, then, date palms and pomegranates were certainly exotic luxury plants, every part of which was useful in some way. While there is today a tendency to view ornamental plants from a purely aesthetic vantage point, and thus as attractive but not useful, the opposite was true in antiquity. Beauty was dependent on usefulness, as the etymology of "ornament"—something functional requiring craftsmanship to produce—

indicates (see Chapter 7 of this volume for further discussion). Importantly, common, humble plants thus were no less ornamental than exotics, and as elaborated below, they were no less sacred or symbolic.

GATHERING IN CREATION

As noted earlier, it was useful plants that were named, and it was also these whose physical characteristics were catalogued, some of the most important catalogues being herbaria in the pre-eighteenth-century meaning of the word: works on medicinal plants, which include portions of the Hippocratic Corpus (fifth and fourth centuries BCE), Theophrastus' *Enquiry into Plants* (*Historia plantarum*, written some time between *c.* 350 BCE and *c.* 287 BCE), Pliny's *Natural History* (*Historia naturalis*), and the second-century CE physician Galen's writings on foodstuffs and simple drugs (*De alimentorum facultatibus* and *De simplicium medicamentorum*) as well as Dioscorides' *De materia medica* (On the Preparation, Properties and Testing of Drugs) (see Chapter 5 of this volume for further discussion).

In a very real sense, the Prima Porta garden is a collection of medicinal plants and, as such, is the visual counterpart of the herbarium. From ivy and violet to oak, all the plants depicted were somewhere documented as possessing healing powers. For instance, of the violet Dioscorides writes:

> The violet has a leaf smaller than cissus, thinner and darker; and little stalks in the midst (from the root) on which is a little flower, very sweet, of a purple. It grows in shady rough places. It is cooling, so that the leaves (applied by themselves or with polenta) help a burning stomach, inflammation of the eyes, and prolapse of the perineum. A decoction of the purple part of the flower helps the abscessed throat and epilepsy of children.
>
> (*De mat. med.* 4.122)

Ivy, for its part, could be used to treat dysentery, heal burns and sores on the head, and serve as an abortifact (*De mat. med.* 2.210). All of an oak's parts were astringent, with the film between the bark and the stock being the most therapeutic for the bowels. Dioscorides reports that "a decoction of this is given for coeliac [intestinal complaints], dysentery, and to blood-spitters, and pounded into small pieces it is put into suppositories for women troubled with excessive discharges of the womb" (*De mat. med.* 1.142).

As an herbarium, the Prima Porta garden is also a botanic garden, gathering useful plants from around the then-known world; as already noted, the plants appearing in the fresco "map" the Roman Empire. In addition, the garden is a collection or assemblage of the Greco-Roman pantheon, for all of the plants were sacred to and/or embodiments of one or more deities. Date palms, for example, were sacred to Apollo—the sun god and also god of prophecy, healing, and light—whose mother clutched a palm while giving birth to him on the island of Delos. The bay laurel, too, was Apollo's tree, a living memorial to his beloved Daphne, the nymph whose arboreal transfiguration was precipitated by her desire to escape god's unwanted advances (see Figure 0.5). Plane trees had a close association with Zeus (Roman Jupiter), king of the gods, who, according to Greek mythology, impregnated the princess Europa in the shade of that tree at Goryton in Crete (*Hist. pl.* 1.9.5; Baumann 1993b: 46). The oak, too, was sacred to this god, his

FIGURE 0.5 *Apollo and Daphne*, sculpted by Giovanni Lorenzo Bernini (1598–1680). Daphne transformed into a tree. Villa Borghese, Rome. © Bettmann. Courtesy of Getty Images.

most important oracle being at Dodona in Epirus where, from the sanctuary's foundation and until the end of the fifth century BCE, worship took place in the open, under the god's sacred oak tree; the sanctuary's priests delivered Zeus' divinations by interpreting the rustling of the sacred oak's leaves. Poppies were sacred to Demeter (Roman Ceres), goddess of grain and the harvest, who availed herself of its palliative qualities to lessen her grief over the abduction of her daughter, Persephone, who would become Hades' queen (Baumann 1993: 69). The strawberry tree had links to the cults of the messenger god Hermes and Hera, Roman Juno, queen of the gods and protector of women as wives and mothers (Murr 1890: 70). Pine was sacred to Poseidon (Roman Neptune), god of the sea; the Eastern mother goddess Cybele; and Dionysus (Roman Bacchus), the shape-shifting god of wine. Violets were sacred to Aphrodite (Roman Venus), goddess of sexual love and procreation—whose powers were also embodied by the fragrant, sensual lily, rose, and myrtle—and to Dionysus. Evergreen and symbolic of everlasting life, ivy, too, was famously associated with Dionysus (see Giesecke 2014, passim, for all of the above). And while each plant could be linked with one or more deities, *all* plants were, in fact, also sacred to Aphrodite as a fertility goddess, and to Dionysus, who was in origin a Near Eastern vegetation deity.

DECODING THE LANGUAGE OF FLOWERS

It remains to consider whether there was an overarching message in the Prima Porta fresco's assemblage of plants. Were there similar monuments in antiquity, and could these provide further clues to unlocking the fresco's cultural code? Two such monuments come to mind, both of them representations of collections of diverse species of plant rendered in polychrome sculptural relief: one is Egyptian and decorates the walls of the Sun Rooms in the Festival Temple at Karnak built at the behest of pharaoh Tuthmosis III (1490/68–1436 BCE), and the other is the Ara Pacis (Altar of Peace) commissioned by the emperor Augustus and dedicated in the year 9 BCE to celebrate the peace that Augustus, through his military victories, brought to the Empire. The Sun Rooms that house the Karnak reliefs are tucked away in a special offering area to the rear of the sanctuary, which itself was a memorial to Amun Re, the Sun God and Tuthmosis' "father." The so-called "botanic garden reliefs" were a departure from earlier Egyptian representations of plants which, generally speaking, took the form of scenes from daily life in which plants

FIGURE 0.6 Botanical Garden relief of Tuthmosis III from his Festival Hall at Karnak, Egypt. Representations of plants and animals brought from Syria by Tuthmosis III in 1465 BCE, Karnak Temple Complex, Luxor, Egypt. Photo by Dan Oldenburg. Courtesy of Alamy Stock Photo.

and animals defined types of landscape (deserts, gardens, fields, exotic lands) or types of human activity (fishing, hunting, gardening), but they also appeared as offerings or tribute conveyed in ceremonial contexts (Beaux 1990: 47–9). No earlier Egyptian painting or relief approximates the extent and diversity of plant species represented here: some fifty different plant species together with at least twenty-five species of birds and several cattle, goats, and gazelles appear (see Figure 0.6). The botanic specimens include plants that are familiar from the Prima Porta fresco—iris, myrtle, oak, and bay laurel, for example—and all are identified as having been brought to Egypt by Tuthmosis on his return from military campaigns in Palestine and Syria: "Plants which His Majesty found in the Land of Retenu. All plants that grow, all flowers that are in God's land." They are offerings to Amun (Wilkinson 1998: 137–9). In other words, the botanic garden is a reflection of empire, conquered lands and control of their wondrous resources. Presumably this living, transplanted booty would thrive bathed in the nourishing light of Amun's Sun; not coincidentally, the Sun Rooms were linked with the pharaoh's perpetual rejuvenation and regeneration.

The very much later Roman Altar of Peace is, more precisely, an altar enclosed by a precinct wall, and it was ornamented, inside and out, with carving in relief. Interior panels bore renderings of garlands, heavy with foliage and various fruits, while exterior panels featured Romulus and Remus, Rome's founders; pious Aeneas, progenitor of the Roman race; the fecund Earth personified; Augustus in the garb of priest, accompanied by his family; and, below these peopled scenes, an exuberant garden in which some ninety species of flora—variously signifying fertility, rebirth, good fortune, and healing and linked to all Rome's gods—have been identified (Caneva 2010). Predominant among this sculpted garden's plants are spiraling acanthus and bay laurel, the god Apollo's favorites,

FIGURE 0.7 Ara Pacis Augustae (Altar of Peace), detail. Scrolling vegetal motif. Dedicated in 9 BCE, Rome. Photo by Lanmas. Courtesy of Alamy Stock Photo.

entwined with Bacchus' life-asserting vine (see Figure 0.7). Here, too, Venus makes her munificent tutelage known through the presence of roses and swans, majestic birds sacred also to Apollo. All three gods had special ties to Augustus and his personal mythology. Apollo was his patron and protector; through his adoptive father Julius Caesar, Augustus was descended from Venus; and Bacchus, once associated with Augustus' rival Mark Antony, was cast in the Augustan period as a god of poetry and prophecy, not drunkenness, and, as Liber/Bacchus, together with the noble goddess Ceres, guarantor of those staple foods (bread and wine) upon which the advancement of humankind depended (Verg., *G.* 1.1–7). The plants most closely linked with these deities figure heavily in Augustan official iconography and self-representation beyond the Ara Pacis. For example, a pair of bay laurel trees flanking the doors of the emperor's house on the Palatine Hill marked the day on which he, born Gaius Octavius, was given the title "Augustus" by the Senate; appearing on Augustan coins, these two trees served as the emperor's metonymn (Kellum 1994: 213). And when a palm, another Apolline plant, sprang up before his house, the emperor reputedly transplanted it to Apollo's temple, conveniently located just next door (Suet., *Aug.* 92.1–2). Livia, Augustus' wife, gifted Rome and its populace with a prodigious grape vine that, growing on the portico that the emperor had built in her honor, provided welcome shade and yielded an abundance of wine (Plin., *HN* 14.3). A crown of Venus' myrtle, meanwhile, features prominently in the opening of Vergil's *Georgics* (1.28) where it graces the head of Octavian, not yet Augustus, who, strikingly, is here invoked as a deity in conjunction with a host of gods called upon to ensure the success of the poet's endeavor.

As in the case of Karnak, the Ara Pacis "garden" is nourished by the sun, in this case the "ray" of the Egyptian obelisk that, like the altar, had been erected on the Campus Martius. Again, Amun Re was Tuthmosis' father and the Sun god Apollo Augustus' personal protector (Frischer et al. 2017; Swetnam-Burland 2010). Also as at Karnak, the Ara Pacis reliefs are certainly to be understood as a "garden of state," a botanical reflection of the Empire and its flourishing under the current regime. The Prima Porta fresco, being from a villa owned by the family of Augustus—for the Livia in the villa's name was none other than the emperor's wife—must be "read" as a companion Ara Pacis vegetal frieze. In both monuments the fertility and prosperity of the Roman state is everywhere manifest.

FLOWER POWER AND GARDENER KINGS

Tuthmosis III was not Augustus' only predecessor as gardener-king; indeed, there had been a long succession of these (Stronach 1990: 171–4). In Mesopotamia were the Babylonian king Adad-shuma-usur (r. *c.* 1218–1189 BCE), a relatively shadowy figure whose palace gardens are attested in inscriptional evidence, and the Assyrian Tiglath-Pilesar I (r. 1115–1077 BCE). The latter's annals boast of his concern for the prosperity of his land and people, which he assured by the annexation of new territories, the expansion of cultivated land, and the importation and planting of exotic trees. "Plows I did harness throughout the whole of the land of Assyria and I heaped up more heaps of grain than my forefathers did," the annals declare, adding:

> Cedar, boxwood, Kanish-oak from the lands over which I gained control—those trees which none of the previous kings my forefathers had planted—I took and I planted them in the orchards of my lands. Rare orchard fruits, which did not exist in my land,

I took and filled the orchards of Assyria [with them].

(Green 2010: 48–54)

Similar sentiments were echoed and amplified by a host of others, among them Assurnasirpal II (r. 883–859 BCE). In his palatial garden at Nimrud, he planted trees, cuttings, and seeds collected from all the lands through which he had traveled in the course of his campaigns. Some forty-one plant species are referenced, including cedar, cypress, box, juniper, myrtle, date palm, ebony, olive, oak, tamarisk, almond, terebinth, ash, fir, pomegranate, apricot, pine, pear, quince, fig, grapes, plum, mint, sycamore, and frankincense (Stronach 1990: 477). Assurnasirpal's successors included Sargon II (r. 722–705 BCE), who constructed at Khorsabad a new capital with "a park like unto Mount Amanus ... laid out by its side"—a marvel in this flat landscape—and in which were gathered "all the spice trees of the Hittite land" and "the fruit trees of every mountain." Amanus, symbol of the king's conquest of Hittite territories, is a mountain range in what is now south-central Turkey. Sargon II's son Sennacherib (r. 704–681 BCE) followed suit, constructing not far from his own palace "a great park like unto Mount Amanus," another garden requiring no small amount of engineering in order to sculpt the land (Green 2010: 59). Later, at Babylon, rose the fabled Hanging Gardens of Nebuchadnezzar II (r. *c.* 604–562 BCE), and not to be outdone, the Persian king Cyrus the Great built the garden city of Pasargadae as capital of his vast, newly forged empire (Giesecke 2015). In the case of Augustus, as with all these Near Eastern kings, his stewardship would ensure the burgeoning of the Empire. In the era of peace introduced by Augustus on the heels of decades of civil war, the Roman state could thrive and enjoy the fruits of a rediscovered Golden Age.

CONCLUDING REMARKS

As just one artifact from a single region has shown, acknowledgment of the human dependence on plants lay at the heart of cultural expression. Plants and their produce served as staple foods and were coveted luxuries conveyed from distant lands, driving both trade and exploration. They yielded medicine and raw materials for any number of endeavors. Plants served as offerings to the gods, ensuring their continued favor, and they were believed also to embody divinity. They were symbols for victory, birth, death, prosperity—in short, for all significant achievements, events, and "natural" milestones encountered in the course of life. As such they informed every medium of human creative expression, from literature to paintings that covered both humble and princely halls.

CHAPTER ONE

Plants as Staple Foods

JENNIFER RAMSAY, SARAH WALSHAW, AND
KARLA HANSEN-SPEER

Plants are an integral part of human evolutionary history. Indeed, we would not exist without them. Plants provide us with the air we breathe, fibers for textiles, building materials, and, of course, the food we eat. As we have evolved and learned to manipulate the world around us we have, consciously or not, changed the course of plant evolution. Humans began as foragers, moving from one location to the next gathering food when it was in season. For hundreds of thousands of years this strategy was effective and provided our ancestors with the nutrients they required while not altering the natural cycles of the plants. Ten thousand years ago something happened that would change the world forever: a shift from foraging to plant domestication, largely brought on by climate change. Arguably the most important development in human history, domestication changed people's occupation from hunter-gatherer-forager to farmer. The words "domesticate" and "domestication" come from the Latin *domus*, house or dwelling. To domesticate therefore means to bring into the household, which can include the house, garden, barn, field, orchard, and fold (Harlan 1992: 63). There is global archaeological evidence recovered for the shift to domestication from Southwest Asia, Africa, and China to the Americas and even to locations as remote as the highlands of New Guinea. The domestication of plants and animals allowed for the exponential growth of our human population, which literally changed, and continues to change, the face of the earth.

While we cannot know for certain how or why plants were brought into cultivation, and eventually domesticated, we can document the changes that plants underwent as a result of human selection. Practical benefits of the initial cultivation of cereals would have been to provide both a reliable source of food and a surplus in environmentally uncertain times. For some plants, an increase in the size of the edible portion is documented (e.g. maize), and for most plants domestication also resulted in traits that made it easier for humans to harvest and process them, including preventing natural dispersal (food portions remain on the plant), and synchronous ripening (heads of grain mature at the same time in one field). Humans selected for reduced toxins (yam, cassava), taste preferences (banana/plantain), and even color (potatoes). The wide variety of selected traits is a testimony to the creativity and innovation of early farmers.

This chapter focuses on domesticated plants that became staple food items. Staple plant foods would have made up a substantial portion of an annual diet, generally comprising plants that could be stored and were widespread in their importance. Below is a look at the culture history of staple plant foods whose origins can be traced to specific geographic regions. The chapter begins with Southwest Asia, which has yielded some of the earliest evidence of the domestication of cereals, such as wheat and barley. Significantly, Mesopotamia, part of the "Fertile Crescent" of that region, would become a "cradle of civilization" in large part due to the proliferation of staple crops. The discussion then shifts to Africa, the heartland of humanity, and an examination of staple resources prior to the medieval (Islamic) period, specifically south of the Mediterranean zone. We conclude with the Americas and prehistoric (pre-European contact) plants used by the indigenous peoples. Although not an exhaustive examination of staple food around the world, this chapter aims to provide an introduction to the great diversity of not only plant species on this planet but also the cultures that have played an integral part in the domestication process of our daily fare.

SOUTHWEST ASIA

In the context of this discussion, Southwest Asia—which has been variously defined depending on discipline—includes the Levant, Mesopotamia, the Arabian Peninsula, Anatolia, the Iranian plateau, Armenian Highlands, and Southern Caucasus. The modern countries the region encompasses are Bahrain, Iran, Iraq, Israel, Jordan, Kuwait, Lebanon, Oman, Qatar, Saudi Arabia, Syria, Turkey, UAE, Yemen, and the West Bank/Gaza. As in antiquity, this region is a crossroads between Africa and Asia and has a rich cultural history. This region saw the growth of societies that provided the modern world with the cultural characteristics that we use to define civilization today, such as cities, social stratification, long-distance trade, writing, monumental architecture, and religion. Fundamental to all these cultural advances is the production of surplus, often "staple" foods, specifically cereals, which enabled the development of these initial civilizations. To date, the earliest known civilization developed along the Tigris and Euphrates rivers in what are now Iraq, Syria, and Turkey. It is here that we see the rise of the first Sumerian city-states during the Ubaid and Uruk periods between 5000 and 3000 BCE. The emergence of these city-states and the hallmarks of civilization led to the rise of the great empires of the Assyrians, Akkadians, and Babylonians.

The end of the Pleistocene Ice Age approximately 11,500 years ago marked the beginning of the Holocene epoch, a time that encompasses all of written history, the growth of world civilizations, and advances in human technology. With the melting of the glaciers, the region had a comparatively moist and fertile environment in an otherwise arid and semi-arid western Asia. The warming of the planet at the end of the Ice Age saw humans influencing the development and reproduction of plants and animals, which led to domestication. Some of the earliest evidence for the domestication of plants comes from the Fertile Crescent. In the Neolithic, a period in the development of human technology beginning about 10,200 BCE, the region of Southwest Asia became home to the eight "founder" crops important in early agriculture as staple foods (i.e. wild ancestors to emmer wheat, barley, flax, chickpea, pea, lentil, bitter vetch). It is with these crops that we can begin to look at the culture history of staple plants. The following is an overview of what the archaeological and genetic evidence has uncovered about the history of groups of staple plants and how culture affected their development into staple foods.

Cereals

Cereals can be stored for long periods and have both a high nutrition value and relatively high yields. The mechanisms by which cereals became domesticated are still a topic of debate, but the results have had a fundamental impact on human society in Southwest Asia. Culturally, communities had to shift from a nomadic or semi-nomadic lifestyle of hunting, gathering, and foraging to a sedentary one. Although there is discussion around why people started to settle down —whether it was related to food acquisition, environment or religion—this move permanently changed human culture. Cultivating cereals required people to be responsible for planting, harvesting, and protecting these plants, which demanded a more settled existence. Once communities became sedentary, pottery production began, there was a rise in social complexity, and populations grew. People went from exploiting a wide range of plants and animals for sustenance to relying on just a few, what we consider staple foods today. Wheat and barley are two of the most significant staple cereals that were domesticated in Southwest Asia (see Figure 1.1).

Arguably, the most basic staple food in Southwest Asia has been bread, which is generally made from cereals, annual grasses grown for their grain, like wheat and barley (Zohary et al. 2012: 20). As a result, many farming cultures in the region today and in antiquity equated food with bread. Also consumed in soups and porridges, cereal grains could be moistened and then shaped or pressed into molds before baking (Allred 2013: 2085; Bottéro 2004: 62), or they could first be ground with hand mills or mortar and pestle into flours of varying fineness. Many different varieties of bread ranging from sprouted and flat breads to leavened and unleavened breads were produced and were baked using various methods, among them direct exposure to flames or glowing embers in fire pits. As still today, bread was also baked in domed ovens or, in the case of flat breads, on the exterior of a cylindrical ceramic *tinûru*. The so-called Mesopotamian Great Encyclopedia—a fragmentary, bilingual text in Sumerian and Akkadian of the early second millennium BCE—alone lists some two hundred varieties of breads distinguished

FIGURE 1.1 A seated vegetation goddess is greeted by three other deities. Stalks of grain sprout from the females, while tree branches grow from the two males, perhaps referring to a specific myth. Mesopotamian cylinder seal impression (Akkadaian), *c.* 2350–2150 BCE. © The Walters Art Museum.

not only by flour type but also by additives (such as spices, fruit, oil, milk, and beer), size, and shape, taking the form of balls, rings, and crescents as well as of body parts such as the heart, head, or ear (Bottéro 2001: 49; Ellison 1984: 91). Importantly, bread was consumed not only for nutrition but also for social cohesion, and as a result, it played a key cultural role in communities (Samuel 2002).

The main cereal involved in the production of bread is wheat, which was a universal cereal of Southwest Asia and currently is tied for second with rice in grain production, whereas corn is ranked first. There were several distinct species of wheat (genus *Triticum*) introduced into domestication, but the two most common today are several cultivars of bread wheat (*Triticum aestivum*) and hard or durum wheat (*Triticum turgidum* subsp. *durum*). Bread wheat is used for producing high-rising bread, and hard wheat was commonly used for pasta or low-rising bread. The first genetically domesticated wheat species appear to have originated from a wild species of emmer wheat (noted variously as *T. araraticum*, *T. turgidum* subsp. *dicoccoides*, or *T. dicoccoides*) that were widely distributed over much of Southwest Asia (Avni et al. 2017)). The earliest evidence for the collection of wild wheat (and barley) comes from the submerged, approximately 23,000-year-old Upper Paleolithic site of Ohalo II on the shores of the Sea of Galilee (Kislev et al. 1992). Wild emmer was likely gathered extensively by Upper Paleolithic peoples before it was cultivated, as attested by finds at the sites of Tell Abu Hureyra and Mureybit in northern Syria dating to approximately 12,700–11,300 years ago (Hillman 1975; 2000: Hillman et al. 1989; Van Zeist and Bakker-Heeres 1986; Van Zeist and Casparie 1968). Remains of domesticated wheat appear shortly thereafter at the sites of Çayönü and Cafer Höyük in southern Turkey (de Moulins 1997; Van Zeist 1972; Van Zeist and de Roller 1991). Wheat would play a major role in the spread of agriculture into Europe, and emmer became the principal cereal of early farming settlement in Southwest Asia.

Like wheat, barley (*Hordeum vulgare*) was a staple cereal crop in Southwest Asia and wild forms would have been collected long before it was domesticated. Barley was often considered an inferior staple to wheat, especially with respect to bread production; however, it withstands drier conditions, poorer soils, and some salinity, which make it a primary grain over a larger geographic distribution (Zohary et al. 2012: 52). Barley was the main cereal used for beer fermentation, and beer was definitely considered a staple food in antiquity. With the increase in population and the rise of urbanization, pathogens spread by contaminated water became a very serious health risk. Since its alcohol killed many detrimental microorganisms, beer was a safer alternative to water. Beer was generally produced by combining malted barley bread and water to yield a sweet liquid (wort) that was allowed to ferment by airborne yeast. Although beer may not have been a replacement for water, it offered the further, valuable benefit of being a more calorie-laden beverage that provided energy and nutrients for laborers.

Evidence for the harvesting of wild barley in the Levant dates back to 50,000 years ago at the Kebara Cave (Lev et al. 2005) and as previously mentioned, 23,000 years ago at Ohalo II (Kislev et al. 1992). Domesticated barley shows up on many archaeological sites in Southwest Asia beginning around 10,000 years ago, and chemical tests of pottery jars indicated that beer was being produced in Iran by at least 5,500 years ago. Literary sources also document the use and production of beer, and therefore barley, as a staple, as early as 5,000 years ago in a cuneiform text that indicated that workers in the ancient city of Uruk were paid in beer (George 2016) and 3,900 years ago in a Sumerian Poem honoring Ninkasi, the patron goddess of brewing (Black et al. 2006) (see Figure 1.2). Wheat and barley continue to hold great influence on our world today, as bread and beer are still some of the most important food items around the globe.

FIGURE 1.2 Administrative account of barley distribution. Mesopotamian Proto-cuneiform clay tablet *c*. 3100–2900 BCE, Sumerian, probably from Uruk. © Purchase, Raymond and Beverly Sackler Gift, 1988. Courtesy of The Metropolitan Museum of Art.

Legumes

Legumes (beans, peas, lentils, etc.) are an excellent complement to cereals as they are rich in proteins, as opposed to cereals, which are rich in starch. Many species of legume have long been considered staple dietary plants, and in traditional agricultural societies (and for modern vegetarians) they still serve as the main substitute for meat. Legumes in Southwest Asia were domesticated together with cereals (Zohary and Hopf 1973) and have the benefit of being able to fix nitrogen in the soil. By rotating legume cultivation with cereals the cultivator would have been able to maintain higher levels of soil fertility (Zohary et al. 2012: 75).

Lentil (*Lens culinaris*) was a companion to cereals and is among the oldest cultivated plants in Southwest Asia. Lentils contain about 25 percent protein and were often used in soup, made into bread, and cooked or boiled. Archaeological evidence points to the collection of wild lentils as early as wild barley and remains show up in similar contexts at both the Kebara Cave and Ohalo II sites in the Levant. Many sites in the region, such as Yiftah'el, Israel, and Jarmo, Iraq, have evidence of quantities of cultivated lentils starting at around 10,000 years ago (Garfinkel et al. 1988). Today lentils continue to be an appreciated food source and are a dietary staple of most countries in South Asia.

The pea (*Pisum sativum*), like lentil, ranks among the oldest legumes domesticated and was also a companion species to wheats and barley. Peas are high in protein and were a staple food item for early agricultural communities. Even today, the pea ranks as one of the world's most important legumes (Davies 1995; Smartt 1990). Like lentil and barley, remains of wild peas were found at the site of Ohalo II, and later domesticated forms were recovered from Tell Aswad in southern Syria, dating to approximately 10,500 years ago (Van Zeist and Bakker-Heeres 1985).

Another legume that was domesticated early on in Southwest Asia was the chickpea (*Cicer arietinum*), which was valued as a meat substitute for its protein content. Although not found in the same quantities as lentils and peas, evidence does point to a similar period of domestication as chickpeas have been found in deposits from the site of Çayönü that date to approximately 10,250 years ago (Van Zeist 1972). In later periods, like the Bronze Age, they are much more abundant. Perhaps the preparation method contributes to the poor recovery of chickpeas from early archaeological sites, since chick peas are often boiled, cooked in stews, mashed into paste (hummus) or ground into flour. Today they are still an important staple in Southwest Asia as well as around the Mediterranean basin and in South Asia.

The final staple legume that belongs to the group of early-domesticated plants in Southwest Asia is the faba (fava) bean (*Vicia faba*). The large seeds of these plants are high in protein and would have constituted a significant food group in antiquity. Faba beans are difficult to differentiate morphologically from some vetch species (e.g. *Vicia narbonensis*), and as a result their identification in the archaeological record is complicated. Although their conclusive identification is problematic, faba bean-like seeds appeared around 13,000 years ago in Iraq ed Dubb, Jordan (Colledge 2001), and 10,700 years ago in Djade el Mughara, Syria (Willcox et al. 2008). Even today in some Mediterranean and Asian countries the faba bean provides a staple protein supply to poor populations.

Fruits

When focusing on food in the Mediterranean it is difficult not to think of wine and olive oil. Today, as in antiquity, the products of fruit-bearing trees and vines, including olive, grape, dates, and fig, constitute dietary staples. These crops represent four of the biblical "seven species" (wheat, barley, and pomegranates are the other three), reflecting their enduring economic importance in Southwest Asia. Indigenous fruit trees and vines were generally domesticated much later in Southwest Asia than cereal grains, with the first definite signs of this transformation occurring approximately 7,000 years ago—although fig is the exception, possibly being the first intentionally cultivated plant on the basis of evidence from Gilgal, Israel, dating to approximately 11,000 years ago. The later domestication of the other fruit-bearing trees and grapes was likely a result of the greater investment required for tree crops than for cereal grains. Unlike cereal crops that can be harvested only a few months after planting, orchards take three to seven years to mature and need to be protected year around, requiring a settled lifestyle.

The olive (*Olea europaea*) is one of the most economically important plants in Southwest Asia (Zohary 1995a). Products yielded by the olive range from edible fruits to oil used for cooking, fuel, and ointment. There are many different olive cultivars that can roughly be divided into those used for oil or for fruit, and they differ in size, shape, color, and taste. Today's olive is thought to have originated from the wild olive (*Olea europaea* subsp. *europaea* var. *sylvestris*), which is indigenous to the Mediterranean and was likely "pre-adapted" to domestication because it can reproduce by simple vegetative propagation. Olives were very probably collected in the wild long before they were domesticated, as indicated by olive stone remains recovered from the Ohalo II site (23,000 years ago) and the Neolithic Nahal Zehora site in Israel that dates to approximately 8,000 years ago. Olive's domestication is complex and may have happened in many different regions around the Mediterranean independently. Definite evidence for domesticated olives comes from several sites in Israel and Jordan that date as far back as 6,800 years ago (Zohary et al. 2012). Bread and olives were, and still are, part of the staple diet of "developing" communities throughout Southwest Asia and the Mediterranean basin (Zohary et al. 2012: 116).

Grapevine (*Vitis vinifera*) makes a significant contribution as a staple plant in Southwest Asia as its fruits are rich in sugar (berries contain 15–25 percent sugar), can be stored and transported easily when dried as raisins, and their juice can be fermented into wine (Zohary et al. 2012: 121). Like the fruit-bearing trees, grapevine requires a higher degree of investment than cereal crops. Wild populations of grapes reproduce from seeds but cultivars are propagated by rooting twigs or by grafting. Fruit production also has to be regulated by regular pruning to keep the plants at a manageable size (Zohary et al. 2012). As with the olive, grape berries would have been gathered from wild plants prior to domestication. It can often be difficult to determine whether a grape pip is wild or domesticated, as there is a good deal of similarity in grape pip (seed) shape between the species (see Figure 1.3). As a result, documenting the first appearance of the domesticated grape is challenging. The earliest archaeological evidence for grape domestication comes from several Early Bronze Age sites in the Levant, such as Jericho, approximately 3,500 years ago. However, cuneiform texts from Mesopotamia record the presence of grapes, raisins, and wine as early as 5,000 years ago (Postgate 1987). Interesting information regarding the early uses of grape is also coming from tartaric acid residue detected on pottery vessels that points to the storage of wine, raisins, or vinegar 8,000 years ago (Cavalieri et al. 2003; McGovern 2003; Miller 2008; Singleton 1994).

The fig (*Ficus carica*), one of the original horticultural domesticates, has acted as a staple in the diet of Southwest Asia at least since the Early Bronze Age. Figs are a relatively fast-growing fruit crop, taking three to four years to produce fruit. Once populations became sedentary in Southwest Asia the fig provided a sugar-rich source of both fresh and dried fruit. Similar to grape and olive, fig was likely collected from wild fig trees, which are native to the region as well as to most of the Mediterranean basin (Zohary et al. 2012). Archaeologically fig can often be overlooked due to the small size of the pips, which makes it challenging to determine not only its presence at a site but also, if present, whether it is wild or domesticated. Fig pips have been recovered from the site of Gesher Benot Ya'akov, Israel, that date to 800,000 years ago (Melamed et al. 2011). Many sites in Israel, Jordan, and Syria attest to the presence of fig in the Early Neolithic, and at some (e.g. Netiv Hagdud and Gilgal, Israel), these finds may represent the first domesticated evidence for the species (Kislev 1997; Kislev et al. 2006).

The last of the first fruit trees in Southwest Asia to be domesticated was the date palm (*Phoenix dactylifera*), which requires a much warmer and drier, low-humidity climate than the others. As a result, most date horticulture takes place in the southern fringe of Southwest Asia. Once under cultivation, a date harvest can be very productive with yields as high as 100–200 kg (200–400 lbs) a tree with each fruit containing between

FIGURE 1.3 Carbonized grape pip (*Vitis vinifera*) from the Hellenistic period (323–32 BCE) at Tall-al-'Umayri, Jordan. Courtesy of Jennifer Ramsay.

60–70 percent sugars. Consequently, dates have been considered a staple food throughout antiquity (Zohary et al. 2012: 131). There is little doubt that the wild form of the date palm was indigenous to the hot, dry parts of Southwest Asia, and like some of the earlier noted species, date remains have also been recovered from Ohalo II dating to 23,000 years ago (Liphschitz and Nadal 1997; Zohary et al. 2012). We find many later archaeological sites in Egypt, Israel, and Iran that have indicated the presence of date stones starting around 8,000 years ago. Dates are also mentioned often in Sumerian and Akkadian cuneiform documents around 5,000 years ago, which lends support to their role as a staple food plant in antiquity (Postgate 1987).

The Spread of Southwest Asian Staple Crops

Methods of plant domestication and farming that had developed by the tenth millennium BCE in the Levant spread into Anatolia and, later, across the Mediterranean Basin, to the Greek mainland, the Danube Valley, Italy, Sicily, Sardinia, Corsica, Spain, and France as well as to central Asia. Inevitably, the Neolithic founder crops—among them wheat, barley, lentils, peas, and chick peas—followed and were intensively cultivated, as were olive, grape, and fig, which could be grown in a wider range geographical range than the date (McGregor 2015: 25–38; Zohary et al. 2012: 3–5). The extensive literary output of the Greco-Roman world, together with abundant artifacts, provides a wealth of information about the various uses and cultivation methods for staple food crops in the historical period, and what follows are necessarily limited remarks. The Greek historian Herodotus, for example, describes the cereals wheat and barley as the foods typifying civilized peoples (*Hist*. 2.26.2). However, in Greece, barley was for a long time more prevalent than wheat; although it was the preferred grain for making bread, wheat did not grow well in central and southern Greece and thus was imported, making it somewhat of a "luxury" staple (Carla 2013: 7099; Dalby 2003: 349). Wheat also did not grow well everywhere in Italy, where it was nonetheless of enormous dietary importance. From at least the second century BCE, the Roman government offered wheat to the urban poor at subsidized prices, and in 58 BCE Publius Clodius Pulcher (famously hated by the orator and politician Cicero) initiated a free distribution of grain via his *Lex Clodia Frumentaria* (Clodian law on grain). Indeed, provisioning the growing city of Rome with wheat became increasingly problematic, with wheat being largely sourced first from Etruria in Central Italy, then from the province of Sicily, and, later, from Egypt when that country fell into Roman hands in 30 BCE. Regarding bread making in Roman Italy, Pompeii's well-preserved bakeries (complete with ovens and mills), frescoes (one of which depicts a baker selling his loaves, see Figure 1.4), and carbonized remains of bread offer invaluable evidence.

As for the olive, source of the principal vegetable oil of the Mediterranean, its importance in Greece was symbolized by its status as a symbol of Athens, having been a gift to the city from its patron goddess Athena. Olives and olive production were no less important to the Romans, whose imperial expansion brought with it widespread production, import/export, and use of olive oil throughout the empire, including North Africa, where new irrigation methods sustained intense olive cultivation. Though also used as fuel, soap, and a base for perfumes, olive oil, available in various grades, found essential applications as a preservative and marinade as well as a medium for cooking in culinary contexts (Dalby 2003: 237–40; Mabberley 2017: 644; Rossiter 2013). As in the case of wheat, Pompeii and the Bay of Naples—where Vesuvius' pyroclastic flow preserved the accouterments of daily life—are among the sites that have yielded abundant evidence of olive processing in the form of stone mills and presses. The Vesuvian region also preserved valuable evidence of viticulture; in the case of the rustic villa of Publius Fannius Synistor at Boscoreale, for example, both wine and olive oil were produced.

In the Greco-Roman world, olive oil and wine were deeply ingrained in daily life and formed the backbone of the Greek and Roman economies. By the Classical period in Greece (fifth and fourth centuries BCE), wine was produced throughout the mainland and the islands, and wine, always diluted with water, was consumed at all levels of society (Dalby 2003: 350–60; Estreicher 2019). Wine consumption was ubiquitous, too, in the Roman world. It is estimated that in the first century CE, when the city of Rome's

FIGURE 1.4 Breadseller, Roman fresco from the House of the Baker, Pompeii. Museo Archeologico Nazionale, Naples, Italy. Photo by Azoor Photo. Courtesy of Alamy Stock Photo.

population had reached a high point, annual consumption of wine in Rome exceeded 85 million liters (22.5 million gallons). Many varieties of wine were produced throughout Italy, but wine was also imported from all corners of the Empire, including Spain, North Africa, Greece, and the northern provinces. In his *Natural History*, Pliny the Elder catalogues ninety-one grape cultivars, fifty varieties of high-quality wine, and some thirty-eight varieties of foreign wine (Unwin 1991: 93).

While the date and fig both were Southwest Asian staple fruits that became embedded in Greek and Roman mythology, agronomic sources reveal that they did not perform equally well when planted in Mediterranean zones other than their places of origin.

Roman author Varro writes of the fig that one could "pass a string through the figs when they are ripe for eating, and after they have dried they are tied in bundles and may be sent where we will; and there they are planted in a nursery and reproduce" (*Rust.* 1.41.3–6). The Roman encyclopedist Pliny the Elder (23–79 CE), on the other hand, notes that the best dates were grown in salty and sandy soils in both Judea and elsewhere and that the tree will bear fruit only in the hottest climates (*HN* 13.9). For the Greeks and Romans, the date thus was more delicacy than staple.

Concluding Remarks

Although not an exhaustive list, the plants covered above represent the most significant staple plants both culturally and nutritionally that were brought under domestication in Southwest Asia. Domesticated during the Neolithic period, these plants and their intensive year-round cultivation were the catalysts for the development of the Bronze Age civilizations in Mesopotamia and the Levant. The predictable and secure food source that these plants provided allowed for the emergence of craft specialization, since for the first time the whole population did not have to be involved in the acquisition of food. From this arose social stratification, the development of writing, government, and laws, as well as literature, art, and religion. Cultivation of Southwest Asian staple species spread throughout the Mediterranean basin in antiquity and, significantly, form a substantial component of the global agricultural economy today.

SUB-SAHARAN AFRICA

Staple plant foods are those starchy foundations that have been prepared for millennia in African kitchens to accompany side dishes and stews of seasonally available vegetables, fish, and meat. Our knowledge of culinary and agricultural practices from the period prior to 500 CE is limited because very few African societies used systems of writing before the introduction of Islam; notable exceptions include the hieroglyphs of ancient Egypt, hieratic script in Nubia, and Aksumite script. More often, scholars look to archaeological evidence such as tools, vessels, and most of all plant remains that document the use and domestication of African staples, among them sorghum, millets, rice, and yam. Evidence of movement of crops and time depth of use can also come from linguistic analysis of terms related to crops and cuisine. And, as is true for much archaeological interpretation, observations of these crops in use today can help infer past practices, particularly when they are produced in non-mechanized systems analogous to ancient agriculture.

It is very likely that sub-Saharan culinary staples today were also important when the plants used to prepare them were domesticated: millet flatbreads (like the Ethiopian *injera*), rice, and stiff porridges made from grains (*ugali* in eastern Africa) and tubers (*fufu* in central and western Africa). In many places staple grains are also brewed into beer with high nutrient value and variable, usually low, alcohol content. Archaeological evidence suggests that we can project beer brewing into at least the first millennium BCE based on vessel forms (e.g. Empire of Kush in present-day Sudan; Edwards 1996). Today, staple plants hold value beyond merely being an ingredient: some people will claim that they "do not feel full" if served a meal lacking in the favored carbohydrate, and staple crops can also serve as key drivers of economic, political, and social status. In his famous novel, *Things Fall Apart*, set in nineteenth-century Nigeria, Chinua Achebe writes: "Yam, the king of crops, is a man's crop." Igbo women, meanwhile, would have been responsible for growing cassava (from the Americas), cocoyam (taro and relatives) and legumes (Achebe 2009: 23). While we cannot be certain that these gendered associations have deep time depth, such observations can serve as hypotheses for understanding the social relations of food production among ancient societies.

This section discusses indigenous African plants used as staple resources prior to the medieval (Islamic) period, focusing on Africa south of the Mediterranean zone and the Sahara desert. The foodways of ancient Egypt and African Mediterranean ports often adopted wheat, barley, lentils, and other crops domesticated in the Near East (and discussed above). However, domesticates important south of the Mediterranean zone include the widely important millets, rice, and yam, as well as regional cultivars enset (a banana relative valued for its underground starch), fonio, and t'ef, both tiny grains. As mentioned, much of the evidence comes from the archaeological record since documentary sources and oral traditions are limited with respect to early food production. Archaeological evidence comes from botanical remains—charred or desiccated seeds and processing waste—as well as tools of agriculture and food preparation. However, it is important to note that plant remains may not be preserved archaeologically due to the soaking and boiling involved in turning millets into beer or stiff mash (Young and Thompson 1999). Where cereal chaff was used as temper in pottery production, archaeobotanists can often find identifying characteristics of plant types and domestication status by looking at ceramic chaff impressions under high magnification. In fact, ceramic chaff impressions currently form our earliest evidence for domesticated sorghum (see below).

Grains, Great and Small

The oldest known African plant domesticates are grains, among them millets and African rice (*Oryza glaberrima*). Millet is a generalized term for a group of small-seeded grasses commonly cultivated for food and/or fodder in African nations, as well as in India and China. These include the Asian domesticates broomcorn millet (*Panicum miliaceum*), foxtail millet (*Setaria italica*), and African domesticates which are described in further detail below. Millets are popular cultivars in the semi-arid tropics because they yield reliably in poor soils and can tolerate low rainfall. We lack direct knowledge of ancient cultivation methods in many areas, and it appears that hand harvesting predominated. Exceptions include use of the plow in Nubia and Aksum, the latter incorporating mixed-crop dryland (non-irrigated), ox-plow agriculture by about 500 BCE (Boardman 1999). Millet grains could be boiled, ground into flour, baked as flatbread, and fried as fritters, among other tasty options, as evidenced by grinding stones, and pottery bowl forms found in archaeological contexts. A drawback of millets is that they can be difficult to process once harvested; in particular, tiny grains like t'ef (*Eragrostis tef*) and fonio (*Digitaria exilis*) are increasingly in danger of becoming "lost crops" due to the arduous task of milling by hand (NRC 1996). Knowing this, we can appreciate how access to labor was a major limiting factor in the success of early farming communities.

The earliest grain domesticated in Africa is pearl millet (*Cenchrus americanus*), so-named because its teardrop-shaped grains are a pearly gray/green hue; it has also been referred to as "bulrush millet" owing to its tall spike that holds hundreds of grains. Pearl millet is found in domesticated form dating to 4,500 years ago in the Tilemsi Valley, Mali (Manning et al. 2011). Previously, the earliest known domesticated pearl millet had been found between 2,500 and 3,500 years ago in the woodland savannah zone of Ghana at the site of Birimi (D'Andrea et al. 2001). Interestingly, domesticated pearl millet is found by 1700 BCE in South Asia (Fuller 2003; Fuller et al. 2004), a pattern shared by sorghum and finger millet (see below), yet these staples are considered African domesticates due to the African distribution of their wild ancestors. Pearl millet plants and the knowledge to grow them likely spread along the Sahel (a zone directly south of the Sahara where rain-fed agriculture is possible but limited) from west to east, and eventually became part of cultural exchanges across the Red Sea and beyond (Manning et al. 2011). Based on its wide distribution throughout the savannah zones of eastern and southern Africa, pearl millet is thought to have spread through the movements of agro-pastoralist peoples speaking Bantu languages between 2,000 and 4,000 years ago.

Sorghum (*Sorghum bicolor*), "great millet," is arguably the most important and widespread crop in Africa, and it appears to have been in the past as well (see Figure 1.5).

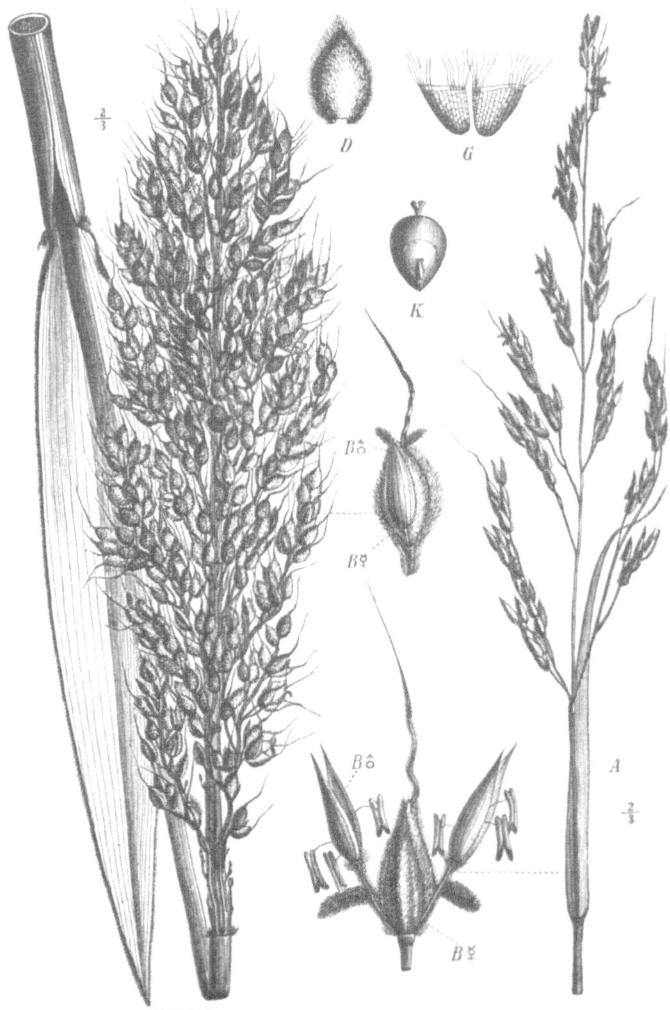

Fig. 5.
Andropogon Sorghum (L.) Brot. — A die Stammform (A. halepensis Brot.). — B eine Kulturform. B♂ männliche Blüthe; B☿ Zwitterblüthe; D Deckspelze; G Schwellschüppchen; K Frucht.

FIGURE 1.5 Great millet, *Sorghum bicolor* (L.) Moench [as *Andropogon sorghum* (L.) Brot.]). Illustration from A. Engler (1895), *Die Pflanzenwelt Ost-Afrikas und der Nachbargebiete*, vol. 1, 35. Courtesy of Missouri Botanical Garden.

It is successfully cultivated in both tropical and temperate environments, and has a tough constitution that allows it to thrive even in waterlogged conditions (NRC 1996: 127–8). Sorghum affords farmers great flexibility due to genetic variations that offer different varieties and ways of growing the plant as well as a wide range of uses as food and fiber—from popped grains, beer, and bread, to forage, hay, silage, and brooms. Wild sorghum is found across sites in the Sahara during the late Pleistocene, when the Sahara was wetter and greener than today, and large grasslands supported abundant herds (Marshall and Hildebrand 2002; Neumann 2003). The earliest form of domesticated sorghum has been found in the eastern Sahara, and impressions in ceramics show the presence of domesticated sorghum in the fourth millennium BCE in the Butana Cultural Group area, which is located just east of the emergence of the Blue and White Nile rivers (Winchell et al. 2017). These chaff impressions come from both wild-type and domesticated sorghum, showing that there was cross-breeding between wild and domestic plants, which suggests a long process of domestication.

Finger millet (*Eleusine coracana*) is small with dark reddish-brown seeds growing on a head sectioned into "fingers," resembling a human hand. It is widespread in eastern and southern Africa where it is ground into flour and used in bread and beer. Finger millet's tiny grains are laborious to process, so many communities are now turning to maize, despite finger millet's considerable nutritional profile (NRC 1996). Finger millet's earliest domesticated identification is from first century CE contexts at Kursakata, Chad (Klee et al. 2000). Early to mid-first-millennium CE grains found in Ethiopia at Ona Nagast (D'Andrea 2008) and in Rwanda (Giblin and Fuller 2011) suggest that finger millet spread south and east following its initial domestication.

A species of rice (*Oryza glaberrima*) was independently domesticated in western Africa, sometimes named "red rice" after the hue of the husk. French ethnobotanist Roland Portères (1970; 1976) hypothesized that it was domesticated between 2000–3000 BCE in the Middle Niger Delta, Mali, but the archaeobotanical evidence needed to verify this has only been collected in recent decades. The earliest archaeological evidence comes from the site of Dia in Mali, where domesticated rice dates to the mid-first millennium BCE (radiocarbon date 719–443 BCE; Murray 2004). Rice continued to be important in the regions of the Senegal River and the Inland Niger Delta, and could be considered the most important food in the early first millennium at the city of Jenné-Jeno, Mali (McIntosh and McIntosh 1979). Today African rice is usually restricted in cultivation, grown in lowland depressions or along rivers; elsewhere it has been widely replaced by Asian rice. In the past, however, it would have been distributed more widely in the forested areas from Senegal to Ghana and Mali (Logan 2012: 43). There are many cultivars of African rice, including floating rice, that can be cultivated in inundated freshwater and harvested by canoe, owing to the elongated stalk that keeps the growing plant head above water (NRC 1996: 26).

Small grains can hold importance beyond their size. T'ef (*Eragrostis tef*) is most famous as the ingredient of the labor-intensive indigenous Ethiopian flatbread, *injera*. T'ef has been found in domesticated form at archaeological sites in what is today the northern regions of Ethiopia and Eritrea prior to and during the Aksumite Empire, roughly 800 BCE–700 CE (Boardman 1999, 2000). It was recovered from the site of Aksum and from first-millennium BCE contexts at sites in the Asmara region of Eritrea (D'Andrea et al. 2007). Interestingly, t'ef domestication does not appear to have resulted in a consistently larger grain size than the wild *E. pilosa*; compared to other plant taxa, such as wheat and barley, priority in t'ef domestication appears to have been given to the production of satisfactory yields in suboptimal conditions, rather than to seed head size or high productivity (D'Andrea 2008: 562).

Fonio (*Digitaria exilis*) is a tiny grain cultivated in the Sahelian region of west Africa valued for its drought-tolerance and food qualities, including taste and wheat-like nutritional profile (NRC 1996: 59–62). Fonio is called "*acha*" in Nigeria, and today is widespread from Senegal to Chad; it remains popular cooked as cous-cous and porridge, brewed into beer, or ground into flour for bread. Black fonio/iburua (*D. iburua*) has a more restricted distribution but is locally important in the Jos Plateau in Nigeria as well as northern Benin and Togo. Fonio was important at some sites in the ancient Empire of Ghana (*c.* 400–1200 CE). It was recovered from the site of Culabel along with dominant grain taxa including pearl millet and a wild grass (*Panicum* sp.), and it was also found in small quantities at Jenné-Jeno (McIntosh 1995).

Tubers and Other Starches

Tropical lowland forest is the dominant environmental zone in equatorial Africa, reaching north to border the grasslands of western African nations from Cameroon to Sierra Leone. Here, cereal cultivation is constrained by overly wet conditions and lack of adequate sunlight, and grain agriculture has limited success. Yam is the indigenous "king of crops" in these parts of central and western Africa, and enset is cultivated in the Ethiopian highlands, with banana as an important newcomer that arrived by the first millennium BCE. Our knowledge of the domestication and expansion of these crops is limited by archaeological invisibility, which is only recently becoming remedied through phytolith (opal silica) and starch analysis of archaeological residues of these non-grain plants. The discussion below focuses on the current and historic importance of these starchy staples, using linguistic evidence where the archaeological evidence to anchor these culinary practices in deep time is lacking.

White yam (*Dioscorea rotundata*) and yellow yam (*D. cayenensis*) are indigenous to western Africa where they still serve as staple carbohydrates in the "yam zone" that today includes Nigeria, Benin, Togo, Cameroon, Ghana, and the Ivory Coast. Yams of all kinds need to be cooked to be palatable due to presence of a bitter and toxic chemical (dioscorene) in raw yam tubers. Yams can be cooked and eaten a variety of ways: roasted whole, fried as chips, or boiled whole and prepared into flour. *Fufu* is a common dish in western and central Africa: yam is pounded and mashed and served alongside a vegetable or meat stew. Yam is thought to have spread south through the tropic forest zone along with early Bantu-speaking fishing communities. Linguistic analyses of northern Bantu languages suggest a long time depth (at least 5,000 years, perhaps longer) of yam use (Philippson and Bahuchet 1994/5); however, linguistic evidence does not differentiate between wild and domestic yams, or between the two cultivated species. Other yam species made their way from Asia, such as the water yam (*D. alata*) by the sixteenth century, which does not seem to have replaced the indigenous species in west Africa in the same way that Asian rice did to African rice (Alpern 1992).

While difficult to establish its antiquity, enset (*Enset ventricosum*) is an indigenous African relative of the banana plant and is a staple in southwestern Ethiopia in the truest sense (see Figure 1.6). Not only is the pulp of the lower leaves and corm (underground storage organ) valued for a variety of food products, but farmers claim that the rest of the plant can be used as "our clothes, our beds, our houses, our cattle feed, and our plates" (NRC 2006: 174). Enset is not widely cultivated outside of Ethiopia. Studies of Ethiopian

FIGURE 1.6 Illustration of *Enset ventricosum* in an Abyssinian setting. Walter Hood Fitch (1817–92), *Curtis's Botanical Magazine* (1861), vol. 87. Photo by The Picture Art Collection. Courtesy of Alamy Stock Photo.

languages suggest that Omotic speakers introduced enset to eastern Cushitic speakers in the southern Ethiopian highlands by 1000 BCE (Ehret 2014: 241), but archaeological evidence is absent. Elisabeth Hildebrand's (2003) ethnoarchaeological work with Sheko expert farmers yielded important insights into the enset domestication process, which was found to have resulted in smaller corms, but more palatable food products.

Bananas and plantains (*Musa* species) are not indigenous to Africa (wild species are found in Southeast Asia to the Pacific), but after introduction they became beloved anchors of many food systems in the central rainforest areas and beyond (e.g. *matoke* in Uganda).

The process of domestication diminished the seeds of the banana, so archaeologists are reliant on its distinctive phytoliths, plant cells that have been naturally filled by silica and can be recovered from archaeological sediments. This makes bananas difficult to find archaeologically, and early evidence for banana is limited. The site of Nkang in central Cameroon appears to yield the best evidence for the earliest presence of *Musa* phytoliths. There banana phytoliths were recovered from an ancient pit (F9) associated with charcoal dated to 840–370 BCE (Mbida Mindzie et al. 2001).

Staples and Their Accompaniments

Cereals, yams, enset, and banana/plantain together formed the staple plants that undergirded African daily cuisine, yielding foods that include fufu, injera, jollof rice, matoke, and ugali. These foods were important in local cultural contexts; without having ingested the regional preferred staple, most people would not consider themselves fed. However, as is still the case today, most meals would have served a sauce or stew as an accompaniment to balls of pounded yam, flaps of injera, or handfuls of rice. These incorporate seasonal vegetables, frequently legumes and leafy greens—many of which are themselves African domesticates. Domesticated African legume taxa include hyacinth bean (*Lablab purpureus*; probably domesticated before its dispersal to India by mid-second millennium BCE), cowpea (*Vigna unguiculata*; by second millennium BCE), and Bambara bean (also called groundnut, *Vigna subterranea*; by 500 BCE) (Fuller and Hildebrand 2013). Various leafy green taxa known and valued today may or may not have been domesticated, but would easily have grown wild in disturbed areas, such as amaranth (*Amaranthus* species). Many trees also yield edible leaves, especially when young, including moringa (*Moringa oleifera*) and baobab (*Adansonia digitata*) (NRC 2006). Green leaves could be cooked in soups or stews, or be served as green accompaniments (similar to spinach). Watermelon was also an early African domesticate, found in sites in Egypt by 4,000 years ago, and possibly earlier in wild or semi-domesticated form in Libya.

Concluding Remarks

African domesticates and foodways have provided innovative examples to archaeologists in recent years. African patterns of plant domestication and the agricultural support for emerging cities and states are different than those original models developed for the ancient Near East. Domestication of grains appears to have occurred well after domesticated cattle appear, the inverse of the Near Eastern model (Marshall and Hildebrand 2002). Plant domestication did not always lead to larger grains or higher yields, but may have enhanced the availability or edibility or ease of cultivation. Social differentiation may have been based on surplus of food rather than the consumption of luxury foods in some cases. And the rise of cities and states may not have been predicated on rural producer sites feeding central consumer sites. These and other observations not only allow for productive thinking about African pathways to food production and the reception of introduced foods but also inspire re-examination of agricultural origins and dispersals around the globe (e.g. Logan 2012), including plant foods brought to Africa with Indian Ocean travel of Polynesians, such as banana, coconut, and Asian rice (Crowther et al. 2016), and through exchange with Near Eastern food cultures (barley, wheat, and peas, etc. noted above). Methodologically, Africanists combine archaeological evidence with linguistic data, and employ ethnohistoric records, ethnographic observations, and ethnoarchaeological research to build interpretive frameworks. African agriculture, past and present, is attracting more research every year, and scholars are pushing back the earliest known domesticated finds as we apply archaeobotanical methods to known and new archaeological contexts.

THE AMERICAS

Humans have lived in the Americas (North, South, and Central America) since before the end of the last Ice Age, *c.* 10,000 BCE. The culture of the region's earliest inhabitants is labeled "Paleo-Indian" by archaeologists and endures until approximately 6000 BCE in some regions. Subsequent cultural stages that vary geographically may be variously categorized as "Archaic" or "Formative" (which includes the Olmec of Mexico at 1200–400 BCE, for example); the time covered in this volume stops short of the well-known cultures of the Inca, Classic Maya, and the Aztec. People were grinding nuts and seeds 14,500 years ago in Monte Verde, Chile (Dillehay et al. 2015), long before they settled down to farm. Throughout the period of antiquity discussed in this volume, wild plants continued to make up a significant portion of people's diet. Even when people began to produce domesticated crops, wild plants were still used. For example, in the Early Agricultural period (2100 BCE to 50 CE) in Arizona cactus fruits and mesquite beans (*Prosopis* spp.) were processed along with maize (Diehl 2009). Wild plants added seasonal diversity to the diet with fruits, greens, tubers, and nuts.

Acorns (*Quercus* spp.), hickory (*Carya* spp.), and walnuts (*Juglans* spp.) are wild plants that could be considered staples because they can be stored and contain a significant source of calories. In California, for example, prehistoric people relied heavily on acorns from the coast to the Central Valley, as least as early as 500 BCE (Schulz and Johnson 1980; Tushingham and Bettinger 2013). Bedrock with holes and basins ground into them, and pestles for pounding nuts are commonly found and attest to prehistoric people's dependence on nuts. While wild plants continued to be widely used throughout the Americas in antiquity, plants that were domesticated became the most enduring and pervasive staples. Some of these plants, such as maize (*Zea mays* subsp. *mays*), were so much a part of daily life that they entered into the symbolic and ritual life of prehistoric people in the Americas.

Stretching from the Arctic Circle to Tierre Del Fuego, the Americas present every type of environment from cold, dry highlands to warm tropical forests. Yet despite this variety of ecosystems, there are a handful of crops that became staple resources from one end of the Americas to the other. Maize, squash (*Cucurbita*), and chenopod (*Chenopodium*) were used prehistorically in North America, through Mexico and Central America, and in South America. They are, of course, not the only food plants to have been domesticated: beans (*Phaseolus*), potatoes (*Solanum*), agave (*Agave*), and sunflower (*Helianthus*) are just a few of the plant foods used by prehistoric Americans. Maize, squash, and beans became known as "the Three Sisters" because they were commonly grown together; however, companion planting of the three crops was widespread (Mt. Pleasant 2006) and certainly stretches back to antiquity since evidence of these three crops are found together in archaeological sites across the Western Hemisphere, for example at Bat Cave in New Mexico (1100 BCE–100 BCE).

The focus here, however, is on maize, squash, and chenopod. These three crops are worthy of note because they are found in so many far-flung ancient sites across the Americas. In North America chenopod is commonly known as a weed called goosefoot, and in South America the domesticated chenopod is quinoa. Both chenopod and squash were independently domesticated several times in different places. They are plants that seem predisposed to domestication as they are weedy and thrive in the scuffed dirt and dump heaps that turn up around human settlements, responding quickly to selection for traits such as color and size. Maize, also known as corn in North America, is the exception: arguably the most important staple crop of the Americas, it was domesticated only once around 9,000 years ago and later became economically and culturally important across two continents. What made maize so successful as a staple was its ability to adapt to different climates. Some modern cultivars grow well in semi-tropical conditions resembling the environment in which it was domesticated. Others thrive in the hot, dry conditions of the Sonoran Desert. Still others grow in the short, cool growing seasons of Canada.

Maize

While the wild ancestors of many domesticated crops are fairly recognizable because they share clear morphological similarities, the wild ancestor of maize (corn) was debated for decades because nothing in the wild matched the large, multi-rowed cob of kernels familiar to us today. Some botanists thought the wild ancestor of maize was extinct; others thought the main contenders might be either of two grasses: *Tripsacum* or teosinte, or perhaps a hybrid of the two (Eubanks 2001; Mangledorf 1954). The debate among academics spanned the better part of the twentieth century. By scouring the Central Balsas Valley of Mexico, which was the presumed region of domestication (Beadle 1972; Iltis 2006; Piperno et al. 2009), and analyzing plant genetics (Benz 2006; Doebley 1990; Matsuoka 2002), researchers confirmed teosinte (*Zea mays* subsp. *parviglumis*) to be the ancestor of maize. Like maize, teosinte is a tall grass, but where maize has one main stalk, teosinte branches widely, and where maize has at least four rows of exposed, soft kernels (and commonly up to eighteen rows), teosinte has a "single" line of kernels encased in a hard fruit case. (Teosinte kernels are arranged distichously, or in two alternating ranks, which give the appearance of a "single" line of kernels.)

Evidence for the antiquity of maize domestication and its spread to regions throughout the Americas comes from a wide variety of sources, such as preserved cobs, kernels, pollen, phytoliths (fossilized plant tissue), starch residue, isotopes, linguistics, and genetics. An international team of researchers sequenced the genome of a 5,310-year-old cob with eight rows that was originally excavated by Richard MacNeish in 1962, the results of which give clues to the relationship between modern maize and teosinte (Ramos-Madrigal et al. 2016). The oldest directly dated maize comes from preserved cobs in Guila Naquitz cave in Mexico and dates to 6,200 years ago (Benz 2001; Piperno and Flannery 2001). Maize in the form of cobs and kernels has also been recovered in archaeological sites from Argentina to Canada. In North America, it reached the Southwest (e.g. Arizona) by 1550 BCE, even before ceramic technology was used in the area (Huckell 2006). By 500 CE, maize was well established in the western half of North America as a staple crop and later supported such groups as the Ancestral Pueblo of Chaco Canyon (800–1100 CE) and Mesa Verde (600–1300 CE). It took a longer time for maize to cross the Mississippi, and although it made its way to Ontario, Canada, as early as 400–600 CE (Crawford et al. 2006), it didn't significantly change agriculture of eastern North America until almost 500 years later, when maize played a significant role in the rise of the Mississippian culture, for example at the site of Cahokia (Illinois) around 1100 CE (VanDerwarker et al. 2013).

Heading south from Mexico into the tropics, direct evidence from archaeological sites for the use of maize is harder to come by because the constant heat and dampness allow microbes to thrive that quickly breakdown kernels and cobs. However, certain microscopic plant parts such as phytoliths, pollen, and starch grains can survive even tropical conditions. Such microbotanical evidence has extended the record of this important crop to places and times where macrobotanical evidence, such as kernels and cobs, has not been preserved. Starch grains and phytolith evidence for maize from nearly 9,000 years ago was recovered in the Balsas River Valley of Mexico, the homeland of teosinte and maize domestication. Likewise, maize starch grains and phytoliths dated to 7,100 years ago have been recovered from central Panama, and maize phytoliths from Ecuador date to 6,600 years ago (Piperno 2009; Piperno et al. 2009). Starch residues that are distinctive of maize have also been recovered from cooking pots in Ecuador from 4,000 years ago (Zarillo et al. 2008). Evidence is quickly accumulating that maize spread south from Mexico earlier than preserved cobs and kernels suggest.

Evidence for the presence of maize without accompanying finds of kernels raises questions about what exactly maize was first used for, and whether it was a staple or a supplement: was it a source of sugar from the stalk, or a fermented drink, or perhaps a grain? If ancient skeletons have been preserved, researchers can address this question by analyzing bone for stable carbon isotopes to discover the proportion of various foods in the diet. C-12, C-13, and C-14 are naturally occurring carbon isotopes, differing in their atomic mass due to different numbers of neutrons. Carbon isotope analysis of diets focuses on the ratio of C-12 to C-13; the isotopic ratio changes with an increase in the diet of certain plants and marine organisms. Through photosynthesis, plants take carbon dioxide out of the air and use light energy to turn it into plant food (carbohydrates). The first stable molecule most plants produce in photosynthesis contains three carbons, and so these plants are called C3 plants. But in hot environments, the C3 method of capturing carbon dioxide can lead to a lot of water loss for the plant. Some plants have a different method of capturing carbon that minimizes water loss; these plants end up with an initial compound containing four carbons, and thus are referred to as C4 plants. C3 plants capture carbon isotopes at a different ratio from C4 plants. Because maize is a C4 plant, stable carbon isotopes can be used to infer whether there was a greater reliance on maize. Interestingly, stable carbon isotopic analysis suggests that few people from Mexico to Peru relied heavily on maize until about 1000 BCE, although it was domesticated thousands of years earlier (Blake 2006: 67). One hypothesis that may explain this delay in carbon isotope signature is that *Zea mays* (either maize or teosinte) may have first been exploited for its sugary stalk and used to make alcohol, facilitating social interactions, celebrations, and ceremonies (69).

Today, tamales are a traditional food in Central America made with much love and labor, especially around Christmas time. *Masa*, a dough made from ground maize, is filled with savory meat, the whole pie is wrapped in maize leaves to form a rectangular package, and then steamed. Tamales, as well as tortillas, have been eaten for thousands of years in the region where corn was domesticated. They depend on an innovation in the preparation of maize flour called nixtamalization. Nixtamalization is the process of soaking kernels in an alkaline solution such as wood ash lye, which makes the kernels easier to grind, adds the vitamin niacin to the flour, and changes the chemistry of the maize flour so it can form dough (*masa*) that can then be made into tamales, or corn tortillas (Bressani et al. 1958; McGee 2004). On the basis of archaeological evidence such as comals (flat stones used to make masa and to cook tortillas), it is believed that people started using the nixtamalization process as early as 3500 BCE (Staller and Carrasco 2009). What researchers believe to be tamales were depicted in the late Preclassic Mayan murals at San Bartolo in Guatemala dating to approximately 100 BCE (Saturno et al. 2005).

With no large domesticated animals suited to pulling plows or carts in the Americas, technology for planting, harvesting, and processing maize was restricted to hand tools. Seeds would have been planted with a digging stick or stone hoe. Cobs would have been harvested by hand and the kernels dried and stored in baskets or clay pots. The dried kernels could be ground into flour and used to prepare a variety of foods. Grinding stones varied through time and by region; for example, large basalt two-handed *manos* (hand-stones) and *metates* (basins) attest to what must have been a daily activity in prehistoric southern Arizona (Cordell 1997). Ground stone tools were used into the time of European contact.

Whatever the reason people first started interacting with maize nearly 9,000 years ago, the plant became indispensable for prehistoric Americans. As a staple food across the Americas, maize did more than provide calories; for many, it was integral to the daily ordering of life. Evidence that maize was a central part of life comes from the widespread use of maize in rituals, stories, and iconography, and can be traced through linguistic evidence (Alcorn et al. 2006; Hill 2006; Stross 2006) (see Figure 1.7). In a creation

FIGURE 1.7 Terracotta funeral urn depicting the God of Maize Pitao Cozobi, grave 104, Monte Alban archaeological site (Unesco World Heritage List, 1987), Mexico. Zapotec civilization, second to sixth century CE. © DEA/G. DAGLI ORTI. Courtesy of Getty Images.

story of the Maya of Mesoamerica (Pre-classic period from 2000 BCE to 250 CE) for example, humans were created from maize dough. Maya divination used maize kernels, and rituals for everything from new house construction to prayers for healing called for maize offerings (Stroll 2006: 586). In antiquity, alcoholic drinks made from maize, such as *chicha* (known as *k'usa* in Amayra, a modern language spoken in parts of the Andes) (Chavez 2006: 628) would have also been used as part of rituals and social bonding.

Squash

Squash is a versatile plant that was domesticated independently several times precisely because of its versatility (see Figure 1.8). The flowers, seeds, and flesh of the fruit can be eaten, and the rind of some types can be used as a ladle, cup, or bowl. As part of the

FIGURE 1.8 Ceramic squash bottle, second to fourth century CE. Peru, Moche Culture. New York, The Metropolitan Museum of Art. © Gift of Judith Riklis, 1983. Courtesy of The Met Museum.

"Three Sisters" companion planting, squash complemented both the growing patterns and nutrition of maize and beans. Nutritionally, the combination of maize and beans provides the diet with complete protein, and squash adds minerals and vitamins such as potassium and vitamin A. While maize grows tall and offers a support for beans to climb, the aggressively sprawling low growth of squash with its broad leaves provides weed control by shading out emerging seedlings in a sort of natural mulch. The intercropping of these three plants is an example of indigenous knowledge that has its roots in the distant past (Mt. Pleasant 2006).

Squash, pumpkins, and gourds are common names for domesticated plants in the squash (Cucurbitaceae) family, which contains several species. The bottle gourd, also known as calabash (*Lageneria siceria*), was one of the earliest domesticates found not only in the Americas, but in Africa and Asia as well (Kistler et al. 2014). The squash and pumpkins we eat come from a range of *Cucurbita* species. The tropical pumpkin (*C. moschata*, which includes butternut squash) is native to South American and depicted on Moche pottery from Peru by 200 CE (Andres 2004). Other types of squash, like Hubbard or turban squash (*C. maxima*) are also native to the region. The green-striped cushaw squash (*C. argyrosperma*) was domesticated in Mexico and traveled to the American Southwest by 500 CE (Cordell and McBrinn 2012). Pumpkins (*Cucurbita pepo* subsp. *pepo*) come from Mexico and the *pepo* lineage from eastern North America (*C. pepo* subsp. *ovifera*) produces summer and acorn squashes (Sanjur et al. 2002; Smith 1998: 192).

C. pepo squash holds the honor as the oldest domesticated plant in the Americas. The earliest indicators of domestication in squash are larger seeds, and later, sweet flesh, a thicker rind, and the shape and color of the fruit were selected for (Smith 1997, 2006b). Seeds, rind, and peduncles (the "stem" on top of the fruit) have been recovered from Guila Naquitz cave in the valley of Oaxaca, Mexico and date to approximately 10,000 years ago (Smith 1997). In Ecuador, evidence of *Cucurbita* in the microbotanical form of phytoliths dates back to 12,000 years ago (Piperno and Stothert 2003). In the tropics phytolith evidence for the domestication of squash generally predates the evidence of larger, fleshy parts of the plant.

Because many cultivars of squash stretch from South America to North America, researchers debated whether there was one independent domestication event in Mexico, similar to maize, or whether there were several instances of independent domestication across the Americas. In particular, scrutiny focused on *C. pepo*, which is found in both Mexico and eastern North America. Subsequent biochemical analysis (Decker-Walters et al. 1993), and genetic analysis confirmed that the Mexican squashes and North American squashes were separately domesticated from different wild progenitors (Smith 2006a). The earliest evidence for domesticated squash from eastern North America comes from Phillips Spring in Missouri, and dates to 3075 BCE (Smith 2006a).

The eastern North American wild ancestor (*C. pepo* subsp. *ovifera* var. *ozarkana*) of domesticated squash has been found along the floodplains of the Buffalo River in Arkansas, in the southeastern United States. Like goosefoot and marsh elder (*Iva annua*), two other Eastern Agricultural Complex plants, the wild squash with its tennis-ball-size fruit is well adapted to colonize the floodplain (Smith 1998: 194), a landscape subject to frequent disturbance. Prior to domestication by humans, squash co-evolved with mega fauna such as mastodon and mammoth that also disturbed the landscape as they ate the baseball-sized bitter fruit and dispersed seeds in their dung. Squash seeds have been recovered from mastodon dung at the Page-Ladson site in Florida (Newsom et al.

2006). When creatures such as the mastodon became extinct, squash lost its "mutualistic dispersal partners" (Kistler et al. 2015). But those squash used by humans found another a source of seed dispersal and thrived while their wild counterparts declined (Kistler et al. 2015).

As part of the Three Sisters, squash was planted along with maize in fields prepared with bone or stone hoes and digging sticks. Because squashes come in such a variety of shapes and sizes, storage and cooking techniques also must have been numerous. A detailed and evocative interview in the early twentieth century with a Hidatsa Indian named Buffalo Bird Woman allows a glimpse into how squash may have been prepared in antiquity. She recounts, "The squash was sliced from side to side, not from stem to blossom. An old woman slicing squash would take up a squash, cut out the stem pit and the blossom, then turn the squash sidewise and slice, beginning on the side nearest her. The cut was made by pressing the bone blade downward into the squash as the latter lay in her palm." The squash rings were then spitted on a willow stick to dry so that they could be stored. Squash was cooked by boiling, steaming, or roasting (Wilson 1917).

Goosefoot (Chenopods)

While squash seeds became larger upon domestication, chenopod showed different markers of domestication. Its seeds remained small, but seed coats became thinner, and more seeds clustered on fewer stalks. It may seem intuitive that maize kernels or squash seeds became larger through the process of domestication: there is more to eat! But why would tiny chenopod seeds that stay the same size (about as big as the head of a pin) have thinner seed coats when domesticated? The answer is that the seed coat, or testa, protects the seed and influences how long it takes to germinate, or sprout. A thin seed coat reduces germination time, and increases success in seedbed competition where fast-sprouting seeds can "shade out their neighbors" (Smith 1995: 187). A thin seed coat as a marker of domestication has been verified by measuring seeds excavated from ancient sites with a scanning electron microscope (SEM).

The thickness or thinness of seed coats may have been *unintentionally* selected for by prehistoric people as they planted and harvested the early sprouting chenopods. However, the *color* of the chenopod seeds was a feature that prehistoric people *intentionally* may have selected for. Pale chenopod seeds are often correlated with thin seed coats, perhaps because the inner part of the seed becomes visible through an almost translucent seed coat (Bruno 2006: 34). Pale-colored chenopod seeds have been recovered from rock shelters in the Eastern US, where cool, dry conditions preserved caches for thousands of years (Smith 2006c: 107). Such preservation conditions are rare, however. Usually, seeds that have survived thousands of years do so because they are burnt, which makes color a difficult trait to trace archaeologically.

Chenopodium is a small grain, but is not in the grass family like maize or wheat. High in protein, this leafy grain was domesticated at least three times: in the Andes of South America, in Mexico, and in eastern North America. Although the domestication events were separated geographically, each time a suite of overlapping changes appeared in the plants: a thin, smooth seed coat and a change in shape where the seed margin becomes truncated (Bruno 2006: 38–42). In South America, the main type of domesticated chenopod is quinoa (*C. quinoa*). Using AMS dating (Accelerator Mass Spectrometry, used primarily in determining carbon 14 content), the earliest directly dated domesticated seeds come from Chiripa, Bolivia in the southern region of Lake Titicaca, and date to around 1500 BCE (Bruno and Whitehead 2003). Quinoa was one of the main crops intensively cultivated around Lake Titicaca, which provided the "agricultural foundations of the Tiwanaku state" (Kolata 1986), a major forerunner of the Inca beginning around 200 CE. Stone hoes, numerous at archaeological sites, were used to prepare raised fields in which to plant quinoa and other crops. Observation of modern native farmers in the Andes provides insight into how people in the past may have planted, harvested, and prepared quinoa. Today, farmers in the Andes sow quinoa by broadcasting the seed into furrows in the field. When the seeds have matured, plants are harvested by pulling up the entire stalk with seed head (Bruno 2008). Quinoa seeds can be eaten whole (toasted or boiled), prepared in soups or ground into flour. Before cooking, seeds should be washed to remove the bitter saponin that coats the grains (Lopez et al. 2011). Leaves of the chenopod plant can also be eaten and provide a nutritious leafy green. Adapted to high altitude, quinoa continued to be of primary importance in native agriculture at European contact and beyond (McCamant 1992).

In eastern North America, the domestic type of chenopod was *C. berlandieri* and was being grown as early as 1800 BCE in the late Archaic period, as evidenced from preserved specimens found at the Riverton site in eastern Illinois (Smith and Yarnell 2009). *C. berlandieri* was not the only domesticated plant from eastern North America. A suite of plants referred to as the Eastern Agricultural Complex includes marsh elder (*Iva annua* var. *macrocarpa*), sunflower (*Helianthus annuus* var. *macrocarpus*), as well as goosefoot (*Chenopodium berlandieri* subspecies *jonesianum*) (Smith 1998: 190). Chenopod, as part of the Eastern Agricultural Complex, was particularly relied upon as a local crop prior to the shift to maize farming in the Eastern US around 1000 CE. Chenopod was a grain crop that fit into a suite of both domestic and wild (acorns, hickories, deer, turkey) resources in a small-scale society. Cultivation of chenopod did not replace gathering of wild plants or substantially change how people moved about the landscape, but was an enrichment of the foodways of people in the late Archaic by 1800 BCE (Smith and Yarnell 2009). The eastern North America domesticated chenopod is now extinct, although it was grown well into the eighteenth century, as evidenced by observations of the French explorer, Le Page du Pratz (Smith 1998: 187). In South America, quinoa is still grown, and has even become a darling crop of the health food industry. In Mexico, chenopods today are a minor crop (Bruno 2006; Wilson and Heiser 1979).

Concluding Remarks

Maize, squash, and chenopods were plants that prehistoric peoples grew and ate with regularity in the Americas. They provided daily sustenance. Each crop began its journey to domestication as a wild plant that was managed by humans. Maize was domesticated once, and from the single instance of domestication, it spread out to the north and south, becoming varied in the size of its cobs and color of its kernels. Maize was arguably the most important plant that was domesticated in the Americas, not only because of its continued economic contribution to the world today, but because prehistorically it became so widespread. Where it went, human populations boomed. Varieties of maize were developed to thrive from the Peruvian highlands and the Sonoran Desert to the Eastern Deciduous Forest. Maize was ensconced in ritual and language. Squash and chenopod tell another story of cultivation. They were domesticated several times. Distinct but related species were grown all across the Americas. They were plants seemingly predisposed to domestication because they grow well in the kinds of disturbed places that humans create, such as clearings for shelter or trash heaps. Wild squash and chenopod were useful and easily accessible for manipulation by human groups.

These three staple plant foods of prehistoric America show the complexity of plant–human relationships: domestication follows a different path for every plant, whether selected domestic traits were for seed size or color or ease of harvesting, whether domestication was achieved once or several times, and how widespread a crop might become. The history of prehistoric America can be traced in the histories of maize, squash, and chenopod, three crops as widespread and adaptable as the people who inhabited the depth and breadth of two continents.

CHAPTER TWO

Plants as Luxury Foods

"Sweet herbs for curry"

ANDREW DALBY

At the end of the period covered in this volume it is easy to point to luxury in certain cultures, and therefore to attempt to identify luxury foods. At the beginning of the period it is more difficult. In the years leading up to 500 CE cultures flourished in several regions of the world that produced literature and art from which a concept of luxury is identifiable: a way of living that a few, rich in money or power, could enjoy without limit, and that the many who were unluckier could observe, envy, and occasionally sample. Whether to those who can command it or to those who cannot, luxury is meaningful only when compared with its absence.

The absence of luxury, in food terms, means eating and drinking to survive, consuming staple foods or any available foods that will keep the consumer alive with the least possible expenditure of resources. If resources permit any selectivity, one may, ascetically, still choose not to select; or one may select even among staple foods for taste and presentation; one may seek a wider range of flavors from food sources that are rarer, more labor-intensive, or fetched from a greater distance. Greater resources allow a more varied diet, one that has the chance to be healthier as well as costlier and more savory. The most resourceful, in the course of acquiring their luxury, have no choice but to share a proportion of it with the producers and traders who supply it, the priests, physicians, cooks, and attendants who mediate it, and even (as further demonstration of power) with all those others around who are best placed to envy it when on some occasion they are invited to partake.

All this can be seen in action, very roughly between 500 BCE and 500 CE, in abundant literary sources from classical Greece and Rome, early India, and China. In those regions, as regards plant foods that may count as luxuries—from fresh fruits to aromatic herbs and spices—the textual evidence is now supported by archaeobotanical finds (associated with the vestiges of staple foods) of rarer and not-so-staple foods, typically in smaller quantities. So, we can make use of similar archaeological finds in regions where there is practically no textual evidence at this date, the Americas, Africa, mainland and island Southeast Asia, Japan, and the Pacific, and class as potential luxuries those foods that were evidently valued but could not have served as staples. We can even look further back, to earlier societies in which social differentiation increased rather in parallel with the development of agriculture, aiming to identify the prehistory of luxury in a continuing use of wild foods in those Neolithic societies and in the early domestication of crops such as tree fruits that can never have been staples.

Can we look back further? Social differentiation, which allows luxury to stand out in relief and be appreciated, is scarcely visible in the profile of human societies before domestication and before the Neolithic revolution. However, even in hunter-gatherer societies there are occasions on which communities come together to enjoy a rare food. Before *Araucaria bidwillii*, the bunya nut tree, became a commercial resource, the bunya nut harvest, which occurs naturally at irregular intervals every three to six years, was the center of a festival that brought together Australian peoples from great distances to what is now eastern Queensland, temporarily ending any warfare (Sked and Macdonald n.d.; Smith and Butler 2002). Such harvests may be encouraged or controlled, as with an Australian tomato-like fruit *Solanum vescum*: it flourished on ground that was periodically cleared by fire by the Gunai of Gippsland, who gathered and enjoyed the fruit about every five years (Gott 2005). The communal enjoyment of such foods that are occasionally or seasonally available, not staples, and therefore luxuries, can be hypothesized for prehistoric as for modern hunter-gatherer societies.

Prehistoric examples can sometimes be identified. In these terms the most ancient luxury food currently known might be the spiny aquatic plant *Euryale ferox* with its meter-wide leaves (several feet-wide, see Figure 2.1). It once grew in the Near East and was of interest

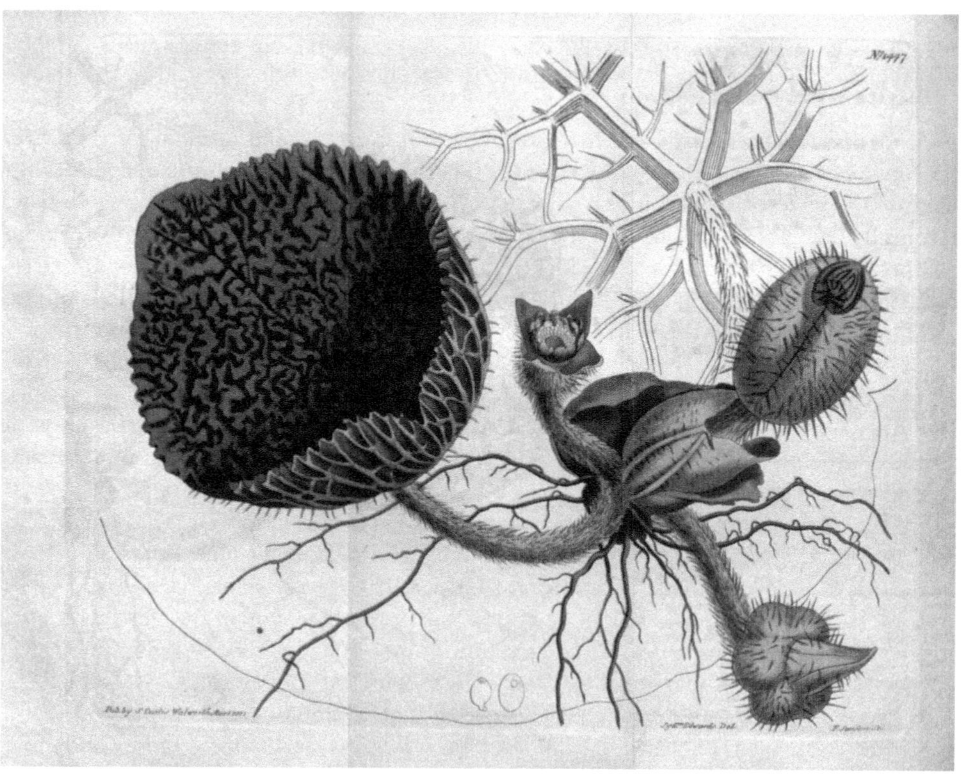

FIGURE 2.1 *Euryale ferox*, from *Curtis's Botanical Magazine* (1811), vol. 35, plate 1447: the leaf, which will open out to a disc of up to one meter across; the flower; the spiny, swollen pod ready to open and eject its seeds. Courtesy of Missouri Botanical Garden.

to the people (*Homo erectus* perhaps) who lived at Jacob's Ford in northern Israel around 800,000 BCE. It is good to eat but hard to harvest: only by observing the flower as it blooms and wilts does one know to dive for the ripe seeds which, when ejected from their spiny capsule, sink to the lakebed. Even at that date, apparently, some knew how to retrieve these useful nuts at the right time of year and would take the trouble to do so (Goren-Inbar et al. 2014). Hesitation is in order, however. The find at Zhoukoudian in the 1930s of charred hackberries (*Celtis barbouri;* cf. modern C. *bungeana*) associated with "Peking Man," dated somewhere between 500,000 and 250,000 BCE, is no longer thought to prove that Peking Man chose to bake and eat them (Binford and Ho 1985; Chaney 1935). The aromatic plants found with the body of "Shanidar 4," a Neanderthal burial of about 60,000 BCE on the western edge of the Iranian plateau, included an Ephedra species, and it was remarked that an Ephedra shrub, for example *E. intermedia*, was in all probability the classic source of the ritual drink, Sanskrit *soma*, Avestan *haoma*, of the Aryans around 1500 BCE when they inhabited the steppes to the north of Iran (e.g. Falk 1989). But there is some doubt that the aromatics are really to be associated with the burial (Leroi-Gourhan 1998; Sommer 1999).

What follows is a survey of ancient plant foods that demand the status of luxury, meaning that ancient peoples could perfectly well live without them, yet wanted them, took unnecessary trouble to get them, savored them and wrote about them. The survey is based on selected sources of information, of quite different kinds, each giving a perspective on plant foods in a particular culture. Needless to say, it cannot be complete.

EASTERN ASIA

Beginning with China, the beautiful and rebarbative *Euryale ferox* is relevant again. Native to eastern Asia in modern times, its seeds are among the plant foods identified at three Chinese archaeological sites of 6000 to 5000 BCE, Tianluoshan, Hemudu, and Kuahuqiao, close to the Yangzi delta, which also offer evidence of the beginnings of rice cultivation (Fuller et al. 2011). People at Tianluoshan additionally appreciated persimmons, peaches, *mei* or Chinese plums, and even, by about 3500 BCE, tea, the relevant find being not of dried leaves but of a row of planted roots of the species *Camellia sinensis* (Owyoung 2013).

The peach (*Prunus persica*) native to inland China is in evidence as human food nowhere earlier than at Kuahuqiao, before 5000 BCE. Peach stones of a domesticated type appear just slightly later in southern Japan, at Ikiriki, from about 4400 BCE, but the trees on which they grew surely had ancestors in China, in the Yangzi valley not far to the west, where domesticated peaches were certainly growing soon afterwards (Zheng et al. 2014). Downy and soft-skinned, the peach attracted the attention of physicians and poets. In the medical manuscripts recovered from tomb 3 at Mawangdui, the tomb of a bookish and health-obsessed nobleman who died young in 168 BCE, various prescriptions call for peach leaf, peach fruit, and even peach down (Harper 1998: 504–5). In Shijing, the "Classic of Songs" (mid-first millennium BCE?) peaches already tempt the reader:

> The peach is soft, fresh,
> Blush red are its flowers.
> This girl goes to marry,
> Fitting her house and home.

The peach is soft, fresh,
Swelling is its fruit.
This girl goes to marry,
Fitting her home and house.
The peach is soft, fresh,
Its leaves are glossy.
This girl goes to marry,
Fitting her home and man.

("The Peach," Shijing 6)[1]

By way of central Asia this fruit of Chinese origin had reached the distant Persian Empire by the late fourth century BCE, the moment of Alexander the Great's conquest. Like some other plants to be mentioned later, it was noted by the Greek scientists accompanying Alexander's expedition. Back in Greece it was duly described by Theophrastos, who collated these scientists' reports. The peach was familiar in Italy by the first century CE: a Pompeian wall painting includes a half-eaten peach.[2]

The mei or Chinese plum (*Prunus mume*) (see Figure 2.2), a versatile fruit in China and Japan, is said to have been domesticated much later than the peach. It is among the most culturally significant of a group of fruits that are both named in the earliest Chinese literature and found among the rich selection of foods placed in Tomb 1 at Mawangdui, the burial place of the lady Xin Zhui who died soon after her young relative (Buck 1975; Chang 1977: 56–7; Lee 2001 with illustrations of the plant finds). This is the tree whose blossom (*meihua*) now defines a spring season in eastern China and Japan. In the "Classic of Songs" the same tree's falling fruit hints that time passes:

Falling are the mei plums;
Seven are the fruits.
Many men seeking me:
Let it be a good one.
Falling are the mei plums;
Three are the fruits.
Many men seeking me:
Let it be now.
Falling are the mei plums;
The slant basket catches it.
Many men seeking me:
Let one speak.

("Mei Plums Falling," Shijing 20)[3]

Other tree fruits preserved in Tomb 1 at Mawangdui were the Chinese pear (*Pyrus pyrifolia*), the Chinese arbutus or strawberry (*Morella rubra*), and the jujube (*Ziziphus jujuba*). Among numerous aromatic plants were five that have been commonly used in food. Lesser galanga (*Alpinia officinarum*) is the least familiar, though still significant locally and important medicinally. Another relative of ginger (*Zingiber mioga*), used not for its root but its shoots and flowers, is mentioned in a poem exactly contemporary with the tombs at Mawangdui, a rhapsody by Sima Xiangru (mid-second century BCE), in a description of the marshy Yunmeng hunting park in Hunan near the mouth of the Yangzi:

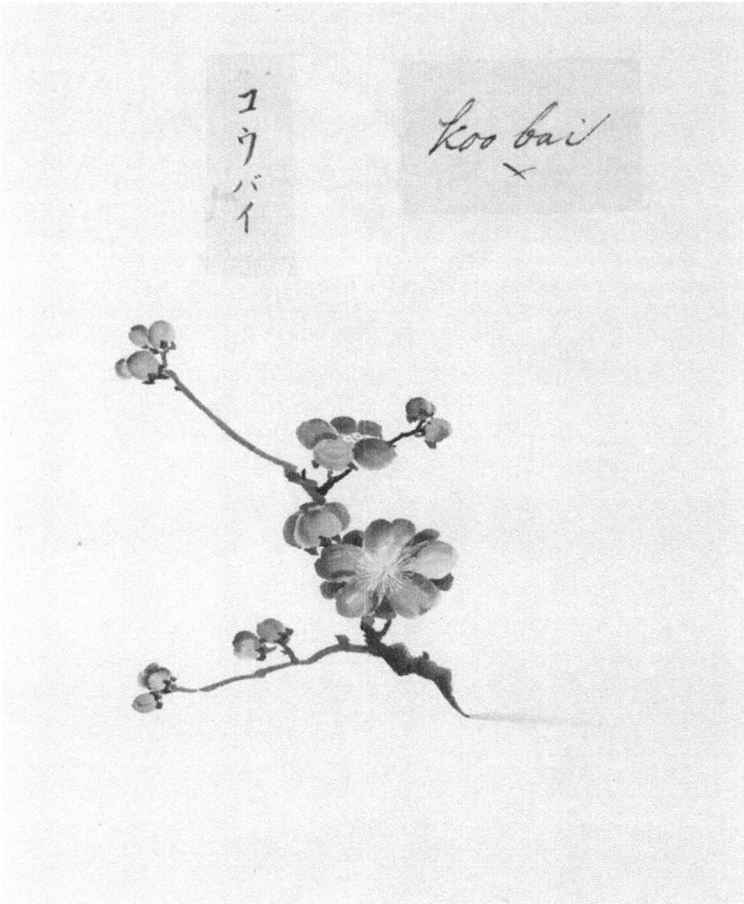

FIGURE 2.2 Mei or Chinese plum (*Prunus mume*). Pencil and watercolor (1823–9), Kawahara Keiga (1786–1860?). Nagasaki, Japan. The Siebold Collection. © Naturalis Biodiversity Center. Courtesy of Wikimedia Commons.

> To the east is the Basil Garden,
> With asarum, thoroughwort, angelica, pollia,
> Hemlock parsley, sweet flag,
> Lovage, selinum,
> Sugar cane, and mioga ginger.
>
> (Sima Xiangru, "Sir Vacuous")[4]

By about 110 CE it was grown still further north, in gardens near Nanyang, according to Zhang Heng in the "Southern Capital Rhapsody." Mioga ginger, "rarely cultivated" according to *Flora of China*, grows wild in moist places in Chinese mountain valleys, though not as far north as Nanyang (Wu and Larsen 2000). Ginger (*Zingiber officinale*) (see Figure 2.3), also found in Tomb 1 at Mawangdui and required in the Mawangdui prescriptions, was not yet familiar further north. Some 250 years later it is listed by Zhang Heng among kitchen supplies at Nanyang:

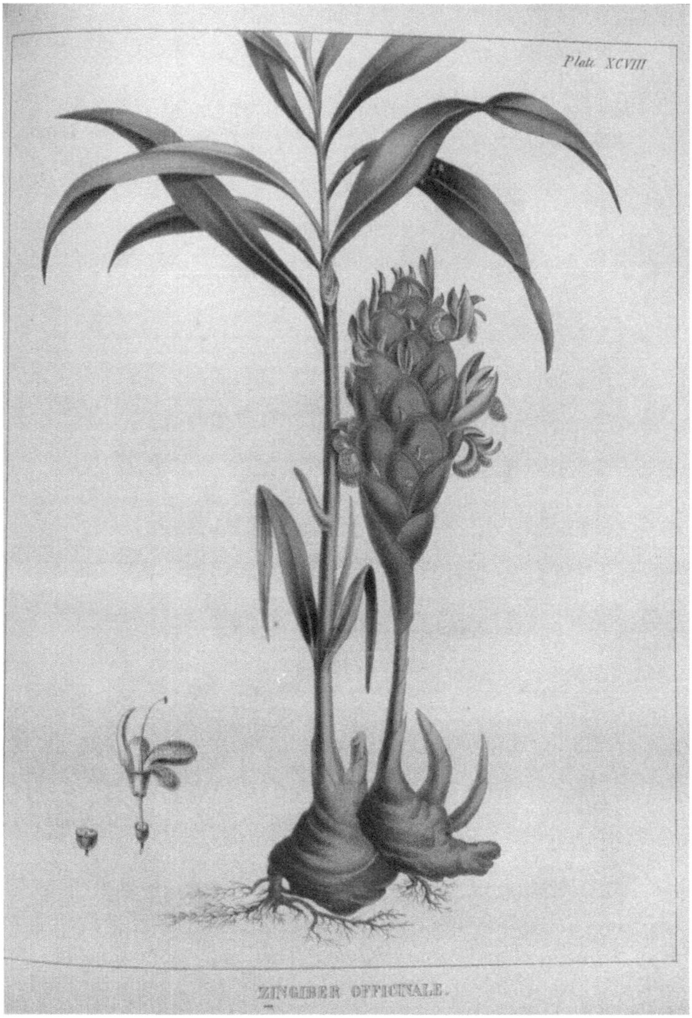

FIGURE 2.3 Ginger (*Zingiber officinale*) from Joseph Carson, *Illustrations of Medical Botany*, vol. 2 (Philadelphia: R. P. Smith, 1847), plate 98. Courtesy of Missouri Botanical Garden.

Sweet and sour flavors,
A hundred kinds, a thousand names:
Spring eggs, summer bamboo shoots,
Autumn leeks, winter rape turnips;
Perilla, evodia, purple ginger.

(Zhang Heng, "Southern Capital Rhapsody")[5]

And yet by that date ginger had spread from southern China far to east and west across tropical seas. Indeed, ginger plants may well have been tended onboard Indian Ocean vessels for their health-giving properties. In the first century CE, far to the west, the Greek pharmacist Dioscorides and the Roman encyclopedist Pliny were both familiar with ginger as a southern import, Dioscorides adding food notes to his medical advice:

> Ginger is a different plant from pepper, grown mostly in Trogodytica [Eritrea] and Arabia, where they make much use of it fresh, as we use leeks, boiling it for soup and including it in stews. It is a small tuber, like galanga, whitish, peppery in flavor and aromatic. Choose roots that are not worm-eaten. Some producers pickle it, to preserve it, and export it in jars to Italy: in this form it is very nice to eat, pickle and all. Its effect is warming, digestive, gently laxative, appetizing; it helps in cases of cataract, and it is an ingredient in antidotes against poison.
>
> (Dsc., *De mat. med.* 2.160)

The two remaining food aromatics from Tomb 1 at Mawangdui, cinnamon (or cassia) and Chinese or Sichuan pepper, are the best known of all in early literature from northern China. They are often paired, as in the early lines of Qu Yuan's great poem of the late fourth century BCE, which relies on the symbolism of aromatic plants throughout:

> The three kings of old were most pure and perfect:
> Then indeed fragrant flowers had their proper place.
> They brought together pepper and cinnamon;
> All the most-prized blossoms were woven in their garlands
>
> (Qu Yuan, "On Encountering Trouble")[6]

Chinese cinnamon (*Cinnamomum tenuifolium* syn. *Chekiangense*) indeed was and still is much used (Li et al. 2008). There are several species of Chinese pepper (see Figure 2.4), all of them typically found on mountain slopes as Qu Yuan suggests in a later reference. *Zanthoxylum bungeanum* is now identified as the best of them medicinally and the most typical of Sichuan cuisine. Certain species were already distinguished in the second century BCE: the Mawangdui medical prescriptions often demand this spice, twice specifying Shu pepper and once Qin pepper. Shu very roughly corresponds to modern Sichuan. Sure enough, Chinese pepper figures in Zuo Si's "Shu Capital Rhapsody," composed *c.* 275 CE, as one among the aromatic and medicinal plants growing on mountain slopes west of Chengdu, the regional capital then and now.[7] These and the many less-edible aromatics found in the tomb were all locally native (with the caveat that ginger is unknown in its wild form but probably native to southern China). The same observation can be made of the Mawangdui medical manuscripts: every item required in the prescriptions could be sourced locally. Noble households in second-century BCE China did not require exotic plants in their food or medicine, a point in which they would have differed entirely from contemporaries in Mediterranean lands.

Early Japan is notable in this context for two green plants that may count as luxuries, whose seeds, found at archaeological sites including Awazu, show that they were gathered from the wild as early as 9000 BCE (Crawford 2011; Matsui and Kanehara 2006). One, unusually versatile, is *Perilla frutescens*, known as *shiso* and by many other names, prized for its leaves fresh, dried, and pickled, its seeds whether ground or pressed for oil. The period at which plant cultivation began to be practiced in Japan is highly controversial, but perilla was perhaps cultivated by about 3800 BCE. It is mentioned in Zhang Heng's poem quoted above as one of the kitchen garden plants in early second century CE Nanyang in China, and has eventually spread in cultivation to Java in the south and India in the west. The second is burdock (*Arctium lappa*), another species with a long history of cultivation in Japan. Its burrs, which cling to hair and clothes, favored its early spread across the Old World. It is widely appreciated for its health-giving roots. Hence its mention in the

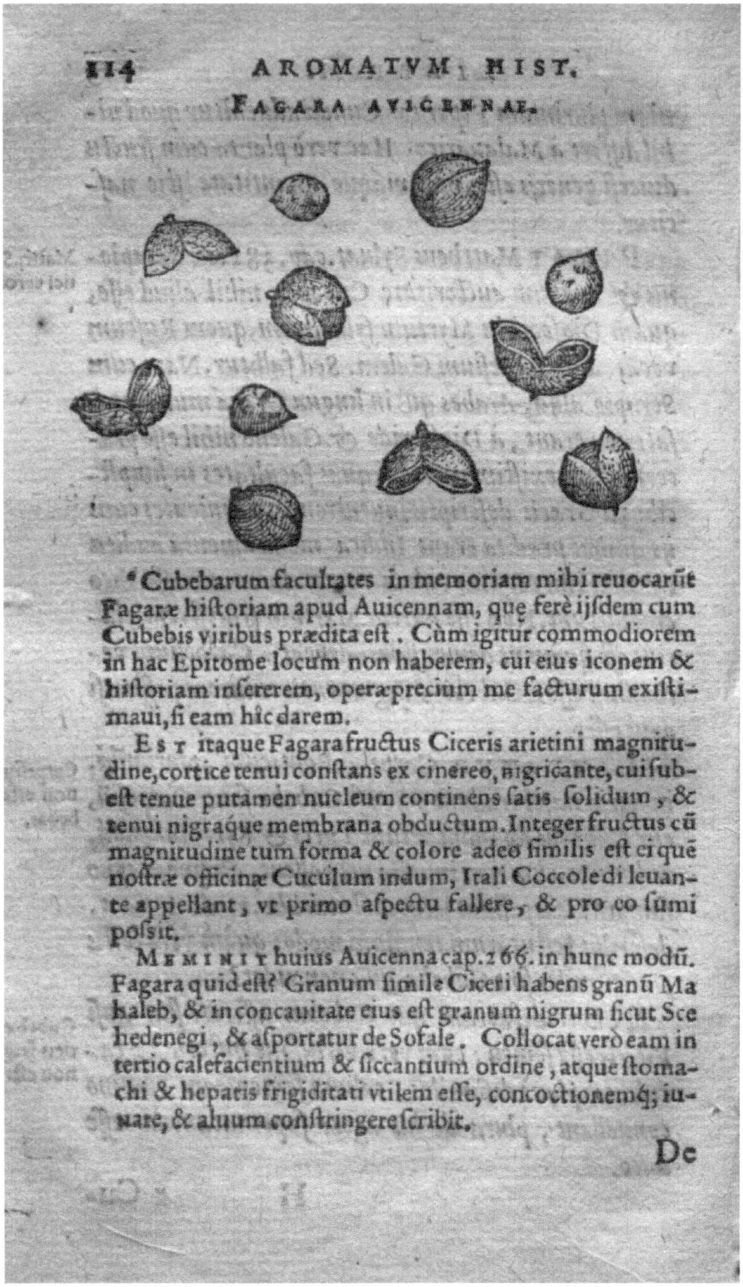

114 AROMATVM HIST.

FAGARA AVICENNAE.

* Cubebarum facultates in memoriam mihi reuocarūt Fagaræ hiſtoriam apud Auicennam, quę ferè ijſdem cum Cubebis viribus prædita eſt. Cùm igitur commodiorem in hac Epitome locum non haberem, cui eius iconem & hiſtoriam inſererem, operæpretium me facturum exiſtimaui, ſi eam hîc darem.

Est itaque Fagara fructus Ciceris arietini magnitudine, cortice tenui conſtans ex cinereo, nigricante, cui ſubeſt tenue putamen nucleum continens ſatis ſolidum, & tenui nigraque membrana obductum. Integer fructus cū magnitudine tum forma & colore adeo ſimilis eſt ei quē noſtræ officinæ Cuculum indum, Itali Coccole di leuante appellant, vt primo aſpectu fallere, & pro eo ſumi poſsit.

Meminit huius Auicenna cap. 266. in hunc modū. Fagara quid eſt? Granum ſimile Ciceri habens granū Mahaleb, & in concauitate eius eſt granum nigrum ſicut Scehedenegi, & aſportatur de Sofale. Collocat verò eam in tertio calefacientium & ſiccantium ordine, atque ſtomachi & hepatis frigiditati vtilem eſſe, concoctionemq; iuuare, & aluum conſtringere ſcribit.

De

FIGURE 2.4 *Fagara Avicennae* (Sichuan pepper) depicted by Carolus Clusius in a side-note to his translation of Garcia de Orta's *Aromatum, et simplicium aliquot medicamentorum apud Indos nascentium historia: ante biennium quidem Lusitanica linqua per dialogos conscripta* (Antwerp, 1567), 114. Courtesy of Missouri Botanical Garden.

Synonyma plantarum barbara, a Roman handbook of medicinal plant names in which local terms for burdock are given in Dacian (Romania) and Gaulish (France). Hence the find of burdock seeds at an archaeological site in Roman northern Gaul (Marinval et al. 2002). Hence, finally, its traditional use in recent England in dandelion-and-burdock root beer.

The best-known citrus fruits are all native to eastern and southern Asia. No fewer than three species are listed among other fine orchard fruits in Sima Xiangru's second century BCE rhapsody on Shanglin hunting park at Xi'an in northern China, where they no doubt received special care during winter frosts:

> And then
> Black kumquats that ripen in summer,
> Yellow mandarins, [bitter] oranges, pomelos,
> Loquats, wild jujubes, persimmons,
> Wild pears, apples, magnolias,
> Date plums, box myrtles,
> Cherries, grapes,
> Dark poplars, dwarf cherries,
> Plums, and litchees
> Are spread among the rear palaces,
> Form rows in the northern orchards,
> Stretch over the hills and mounds,
> Descend to the level plain.
>
> (Sima Xiangru, "The Imperial Park")[8]

Of these three citrus fruits, then, the smallest and the largest are both pleasantly sweet fruits, easy to eat, produced by natural species of genus *Citrus*. The mandarins or tangerines (*C. reticulata*) native to Japan and eastern China, have always been highly popular locally and did not move far beyond this orbit until recent times (see Figure 2.5). The big pomelo (*C. maxima*) comes from much further south, in mainland Southeast Asia. Almost contemporary with this poem are the two earliest archaeological finds of pomelo seeds, from two ancient trading cities close together on the Malay peninsula in southern Thailand near the Burmese border, Phu Khao Thong looking westwards across the Bay of Bengal and Khao Sam Kaeo across the hills close to the shore of the Gulf of Siam. The pomelo was among the fruits carried eastwards across the Pacific by Austronesian peoples (Blench 2004; Castillo et al. 2016). There is one other widespread natural *Citrus* species, the aromatic but not very edible citron, *C. medica*. Native to Burma and Assam, known in early southern China (Li 1979: 127–8), this was the first of the genus to be carried westwards by gardeners. It had reached Media (modern Kurdistan) by Alexander the Great's time (late third century BCE) and is fully described by Theophrastos as "the so-called Median apple":

> The apple is not eaten, but it and the leaves of the tree are powerfully scented. If it is put with clothes it keeps moths off; is useful too if one has drunk poison—because taken in wine it turns the stomach and brings the poison up—and to sweeten the mouth—because the inside of the fruit, cooked in broth etc. and squeezed into the mouth and sucked, makes the breath sweet.
>
> (Theophr., *Hist. pl.* 4.4.2)

FIGURE 2.5 *Background*: a tangerine and a lemon in pots; *foreground*: grafting knives as made in the 1640s (and in almost identical form today). From Ioannes Baptista Ferrarius, *Hesperides, sive De malorum aureorum cultura et usu libri quatuor* (Rome: Hermannus Scheus, 1646), 111. © Biodiversity Heritage Library, Courtesy of Getty Research Institute.

It was introduced to Mediterranean lands soon afterwards: this was to be the only citrus fruit known in the Roman Empire, described in several literary texts and seen in wall paintings and pollen from Pompeii (Andrews 1961; Dalby 2003: 88; Mabberley 2004: 482).

The other well-known citrus fruits originate as hybrids, and the oldest of these is the bitter orange (*Citrus x aurantium*). Again, almost contemporary with its mention in Sima Xiangru's poem is the first archaeological evidence, for a bitter orange was found in Tomb 2 at Mawangdui. Was it a special favorite of Marquess Li Cang (d. 186 BCE) but not of his widow, whose tomb contained so many other luxury foods? Not far south of Mawangdui are the southern Chinese coastal lands where the parent species, the pomelo and tangerine, might have grown side by side.[9]

Looking south and west from China we encounter the sugar cane, source of a traditional spice that in most people's eyes is one of the most obviously luxurious of foods. *Saccharum robustum* was perhaps domesticated in New Guinea; there and in Melanesia its sap was something approaching a staple food. Gradually it was taken west to Southeast Asia, and there perhaps was crossed with *S. spontaneum* to produce *S. officinarum*, the species now cultivated around the world. The sugar cane was known in southern and central China by the second century BCE and had been carried west to India by the same period (Dalby 2000a: 21–6).[10] That was when the Greek geographer Eratosthenes (relying on reports from Alexander's expedition or from a later exploration) wrote of "a big reed, sweet both by nature and by the sun's heat" (quoted by Strabo, *Geography* 15.1.20). By the first century CE something like granulated sugar had been invented, as described by the pharmacist Dioscorides:

> There is also a substance called sakcharon, a sort of crystallized honey, in India and Arabia. It is found in reeds. It is not unlike salt in its texture, and can be crunched between the teeth like salt. It is laxative, good to drink dissolved in water, beneeficial in bladder disorders and for the kidneys.
>
> (Dsc., *De mat. med.* 2.82.5)

Is the betel quid, the combination of *Piper betle* leaf and *Areca catechu* nut, incorporating slaked lime, to be regarded as a luxury food? It is an integral part of hospitality and its psychoactive constituents are swallowed, even if the leaf and nut are not. It is addictive, and the coloring effect on the teeth of those who chew it is sufficiently admired to have been imitated in ancient and modern times. *Areca catechu* was perhaps native to the Philippines, *Piper betle* possibly to New Guinea. There is much conjecture but no reliable archaeobotanical evidence of where and when the two species came together or of their early use. The best clue to the historical origin of this "food" is therefore linguistic: it was in all probability known to the early speakers of Austronesian languages when their migrations had extended no further than the Philippines, around 4000 BCE. In the second millennium BCE they were spreading the two plants and the betel habit eastward across the Pacific, while others were spreading it westward into mainland Southeast Asia (Zumbroich 2007). Legendary narrative sources for the life of the Buddha in northern India in the sixth century BCE describe betel chewing, but these cannot reliably give us more than a terminus ante quem: the habit had become known in Sri Lanka by about the fifth century CE when the texts were written. There is also a quite different textual source from India, the medical compilation *Carakasamhita*, but the version that is now known did not reach its final form until about the same period. We can be reasonably confident that by the fifth century some Indians were reading the following advice, which in essence suggests the use of a betel quid without the slaked lime but with additional aromatics:

> One desiring clarity, taste and good smell should keep in his mouth the fruits of nutmeg, musk seed, areca nut, cubeb, small cardamom and clove, fresh betel leaf and exudate of camphor.[11]
>
> (*Carakasaṃhitā*, "Sūtrasthāna" 5.76cd–77)

SOUTHERN ASIA

The exacavation of Belilena cave in Sri Lanka showed that around 30,000 BCE, at a period long before human encouragement of useful plants had turned into cultivation, foragers were interested in breadfruit (perhaps *Artocarpus nobilis*), bananas, and candlenuts among many other food sources. These tropical fruits, however, with ancient and complex domestication histories, may legitimately be counted as staples, at least in their tropical heartlands. Later archaeobotanical evidence for prehistoric southern Asia is scanty. It is, as with China, reinforced by literary evidence, but the dating of early Indian literature is difficult in the extreme (Achaya 1994 on literary sources; Fuller 2002 on archaeobotany).

If any region of the ancient world excelled for the variety of its foods that region is surely south Asia, unrivalled in its range of fruits and herbs, enviably placed astride the Indian Ocean trading routes. Yet popular literature suggests that for most people in ancient India daily food was simple enough, as it might be today. Two brief quotations from a collection of religious tales demonstrate the claim, one of them supplying the subtitle for this chapter. The Jātaka stories of the Buddha's former lives were written in Pali or translated into that language in the early fifth century CE. In the 42nd tale in the series a cook prepares an everyday pickle or marinade, taking salt, ginger, and cumin (*Jātakaṭṭhakathā* 42 "Kapotajātaka"). The two aromatics are indeed not native to India—ginger had arrived from the east as we have already seen, cumin was introduced from the west—but they had evidently, even at that early period, become naturalized constituents of Indian cookery. In the first tale of the same collection the Bodhisatta, a merchant in Benares, skilfully arranges that his well-manned caravan of trade goods will reach its destination safely, ensuring that among other incidental amenities "my men will find a fresh growth of sweet herbs for curry where the old ones have been picked" (1 "Apaṇṇakajātaka"). He is wise but not wasteful, so why does he care about the herbs? Because *sūpa*, the "curry" of this translation, far from being a superfluous sauce, is the laborer's daily staple, a bean or chickpea soup. The men must forage for fresh greens to complete their daily meals, remain healthy and strong, and deliver the trade goods in due time. "Luxury foods" as well as staples are needed for a proper diet.[12]

The Indian date (*Phoenix sylvestris*) was harvested at Harappa, the principal site of the Indus Valley civilization, which flourished in the late third millennium BCE (Ahmed 2014). The fruit and especially the sap are important sugar sources—when tapped the sap can continue to flow for several months—and there are many mentions in early Sanskrit texts associated with northern India in the first millennium. That is true, too, of three other fruits little known beyond the subcontinent, the Indian fig (*Ficus benghalensis*), the bael (*Aegle marmelos*), and the emblic myrobalan (*Phyllanthus emblica*). The first three recur in the epic *Mahabharata* and in classical Sanskrit literature; all four are noticed in Pali Buddhist texts of the early centuries CE. This myrobalan, important medicinally, is one of the plant foods for which evidence was found at the archaeological site of Navdatoli near the northwestern coast, dating to the mid-second millennium BCE. Seeds of the Indian jujube (*Ziziphus mauritiana*) were found there too (Blench 2004: 44).

Several fruits find their first mention in the *Mahabharata* and literature contemporary with it, of which the best known and most luxurious is the mango (*Mangifera indica*). Alongside several wild and locally cultivated relatives, it originates in the mountainous country between Assam and northern Burma. How long ago it was domesticated is not known, but mango charcoal (which suggests carefully tended trees) was recognized with probability, alongside the charcoal of a citrus species, at the archaeological site of

Sannarachamma in southern India and dated to about 1300 BCE (Blench 2004: 43; Fuller et al. 2017). At Senuwar in northern India, a site of the early first millennium BCE, there was mango charcoal again and also charcoal of the jackfruit, Artocarpus heterophyllus (Fuller 2002: 322), a species related to the breadfruit but native to southern India and perhaps Bengal.

Jackfruit, bananas, mangoes, Indian jujubes, and coconuts, all familiar in eastern India and Sri Lanka, naturally figure in Buddhist texts from that region. More surprisingly, although specimens may never have reached the ancient Mediterranean, all five are described in ancient Greek, the first four very briefly by Theophrastos in a summary of reports from Alexander's expedition:

> There is another tree, of great height, with fruit that is remarkably good to eat and very large: the Indians who are philosophers and go naked use it for their food. There is another whose leaf, about two cubits long, oblong in shape, like an ostrich feather, they fasten on their helmets. Another also has fruit that is long, not straight but twisted, and sweet to the taste. This causes colic and dysentery, hence Alexander ordered that it was not to be eaten. There is another whose fruit is like that of the cornel.
>
> (Theophr., *Hist. pl.* 4.4.5)

In translating into Latin this passage on "trees known from Alexander's conquests," Pliny expands the sentence on bananas, presumably from a version unknown to us: "the fruit grows out of the bark, and is remarkable for the sweetness of its juice: each one is enough for four people" (Plin., *HN* 12.21–4).[13]

More widespread than these other fruits, even in ancient times, was the coconut, which, since it floats, was able to transplant itself using ocean currents, and was probably domesticated on two separate occasions, once in Indonesia, once in southern India or Sri Lanka (Blench 2004: 45; Gunn et al. 2011). Northwards from Indonesia it was known in early China (where, planted far beyond its comfort zone, it is listed in Sima Xiangru's rhapsody "The Imperial Park" (cf. Li 1979: 115–17). In Sri Lanka and India, familiar in Buddhist texts, it was noticed in the early sixth century CE by the Greek trader Kosmas, who drew a sketch of the pepper vine and the coconut palm and described both plants:

> This is the tree that bears pepper. Each one climbs some other tall tree that is not fruit-bearing, because owing to its slenderness it is very weak, like the slender branches of the vine. Every bunch of the fruit is protected by a double leaf. It is deep green, the color of rue.
>
> The other tree bears the so-called nargellia, that is, the large Indian nuts. It is no different from the date-palm, except that it is larger in height, in girth and in the length of its fronds. It produces only two or three spathes, each of three nuts. Their taste is sweet and very pleasant, like that of green walnuts. The nut at first fills with very sweet water which the Indians drink in place of wine … If the fruit is gathered ripe and kept, its water gradually solidifies on the shell, while the water in the middle remains fluid until there is none left. If it keeps too long the fruit thus solidified becomes rancid and can no longer be eaten.[14]
>
> (Kosmas Indikopleustes, *Christian Topography* 11.10–11)

Black pepper (*Piper nigrum*), the subject of Kosmas' first paragraph, is of all Indian luxury food plants the one that has been most sought after in distant markets. This was the case even in ancient times, but the names it was given betray the fact that when long-distance trade began to make black pepper available it was seen as an alternative to something already known. It is, as Kosmas says, a small fruit growing in bunches on a vine, and is most useful and profitable when fully ripe and sun-dried. In southwestern China, when first encountered as a trade item that had crossed the mountains from India, black pepper was correctly understood to be a seasoning from a plant related to the betel vine and was therefore called "sauce-betel" (Li 1979: 46–53). In classical Greece and Rome, at the other end of the Old World, black pepper became familiar in about the second century BCE as the most obvious result of the beginning of regular monsoon sailings across the Indian Ocean. The monsoon was known as *hippalos* in Greek from the name of the Greek mariner who was credited with opening the route. Trade by sea being so much faster and cheaper than caravans of camels or oxen, black pepper immediately began to eat into the market for long pepper (*Piper longum*), whose tight bunches of tiny fruits are stronger, better according to some, and native to northeastern India from which they had previously reached Europe in small quantities at high cost.[15] Long pepper was known in Greece as *piperi* from its Prakrit name *pipali*. In India black pepper has a different name, *marica*, but in Europe the name of the old, expensive spice was extended to its replacement. Pepper was the one indispensable exotic condiment of Roman cuisine, stored in great quantities in Imperial warehouses, required for almost every dish in the fourth-century cookery book *Apicius*.

"They arrive with gold and they depart with pepper," wrote the Tamil poet Tayan-Kannanar (*Agam* 149.7–11) of the Roman merchants in the south Indian port of Muziris, in unknowing agreement with Pliny's bitter calculation of the quantity of Roman gold that was annually lost to India in the purchase of condiments and cosmetics (Dalby 2000b: 194–7). Two forms of cinnamon were part of this ancient trade. One, fine cinnamon bark, was either *Cinnamomum verum* syn. *zeylanicum* of Sri Lanka or else *C. cassia* and other Southeast Asian species: the source of the ancient supply is not known for certain (198–9). It was so expensive in Rome that it was hardly used for food. Romans could, however, afford to flavor their sauces with cinnamon leaves of the species *C. tamala*. As described in the ancient Greek sailing guide to the Indian Ocean shores, local traders acquired this now-forgotten spice, *malabathron*, somewhere in the mountains between Assam and Yunnan, on the very edge of the world known to the Greeks and Romans:

> Every year there turns up at the border of Thina a certain people, short in body and very flat-faced ... called Sesatai ... They come with their women and children, carrying great packs very like mats of green leaves, and settle at a certain place on the border between them and the Thina people and hold a festival for several days, spreading out the mats for themselves, and then take off for their own homes in the interior. Those who know about this go to the place at exactly that time and collect their mats. Taking out the fibers from the reeds ... and lightly doubling over the leaves to make them into ball shapes, they string them on the fibers from the reeds. There are three grades: big-ball malabathron from the larger leaf, middle-ball from the less large, small-ball from the smaller; this is the origin of the three grades of malabathron. Then they are carried down to India by the people who make them.
>
> What lies beyond this area, because of extremes of storm, bitter cold, and difficult terrain and also because of some divine power of the gods, remains undiscovered.[16]
>
> (*Periplus Maris Erythraei* 65–6)

FROM ASIA TO EUROPE

The Qur'an, though slightly late in date, offers good evidence for this chapter. Alongside the olives, grapes, and figs typical of the eastern Mediterranean it names major fruits of the more arid shores of the Indian Ocean, dates (*Phoenix dactylifera*), pomegranates (*Punica granatum*), and two other species that are to be found in Paradise:

> Thornless jujubes, and banana trees with fruits one above another, and extended shade, and water flowing constantly, and abundant fruit, neither intercepted nor forbidden.
> (Qur'an 56:28–33)

Both fruits, dry-fleshed but sweet, flourish in Arabian lands. The banana, Arabic *talh,* would have been a fairly recent introduction, but the jujube of the Near East (Arabic *sidr*, *Ziziphus spina-christi*, the third species of this adaptable genus that we have encountered) is a native wild fruit, and not very welcoming to the fruit-picker unless—as is the case in Paradise—it grows without its thorns.

Composed in a less-arid environment somewhat further north, the books of the Bible name a wider range of fruits than does the Qur'an. A full list would include some that were domesticated long ago in the Near East and were to become typical in Europe, almond (*Prunus dulcis*) and pomegranate; some that originated further north, apple (*Malus domestica*), walnut (*Juglans regia*), and grape; and one other, the terebinth (*Pistacia atlantica*), that by its resistance to aridity can flourish in landscapes more typical of the Qur'an. Terebinth seeds or fruits, tiny and oily, are no luxury, yet this tree makes its contribution to luxury food. Among its other discoveries in central Asia, the expedition of Alexander the Great found a species previously unknown to Greeks, related to the terebinth but with larger, equally nourishing, and much tastier fruits. This was the pistachio (*Pistacia vera*). Once noticed, it was transplanted westwards, a story that is clearly reflected in Greek sources, but plants of this genus are not easy to grow. Success came when pistachio was grafted on terebinth roots, a method that has made the pistachio a Mediterranean tree and is still used (Dalby 2003).[17]

One last discovery of Alexander's expedition, found in the hills of western Afghanistan, was the big fennel-like plant *Ferula assa-foetida*. Greek soldiers immediately noticed its sap or resin, asafoetida, with a powerful aroma that reminded them of the favorite culinary spice of classical Greece, silphion. This came only from wild plants in the hinterland of the Greek colonies in north Africa (modern Libya). With something of garlic, something of onion, and a good deal that was even worse, silphion was an acquired taste, but one with many health benefits. This plant's decline and extinction under the early Roman Empire gave unexpected importance to its central Asian relative, which was perhaps already used in Indian food and medicine as it certainly has been later. Central Asian asafoetida is the laser or silphium frequently called for in the fourth century Roman recipes of Apicius (Dalby 2000a: 110–12; Grocock and Grainger 2006: 331–4).

From recent genetic research central Asia is now known to be the region where the apple was first domesticated. The principal ancestral wild species is *Malus sieversii* syn. *pumila* of the mountains of Kazakhstan and Xinjiang, but the date at which cultivation began and apple cultivars began to multiply is not clear. In the fictional orchard of Alcinous, imagined to be on an island west of Greece and described in the *Odyssey*—written down, it is thought, around 650 BCE—apples figure prominently:

> Outside the yard is a big orchard on both sides of the gates,
> Of four acres, and a hedge runs along each side of it,
> And there tall trees have grown, lush with leaves,
> Pears and pomegranates and shiny-fruited apples
> And sweet figs and olives lush with leaves,
> Whose fruit never fails or falls short,
> Winter or summer, all the year, but forever
> The West Wind, blowing, engenders some and ripens others.
> Pear upon pear grows old and apple upon apple,
> Grapes upon grapes and fig upon fig.
>
> (*Od.* 7.112–21)

Eight hundred years later, Roman authors were celebrating luxury fruit, apples prominent in the catalogue. A poetic dinner invitation in Latin ends with "grapes preserved half a year but as good as when they were fresh, and Signine pears and Syrian pears, and apples fresh-scented from their harvest baskets, rivals of the Picene and nothing for you to fear, cured by the cold of autumn of the dangerous roughness of their juice" (Juv., *Satires* 11.56–76). Signine pears and Picene apples were from central Italy. In the Greek romance of Daphnis and Chloe, written possibly a hundred years after Juvenal, an orchard in autumn is the setting for lovemaking, "the time of year when everything is ripe. There were lots of wild pears, lots of cultivated pears, lots of apples. Some of these had already fallen, some were still on the trees. Those on the ground were more fragrant, and smelt like wine; those on the branches were fresher in color and shone like gold" (Longus, *Daphnis and Chloe* 3.33–4). The smell of fully ripe apples and of stored apples was appreciated by the emperor Augustus, whose favorite lunch might be "perhaps an apple, freshly picked or from store, with a wine-like flavor" (Suet., *Aug.* 77–8) and by the poet Martial, a younger contemporary of Juvenal (both poets came from Roman Spain). Martial noticed "the smell of apples ripening in the winter chest" (*Epigr.* 11.8.1–12). In a poem supposed to accompany a gift of fruit he alluded to the passage of the *Odyssey* already quoted, claiming that the fruit he bought in the street market of the crowded Suburra district of Rome was just as good:

> I have not the royal orchard of an Alcinous. My garden is both safe and fertile in its Nomentan fruit; my leaden apples fear no thief. Accept this waxy fruit of my own harvest, grown for me in the heart of the Suburra.
>
> (Mart., *Epigr.* 10.94)

There is not much archaeobotanical evidence for the apple, though textual sources confirm that it has been the most important tree fruit of temperate Europe and western Asia. There is an unresolved problem about how it spread westwards from central Asia, because, although a few Near Eastern and a very few European apple cultivars will reproduce reliably from seed, the vast majority will not: the result of growing from seed is endless and mostly unwanted variation. Hence the universal method of propagating apples is by grafting, but no one knows when grafting was invented. The strings of dried apple found in the Royal Cemetery at Ur in Mesopotamia, dated to the mid-third millennium BCE, and the tenth century BCE carbonized apples from Kadesh Barne'a in southern Palestine, and the orchard apples of the *Odyssey*, do not necessarily imply domesticated, grafted cultivars, since even the parent wild species is already "shiny-fruited" and worth growing in its own right (Juniper 2001; Velasco et al. 2010; Zohary and Hopf 2000: 171–5).

Grafting is mentioned very briefly by the Greek scientist Theophrastos about 310 BCE: "… to graft suckers that will respond to it on trees of the same, such as olive, pear, apple, fig …" (*Hist. pl.* 2.5.3) and it is described fully by Roman authors on farming, Cato in the second century BCE and Columella in the mid-first century CE. So we can at least say that grafting was commonly practiced by the date of the Latin authors quoted above, Juvenal, Martial, and Suetonius. Their older contemporary, the encyclopedist Pliny (who, incidentally, mentions the fine apple orchards of Ctesiphon, metropolis of Mesopotamia in his time) also gives a long list of apple cultivars known in Italy, many of them named after Roman growers. "Unless I am mistaken," he observes, "the following list is evidence of ingenuity in grafting," but he ends with a reminder that cultivated varieties are not everything: "there are also wild apples, with poor flavor and even sharper smell. Their special defects are fierce bitterness and enough acidity to blunt the edge of a sword" (Plin., *HN* 15.49, 52).

Pears (*Pyrus communis*), more luxurious because most cultivars were fleetingly ripe and could not be stored, accompanied apples into the orchards of prehistoric and classical Europe. Alongside them came plums, a variable group of species and cultivars, and cherries both sweet and sour (*Prunus* spp.). All of these as they spread westwards were, like apples, taking over a space where their wild relatives already flourished. These local relatives provided useful if not essential grafting stock as well as blossom for cross-pollination and possible hybridization. Hence grafting was available as a skill worth trying when, as described above, the pistachio spread westwards into Syria, Anatolia, and Greece.

Figs are a versatile fruit, as fleeting and luxurious as pears in their brief ripeness, a valuable store of sugar, and an excellent aid to digestion when dried and stored, surely to be classed as "luxury" in either form. When fruits of the eastern Mediterranean are listed, figs will be in the list, whether it is written around 520 BCE ("Is the seed yet in the barn? yea, as yet the vine, and the fig tree, and the pomegranate, and the olive tree, hath not brought forth: from this day will I bless you": Haggai 2:19) or in the seventh century CE ("By the fig and the olive and Mount Sinai and this safe country": Qur'an 95.1–4). In the case of figs, the relationship between cultivated and wild is unusually close. If figs are to be planted for fruit in a region where they do not grow wild (which is how they spread around the Mediterranean in ancient times) suckers and cuttings will take root—grafting is not essential—but that is not enough. It is female trees that bear fine fruit, and most cultivars will not ripen their fruit unless pollinated from a male tree. Pollination takes place inside the immature fruit and can only be done by the fig wasp (*Blastophaga psenes*). Male trees are therefore encouraged to grow in wasteland or pasture near an orchard of female trees. At fruiting time growers often place a branch of male figs in a female tree to encourage pollination. They need not fully understand the symbiosis—until the twentieth century none did—but they must not overlook the need for wild figs near an orchard of domesticated figs. It is sufficiently remarkable, considering the knowledge that was needed, that the fig was certainly domesticated in the Near East by about 4500 BCE. Recent evidence from the site of Gilgal in Palestine suggests that the first steps had been taken at a much more ancient period, around 9300 BCE, for the large number of carbonized figs found there, which were female figs of small but edible size, appears to demonstrate that people were already planting cuttings of wild female fig trees and were drying the fruits. It is suggested that the community's barley, oats, and acorns were being gathered from the wild, but meanwhile its figs, the luxury fruits of the future, were being purposefully grown (Kislev et al. 2006).

Among the many other plant foods of Europe that might be labeled luxurious three very different examples must suffice. The bulb of the grape-hyacinth (*Leopoldia comosa* [*Muscari comosum*]) has been a favored food for more than two millennia. "I say goodbye to little dishes of bulbs and silphion, and all other appetizers," wrote the gastronomic poet Archestratos, but other luxury-seekers evidently embraced them. His older contemporary, the comedy author Philemon, wrote: "Look, if you please, at how highly the bulb is regarded for its extravagance, demanding cheese, honey, sesame, olive oil, onion, vinegar, silphion. All on its own it is mean and sour." Then why was so much trouble taken with it? At least partly because it was considered an aphrodisiac, ignoring the wise proverb, "The bulb won't help you if you lack the nerve" (quotations from Greek authors of the fourth and third centuries BCE: Athenaios, *Deipnosophistai* 63e–64e). Why, then, has it remained a favored appetizer in southern Italy and Greece until today, though unknown elsewhere even in Europe, and though no one for many centuries has spoken of its aphrodisiac quality? A writer on ancient luxury can only say that its favored status goes back a long time: grape-hyacinth was one of the flowers found associated with the burial of "Shanidar 4", about 60,000 BCE in Iraqi Kurdistan (Leroi-Gourhan 1998; Sommer 1999).

Poppy and saffron are two spices that originated in Europe, spreading eastwards in ancient or medieval times. "Sow poppy seed where you had a bonfire," Cato the Censor wisely instructed in his Latin farming handbook of the second century BCE. Opium poppy (*Papaver somniferum*) is more important medicinally than as a food, but its aromatic seeds, which are not narcotic, were sprinkled on bread before baking in ancient Greece and Rome as they are today (Dalby 2000a: 134–6; Zohary and Hopf 2000: 135–8). Saffron, the laboriously picked stigmas of *Crocus sativus*, is said to be the most expensive spice in the modern world, having supplanted cinnamon in that rôle because transport is now cheap and labor expensive. A species now known only in cultivation, saffron may have been domesticated in the Greek islands from the wild species *Crocus cartwrightianus* (see Figure 2.6). It was important in religious ritual, though perhaps not in food, at the moment when the Minoan culture of Thera (Santorini) was overwhelmed in the volcanic eruption of *c*. 1628 BCE: a wall painting shows girls and women gathering saffron, apparently for a goddess, from lovingly depicted flowers scattered in a landscape. Almost two millennia later it was an affectation of Roman luxury to serve wine colored and aromatized with saffron, and even (if one were the emperor Domitian) to spray saffron wine into the audience at the amphitheater (Dalby 2003: 289–90).

BEYOND LITERARY EVIDENCE

Of luxury food plants originating in the Pacific and in island Southeast Asia three species have been named above because they found their way westwards into literature in ancient times, *Piper betle*, *Areca catechu*, and *Saccharum officinarum*. Of African food plants the watermelon (*Citrullus lanatus*) had reached Mediterranean lands at the same period (Zohary and Hopf 2000: 193–4), but, in spite of its early domestication and wide distribution, it found very little mention in ancient literature. As to the luxury food plants of the Americas, only one can be named from a written text of the relevant period: it is chocolate (*Theobroma cacao*), famous from the single-word inscription *ka-ka-w* on a chocolate drinking bowl from the Maya Classic period (*c*. 500 CE), an inscription that formed a step in the demonstration that the Maya writing system was phonetic and had been at last successfully deciphered (Coe and Coe 1996). A bowl labeled for chocolate, and the absence of such labeling for any other food, demonstrates unusual status for this highly nourishing but non-staple beverage.

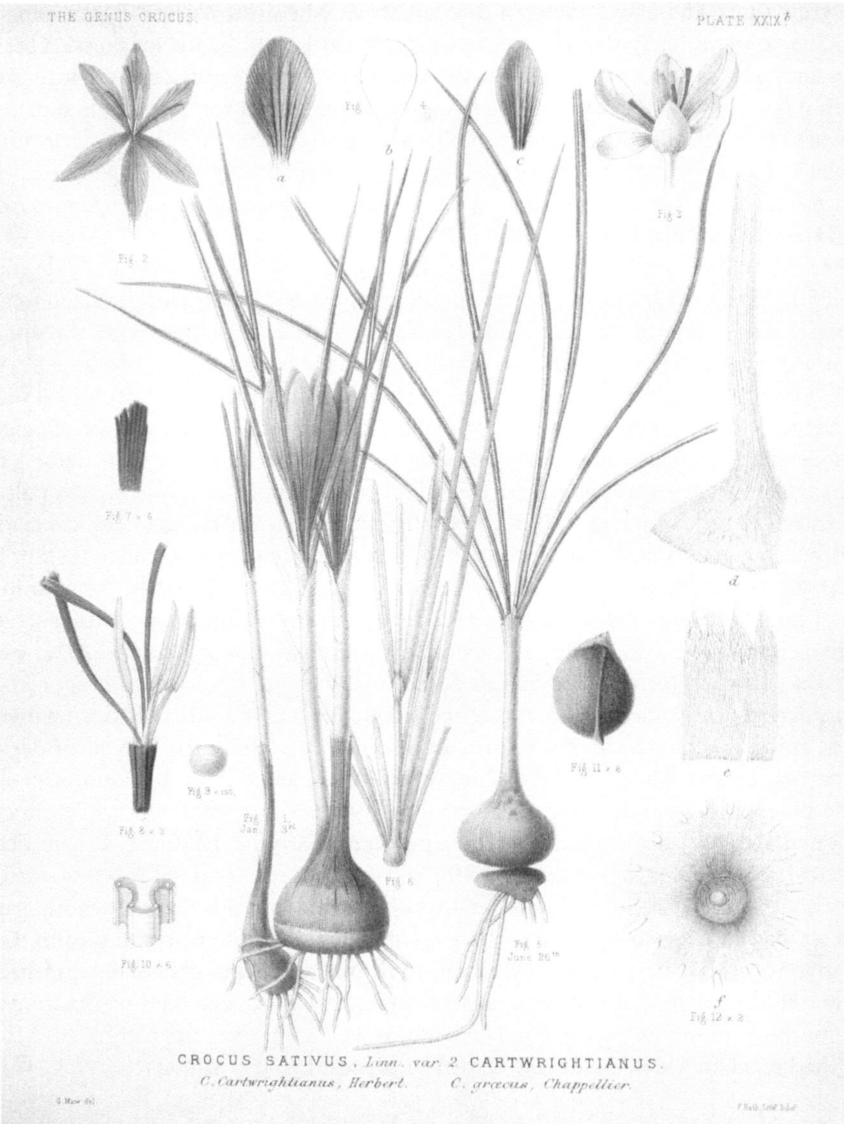

FIGURE 2.6 *Crocus sativus* L. var. *cartwrightianus*, from G. Maw, *A Monograph of the Genus Crocus*, t. 29b (London: Dulau and Co., 1886). The New York Botanical Gardens.

Beyond literature, luxury status of foods can be easily conjectured but with difficulty proved. Stimulants analogous to chocolate such as the kola nuts of tropical Africa (*Cola nitida* and *C. acuminata*) and the coca leaves of South America (*Erythroxylum coca* and *E. novogranatense*) were certainly of interest in the ancient period. For the coca species this is demonstrated by their early domestication in the Andes and the upper Amazon valley (Piperno 2011; Plowman 1984). For the kola nut the evidence is linguistic.[18] A name for this tree can be confidently reconstructed in proto-Bantu, which was spoken around 2000 BCE in southeastern Nigeria. It will have been applied to *Cola acuminata*, which grew there and across western central Africa where the earliest Bantu expansion took place: it seems likely that the tree spread eastwards with Bantu speakers. A series of other names for kola nuts is traceable historically in several west African language groups, extending into areas where *Cola nitida* is native, the main source of the kola nuts traded northwards into the western Sudan in medieval times. Another culturally important fruit tree whose name is traceable to proto-Bantu, and which may well have spread from southern Nigeria across central Africa with the Bantu expansion, is the *safou* or African plum (*Dacryodes edulis*) (Bostoen 2014).

In Mexico and central America some similar linguistic deductions can be made (Brown 2010 for linguistic data below), corresponding, in certain cases, with archaeobotanical evidence.[19] The name of the avocado, *Persea americana*, was known to the speakers of proto-Otomanguean, probably in southwestern Mexico, around 5000 BCE, while archaeobotanical evidence suggests domestication in the same region around 4500 BCE (C. E. Smith 1966). A name for agave, probably *Agave americana*, the source of *aguamiel* for *pulque* and with many other uses (Stix 2015), can be traced to the same date. A name for the species *Capsicum annuum*, source both of sweet pepper and chilli pepper cultivars, can be traced to proto-Amuzgo-Mixtecan soon after 3000 BCE. Archaeobotanical evidence from Romero and Coxcatlán caves takes the human use of this species much further back, to 5000 BCE or before, but domestication came later (Kraft et al. 2014). A name for the jocote or hog plum, *Spondias purpurea*, goes back to proto-Chinantecan *c.* 400 BCE: it was attractive enough to be domesticated twice, in western Mexico and central America, at a date as yet uncertain (Allison and Schaal 2005).

Considering now some fruits that reached Mexico from the south, Mexican names can again be traced to quite early dates, providing evidence of the early spread of desirable foods across regions. The guava, *Psidium guajava*, was named in proto-Popolocan *c.* 600 BCE, the pineapple, *Ananas comosus*, in proto-Chinantecan *c.* 400 BCE, while chocolate (mentioned above) had a name in proto-Zapotecan of around 1250 BCE. It is noticeable that chocolate is the least immediately useful of these four transplanted species—the other three produce fruits that are edible without preparation—but it is the one whose product can be stored. Sure enough, in later Mexico chocolate beans represented wealth. Guava was native to northwestern South America, both pineapple and chocolate to Amazonia (Clement et al. 2010). A third Amazonian resource, *annatto*, the seed of *Bixa orellana*, very important culturally, used as food coloring as well as in cosmetics (Ambrósio Moreira 2015), had reached Mexico and was named in proto-Mayan *c.* 400 BCE.[20]

CONCLUDING REMARKS

Archaeobotanical evidence from the Palaeolithic and Mesolithic periods is rare. Approaching the beginnings of cultivation and domestication, such evidence becomes commoner, and shows the story to be complex. In several regions increasing population and climatic variation impelled change. A wider spectrum of plants was drawn into use, and at the same time plant encouragement was slowly turning into cultivation and leaning towards domestication (e.g. Belfer-Cohen and Goring-Morris 2011). The most widely useful plants and the most reliable staple foods were everywhere the first to be domesticated, the former group including the bottle gourd (*Lagenaria siceraria*), and the fig (*Ficus carica*).

Nearly all luxury food plants were domesticated significantly later than the early staple foods: if reasons are wanted it may well be that the staples were more urgently needed and it may also be that the management of such plants, tree fruits for example, demanded skills that were more slowly acquired. It may remain true for a long time that there is no need to cultivate. There was perhaps always enough *Ephedra intermedia* in the wild to be pressed for the *haoma* or *soma* of northeastern Iran. There were perhaps enough grapes of the Chinese wild vine, *Vitis heyneana*, to make the wine that early neolithic China enjoyed (McGovern et al. 2004). The best-known kola nut sources, *Cola nitida* and *C. acuminata*, were encouraged to spread alongside the people who wanted them (recently, for example, when enslaved west Africans crossed the Atlantic and sowed cola nuts in tropical America). *Piper betle* and *Areca catechu*, having once coincided, thereafter accompanied human migrations and voyages in tandem, not domesticated, barely even cultivated, but welcomed.

Through the whole period of domestication, extending to the present day, certain plants with luxury status could not be or have never yet been domesticated. Samphire, *Crithmum maritimum*, does perfectly well on the salt-sprayed rocks from which it is harvested. An Englishman knows that jams, jellies, and tarts made with wild blackberries (*Rubus fruticosus*) are better, more luxurious, than those from any cultivated blackberry. A Greek knows that caper bushes (*Capparis spinosa*), whose leaves, buds, and fruits are so unique and so luxurious in their flavor "are wild by nature," "do not respond to cultivation but deteriorate" (Theophra., *Hist. pl.* 3.2.1, 1.2.6). But capers choose to grow on the craggy sides of ruined and untended buildings. Wild blackberries flourish nowhere better than in abandoned fields and hospitable hedgerows.

Already in the ancient period plants that produced luxury foods were being transplanted gradually over vast distances along routes that cannot be mapped, watermelons having traveled from south Africa to Europe, chocolate and pineapple from Amazonia to Mexico, poppy and coriander (*Coriandrum sativum*) from Europe to India. Aromatics of plant origin were also commodities of trade, such as black pepper and cane sugar from India, asafoetida from central Asia, but this "spice trade" was as yet relatively little concerned with food. Cinnamon and cardamom of South and Southeast Asia, alongside several even rarer spices (costus, spikenard, zedoary, camphor, benzoin, cubebs, cloves, nutmeg) make no mark on this chapter because when they reached their distant markets they were used in perfumes and cosmetics, in medicine and religious ritual, perhaps in wine, hardly ever in food. The quotation already given from *Carakasaṃhitā* is typical of ancient uses of spices in personal hygiene. Medical prescriptions from India and the Roman Mediterranean also prescribe exotic spices for patients who could afford them.

As far as the Mediterranean region is concerned, medicine was possibly the catalyst for change. Galen, a Roman Imperial physician, writing in Greek in the late second century CE, listed the medicinal effects of fruits and culinary herbs, naturally, in his manual *On the Properties of Foods*. Spices are not there: they are found among other pharmaceuticals in the companion work *On the Properties of Simples*. But, after all, they are evaluated on the same scales. Already in the fourth century CE, if that is the date of the Latin recipes of Apicius, exotic spices beyond pepper and asafoetida are beginning to make their way into recipes for food. In medieval sources, from the Byzantine empire and later from western Europe, exotic spices multiply, for their rare flavors and for their health benefits, if, indeed, it makes sense to separate the two.

CHAPTER THREE

Trade and Exploration

LAURENCE M. V. TOTELIN

CINNAMON: AN INTRODUCTION TO THE TRADE IN PLANTS AND THEIR PRODUCTS

> Regarding cinnamon and cassia, antiquity, led by Herodotus, has told a fabulous story. It is said that these are collected from the nests of birds, and in particular from the phoenix, in the region where Pater Liber was raised, and that they are knocked down from inaccessible rocks and trees by the weight of meat that the birds themselves bring there, or by leaden arrows. Similarly, they say that cassia grows around marshes, protected by a dreadful type of bats with their claws, and by winged snakes. They increase the price of these goods with these lies.
>
> (*HN* 12.85)[1]

Sensational stories contribute to raising the price of spices: this is the damning conclusion drawn by Pliny the Elder, the first-century CE Roman encylopaedist. These stories function as a form of advertising, tempting clients with promises of rarity and exoticism. Pliny did not specify who invented the tales about *cinnamomum* and *casia* (which will be referred to as "cinnamon" and "cassia" for the sake of simplicity). The likely candidates are either the inhabitants of Arabia or Ethiopia, the places where the spices allegedly grew according to Greek and Roman authorities, or the Phoenician merchants who, according to the fifth-century BCE historian Herodotus (Pliny's source), brought the spices to Greece (Hdt. 3.110–11). Other legends circulated about cinnamon in the Greco-Roman world. According to Aristotle, the cinnamon-birds made their nests in high trees, not on rocks (*Hist. an.* 8.13, 616a7–13); and his student Theophrastus, the so-called "father of botany," reported that the plant grew in ravines inhabited by dangerous snakes (*Hist. pl.* 9.5.2).

There was legitimate confusion about the geographical origins of cinnamon and cassia among Greek and Latin authors as well. Theophrastus (*Hist. pl.* 9.4.2), Herodotus (3.107), and Arrian (*Anab.* 7.2, see below) suggested the plants grew in southern Arabia, the land of spices par excellence (see the articles collected in Avanzini 1997), while Pliny was adamant they came from Ethiopia (*HN* 12.87). The spices that today we call cassia (the bark of *Cinnamomum aromaticum*) and cinnamon (*Cinnamomum verum*), for their part, grow in Southeastern Asia, in Ceylon, southern India, and southern China.[2] Modern scholars who work with ancient texts therefore debate whether the Greeks and Romans knew the southeastern spices (see e.g. Amigues 1996a, 2006: 295 and 297; Casson 1984: 239; Dalby 2003: 87–8) or not (see De Romanis 1996: 109; Raschke 1978: 655;

Young 2001: 20). If they did know them, as I am inclined to think, southern Arabia and the Horn of Africa, believed as the actual sources of these spices, would instead have been stopovers for them on their long journeys from the East.

Wherever they came from, these spices were expensive. Pliny stated that cinnamon used to cost a thousand denarii per pound, but that the price rose by half as much after a fire in the cinnamon forests (*HN* 12.93). The price of cassia varied much more, with the best grades costing fifty denarii per pound, and the others five per pound (*HN* 12.97). There was, however, also a particular variety of cassia, called "*daphnitis*," which fetched three hundred denarii per pound (*HN* 12.98); Pliny asserted that this particular variety also grew on the Rhine frontier. For comparison, Pliny gave the prices of types of peppers as fifteen denarii a pound for long pepper; seven denarii a pound for white pepper; and four denarii a pound for black pepper (*HN* 12.28; see Young 2001: 222).

One of the main uses of cinnamon and cassia in the first centuries of the Common Era was in the preparation of antidotes. These were panaceas that supposedly protect against poisons, poisonous bites, and many illnesses beside and that elites conspicuously consumed in emulation of the Roman emperors (see Gal., *De antid.* 1.4, Kühn 14.24).[3] The most famous of antidotes was theriac, which the physician Galen prepared for Roman several emperors (on theriac, see e.g. Boudon 2002; Stein 1997; Totelin 2016). Galen gives us invaluable information on how the precious ingredient cinnamon was kept in imperial storehouses in the city of Rome:

> When I was preparing theriac for Antoninus [the emperor Marcus Aurelius], I observed many wooden vessels filled with cinnamon, some placed in the storehouses under Trajan [98–117 CE], others under Hadrian [117–138 CE], and some under Antoninus [Pius, 138–161 CE], who ruled after Hadrian.
>
> (Gal., *De antid.* 13, Kühn 14.64)

In contrast to his father Marcus Aurelius, Commodus apparently had no interest in theriac, to the extent that he sold off the Roman stocks of cinnamon, so needed in the production of antidotes:

> Commodus having succeeded him, as he had regard neither for theriac nor for cinnamon, sold the surplus of the plant, which had been brought [to Rome] after the time of Hadrian. As a result, nowadays, when the Emperor Severus [193–211 CE] orders me to prepare the antidote in the way it was prepared under Antoninus [Marcus Aurelius], I am forced to choose the stocks from the times of Trajan and Hadrian, which clearly have become weaker [with the passage of time].
>
> (Gal., *De antid.* 13, Kühn 14.65)

Apparently, Galen did not seek fresh cinnamon from sources other than the imperial storehouses. This perhaps indicates that there was very little high-quality cinnamon in Rome under the rule of Severus. In another work, Galen reported that he had lost his own stocks of cinnamon, "more than all the shops put together," in the great fire of Rome in 192 CE (*De indol.* 6). He was therefore almost entirely dependent on the whims of a particular emperor, who could sell the spice to fill his coffers.

To Galen, Commodus' disdain for the precious cinnamon must have been abhorrent and altogether rather odd, for many ancient rulers had famously shown interest in precious spices and led explorations to far-away lands where they grew—an early example being the Egyptian queen Hatshepsut (r. 1473–1458 BCE), who launched an expedition to the land of Punt, perhaps on the northeastern coast of Africa, to fetch the precious resins frankincense (*Boswellia*) and myrrh (*Commiphora*). Some authors went as far as identifying plants as a reason for a ruler's trying to conquer a land. Thus, the historian Arrian (first–second century CE) suggested that Alexander the Great considered conquering Arabia because of its wondrous plants, which allegedly included cinnamon:

> Then the prosperity of the country [Arabia] incited him, since he heard that in their oases cassia grew, and from the trees came myrrh and frankincense; and from the bushes, cinnamon was cut; and that from their meadows nard grew self-sown. Then there was also the size of their territory, since the sea-coast of Arabia was reported to him to be not less long than that of India, and that there were several islands adjacent.
> (Arr., *Anab.* 7.20; translation in Hardy and Totelin 2016: 54)

It is impossible to determine whether Alexander truly considered spices so important that they would warrant a campaign, but we know that natural scientists accompanied him on his expeditions. They described and mapped plants that the Greeks had never encountered in their live form, although they might have seen parts of them dried. Their writings are lost, but fragments are preserved in Theophrastus' work (see Amigues 1996b; Bretzl 1903). Alexander never had a chance to conquer southern Arabia, however, as he died before putting his plan into action. Nor did later Greek and Roman generals and rulers manage the feat of annexing Arabia Felix, which remained independent.

There is much more that could be said about cinnamon and its trade. Certainly, these stories are of great interest, but to focus on them distorts our view of ancient plant trade. A focus on luxury and/or exotic trade can cloud our vision (Vivian Nutton made this point in relation to drug trade in particular [1985: 141]). Thus, while the renewed interest in Rome's trade with the East, and in particular with Arabia and India, in the last twenty years, is to be welcomed (see e.g. Matthew 2017; Parker 2008; Young 2001), one should stress that most plant products traded in the ancient world were not exotic, expensive, or luxurious. This last statement, however, must be qualified in several ways.

First, our literary sources are slanted towards luxury items. Pliny complained repeatedly about *luxuria*, that plague of his times, but that did not prevent him from writing at length about it, for the simple reason that it provided entertaining material for his readers. To find relevant material for the study of trade in non-luxury or staple commodity plants, one must turn to sources that can be more difficult to interpret, such as documentary papyri, epigraphical sources, or archaeobotanical evidence (see Macauley-Lewis 2010 for the types of evidence available on plant trade).

Second, one must reflect on the definition of "luxury" (see van der Veen 2003 on when an ancient food counts as a "luxury"). In their seminal studies, Arjun Appadurai (1986) and Christopher Berry (1994) have shown that luxury is not an intrinsic quality of goods. Berry (1994: 41) defines a luxury good as follows: "a widely desired (because not yet widely attained) good that is believed to be 'pleasing,' and the general desirability of which is explained by it being a specific refinement, or qualitative aspect, of some universal need." Societies define what goods they consider as luxuries; any good can potentially become a luxury, which people conspicuously consume as status symbols (on consumption and consumerism in the Greco-Roman world, see in particular Foxhall 1998, 2005; Greene 2008). In this chapter, we will encounter examples of plants, such as apples, that we now consider ordinary but that take on luxury status.

Third, "luxury" status is not fixed; it can fluctuate over time. The trickle-down effect, whereby interest in a product originates in the upper classes of a society and then spreads to the lower classes, can at times be identified in ancient consumption patterns (see van der Veen 2003: 409–10). While the best cinnamon and cassia remained out of reach of many, cheaper alternatives could be found from less geographically remote sources. Indeed, the pharmacological writer Dioscorides listed several grades of cinnamon and cassia, to which he added products called "false cinnamon" (*pseudokinnamomon*, Dsc. 1.14.3) and "false cassia" (*pseudokassia*, Dsc. 1.13.2).

In this chapter, I examine trade in ordinary as well as luxury plants, and trade over short, medium, and long distances, the particular focus being on examples of ancient trade in various plant products: fruits, flowers, dried herbs, and timber. These examples come for the majority from the Greco-Roman world in the first centuries of the Common Era. I pay attention not only to the plants that are traded, and the value (both monetary and symbolic) that is assigned to them, but also to the people involved in trading them, whatever their social status. I will also examine the links between trade and exploration in the ancient world. The word "exploration"—perhaps unlike "trade"—suggests selfless scientific enterprise, the search for knowledge. However, as already seen with the example of Alexander the Great, plant exploration in the ancient world accompanied military conquest, colonization, and the building of empires. Two of the reasons for expanding a territory in the ancient world were access to resources, including plant resources, and facilitating trade.

TO MARKET ... TO MARKET: TRADING FRUITS IN ANTIQUITY

A Gallo-Roman funerary stele from the Belgium city of Arlon offers a fascinating insight into rural life in the Roman province of Gallia Belgica in the early centuries of the Common Era (Figure 3.1). The deceased couple are represented on the main face. The wife carries a jewel box—a sign of the couple's prosperity. On the left lateral face, we find two scenes. The top one is a market scene. Two vendors serve a client who has put apples in the fold of his cloak. Their produce is placed on a trestle table under which we see three wicker baskets. Above the merchants, we see bundles of plants. The bottom scene shows two men working the land, one with a spade, the other with a rake. The right lateral face also bears two scenes. On the top one, a man travels in an empty cart. On the bottom, a man places fruits in a wicker basket, while another looks on (Esperdandieu 1913: n4044; see also Holleran 2012: 207).

The Arlon stele is that of a relatively wealthy couple; they made their living from the product of the earth, which they sold at a not-too-distant market. The fruits that are handled and sold in the market scene are usually identified as apples, a fruit that grows abundantly in the south of Belgium, and which—alongside the pear—is the northern European fruit par excellence (Zohary et al. 2012: 135–40). Domesticated apples are propagated by grafting, a horticultural technique which consists of inserting the part of a plant (the scion) into that of another which is rooted (the rootstock) (Zohary et al. 2012: 5). This method of vegetative propagation (a cloning method) allows for the properties of a specific cultivar to be preserved. Since apples do not reproduce true to type, an apple grown from seed will be different from its parent, and its fruit may well be unpalatable or inedible. Through grafting, the Greeks and Romans developed numerous

FIGURE 3.1a and b Gallo-Roman stele, market scene (left lateral side, top and bottom); Musée Archéologique d'Arlon, GR/S 049. Courtesy of Musée Archéologique d'Arlon.

varieties of apples, of which Pliny named several in book fifteen of his encyclopaedia (*HN* 15.49–52). He noted that people gave their name to cultivars they had created—a practice of which the encyclopaedist did not approve: "nothing is too small as to be incapable of imparting glory."

The food items on the Arlon stele, however, might actually be something other than apples. The Arlon farmers might instead have sold common, indigenous, round vegetables, such as cabbages, which would explain the presence of the land-working scene on their stele (on cabbages, see Dalby 2003: 66; Zohary et al. 2012: 158–9). It is also possible that the couple grew and, considering their wealth, successfully traded, the fruits of trees newly introduced into the provinces of the Roman Empire in the first centuries of the Common Era. Archaeobotanical evidence shows that, through imperial expansion, the Romans played a crucial role in the spread of numerous species within Europe (see e.g. Bakels and Jacomet 2003; van der Veen et al. 2008). Where the Romans posted their armies, plant species tended to follow. Pollen finds indicate that some plants, such as the sweet cherry, plum, and walnut, were now grown throughout the empire, and became part of the normal diet. As Bakels and Jacomet argue, these plants were considered luxuries at the beginning of Roman occupation in the western and northern provinces, but they became staple commodities with the passage of time. The Roman army perhaps did not have trade as its main aim when importing species to the provinces of the empire; rather it attempted to respond to the wishes of its soldiers (especially those high-ranking ones) who were used to a particular diet. Nevertheless, Roman imperialism created the opportunity for trade, first by selling surplus to the local populations, then presumably by teaching the techniques involved in growing "new" plant species when the environment allowed it.

Literary evidence also testifies to the spread of fruit trees in the first centuries of the Common Era. Pliny wrote that there was no cherry in Rome until 74 BCE, when the general Lucullus brought it back from Pontus, where he had defeated the formidable king Mithradates VI. By Pliny's time, the cherry had "crossed the ocean and reached as far as Britain," although it was impossible to grow it in Egypt (*HN* 15.102). The encyclopaedist named several varieties of cherries, including a "Lusitanian" variety growing in Belgium—an interesting geographical mix. It is likely that the species of cherries Lucullus allegedly imported from Pontus to Rome was the domesticated sweet cherry, *Prunus avium*. Wild cherries were gathered well before their domestication throughout Europe (Zohary et al. 2012: 144). A rather late source, Isidore of Seville (seventh century CE) argued that there were indeed cherries in Italy before Lucullus, while also giving the etymology of the word *cerasium* (cherry tree in Latin):

> The cherry is named from the city Cerasum in Pontus, for when Lucullus destroyed the Pontic city Cerasum he imported this kind of fruit from there and named it "cerasium" from the city's name. The tree is called "cerasus," the fruit "cerasium." Before Lucullus, these were in Italy, but only a hard variety, and hence it was also called the cornel-cherry.
>
> (*Etym.* 17.7.16; on the etymology of the name, see André 2010: 57)

Pliny also reported that Vitellius (Lucius Vitellius the Elder, consul in 34 CE) introduced the pistachio into Italy (*HN* 15.91); and Sextus Papinius (Sextus Papinius Allienus, consul in 36 CE) introduced the jujube tree and the tuber apple into Italy, respectively from Africa and Syria at the end of Augustus' rule (*HN* 15.47). Papinius apparently grew these plants in his camp from seeds (*sata*). From those camp nurseries, he could have transported his saplings in a basket or in a perforated clay vase, known as *olla perforata* among archaeologists (see Kenawi et al. 2012; Macauley-Lewis 2006a, 2006b). While ancient political figures took interest in plants, one may doubt whether Vitellius and Sextus Papinius themselves elected to transplant these fruit trees. It is much more likely that people in their entourage made those decisions and carried out the work—at times delicate—of transporting plants over long distances and replanting them in a new environment.

Some transplants, however, were destined to fail or only be partially successful. The date palm is a case in point. By the fourth century BCE, date palms were growing in the Aegean world, having been transplanted from the Middle East, but they were barren. As Theophrastus noted (*Hist. pl.* 2.2.8), a barren Greek palm could be transplanted back to Babylon where it would then bear fruits; he concluded that "location is more important than cultivation and care" (on the date palm, see Georgi 1982; Hardy and Totelin 2016: 132–3). Pliny, for his part, blamed trees for their pride and obstinacy in refusing to grow wherever they are transplanted or to bear fruit when transplanted (*HN* 16.134–5).

While not all fruit trees can grow anywhere, their fruits can still be traded to other parts of the world. Famously, the Elder Cato played on the Romans' fear of the Carthaginian by producing a fresh fig that had allegedly been imported from Carthage in three days:

> In addition to these things, they say that Cato deceitfully dropped some figs of the Lybian type on the floor of the senate, as he threw back his toga, and then, as the senators admired their size and beauty, he said that the region that bore them was only three days' sail from Rome.
>
> (*Cat. Mai.* 27.1)

This was of course a trick, and Cato probably had plucked the fig from his own garden. Still, it remains the case that fruits were transported over large distances in the Roman world. Macroscopic carbonized remains of plants such as peaches, pomegranates, dates, and figs have been found as far as Roman Britain (van der Veen et al. 2008). These fruits were not grown locally (with perhaps the exception of the fig), but rather imported in a dry state (in the case of figs and dates) or in barrels which provided some refrigeration (in the case of peaches and pomegranates). Remains of two carbonized wooden barrels filled with seeds and pericarp fragments of pomegranates have been discovered in a building dating to the first century CE at Windish, Switzerland, the site of the legionary camp of Vindonissa (Jacomet et al. 2002).

To sum up, the fruits on the Arlon stele are probably simply apples, but the Gallo-Roman cultivators commemorated here lived in an environment that was rapidly changing because of Roman influence. The market at which they sold their goods was one of the thousands of local markets held in the ancient world. We now turn to a market in one of the largest cities of the Hellenistic world, Alexandria, and the flowers there sold.

THE ELEPHANT AND THE FLOWER SELLER: SELLING CUT FLOWERS IN ANTIQUITY

Plutarch (first century CE) told the entertaining story of an elephant who fell in love with girl who sold flower-garlands for a living:

> Such was [the love] of an elephant in Alexandria, who became the rival in love of the grammarian Aristophanes. Indeed, they were both in love with the same garland-seller; nor was the elephant's passion less manifest. For, when he passed through the market, he always brought her fruit and he stayed by her side for a long time and, inserting his trunk into the folds of her garments as if it were a hand, he gently caressed her breasts.
>
> (Plut., *De soll. an.* 18, 972D)[4]

This girl was a *stephanopōlis*, literally a seller (*pōlis*) of garlands of flowers (*stephanos*; see Figure 3.2). She plied her trade on the market, where other vegetable or vegetable-derived products were also on sale. Women who plaited and sold flower-garlands had been a feature of ancient markets since at least the fifth century BCE, and likely much earlier. The comic

FIGURE 3.2 Roman glass bowl with flower garlands, late first century BCE. New York, The Metropolitan Museum of Art. © Edward C. Moore Collection, Bequest of Edward C. Moore, 1891. Courtesy of The Met Museum.

playwright Aristophanes depicted a widow who struggled to feed her five children on the money she made from her flower business (*Thesm.* 446–56). Another comedian, Eubulus, entitled one of his plays (which is now lost save for fragments) *The Female Garland-Vendors* (*Stephanopolides*). There are indications in the fragments of that play (fr. 97), and indeed in the story of the elephant, that flower girls might have traded sexual favours (for discussion, see Hunter 1983: 191). An epigram by Strato (second century CE) hints at yet another story of seduction involving a garland-seller, male this time (*Anth. Pal.* 12.8). These retailers might have led a hand-to-mouth existence, which they needed to supplement with other sources of income, including prostitution.

Quite apart from the implication in some sources that selling flowers might yield marginal returns, there was certainly much money to be made in the trade of flowers in the classical world, for fresh flowers were in high demand for use in religious rituals and at banquets (see Farrar 1998: 135–42). Those who profited the most were the landowners with estates near big cities, as the agronomical author Varro (first century CE) explained:

> It is profitable to cultivate large gardens near a city, for instance gardens of roses or violets and many other plants of which the city has need. It would not, however, be profitable to cultivate the same product on a distant estate, where there is nowhere to bring its products for sale.
>
> (Varro, *Rust.* 1.16.3; see also Cato, *Agr.* 8.2)

There are numerous stories of elites conspicuously consuming flowers in the Greco-Roman world (see Géczi 2008). For instance, Cicero recounted how Verres (governor of Sicily from 73 to 70 BCE) traveled in a litter, resting on pillow from Malta stuffed with rose, wearing a chaplet of roses on his neck, and sniffing a bag of roses (*Verr.* 2.5.27). Suetonius noted that the emperor Nero (54–68 CE) had a false ceiling in the dining room of his palace; it would open up to scatter flowers and spray perfume (*Ner.* 31). Such a ceiling was also to be found in the palace of the emperor Heliogabalus, who allegedly once smothered to death some of his guests under violets and other flowers (*Hist. Aug.* 21.5).

High demand for flowers led to the development of techniques to make them bloom early. Thus, the moralist Seneca the Younger (*Ep.* 122.8) and the agronomist Palladius (*Rust.* 3.21.2) recommended watering flowers with lukewarm water. Winter roses cost much more than their spring sisters, as the Latin satirist Martial noted: “rare things are pleasing: thus, early apples are more esteemed; thus, winter roses fetch a higher price” (4.29.3–4). Winter roses could also be imported from warmer regions, such as Egypt. In another of his epigrams (6.80), Martial wrote that the Emperor Domitian received a gift of Egyptian winter roses. The sailor who brought them from Memphis, however, realized that Egyptian roses paled in comparison with the glorious Paestan roses growing near Rome. Martial closed his epigram by exhorting the Egyptian land: “But you, O Nile, since you are forced to submit to Roman winters, send us your crops, and receive our roses.” Egypt was to be praised much more for its wheat than for its roses!

Even in Egypt, however, flowers could not grow all year round, as an extraordinary second-century CE letter preserved on papyrus testifies. A certain Apollonios and his wife Sarapias wrote to decline an invitation to the wedding of Sarapion, son of Alexandros and Dionysia. They gave illness as an excuse. One suspects, however, that the real issue had been the inability of Apollonios and Sarapias to deliver the flowers necessary for the wedding:

> Apollonios and Sarapais to Dionysia, greetings ... There are not yet many roses here—rather a shortage—and from all the farms and all the garland makers we had difficulty in putting together the thousand which we sent you via Sarapas, even by picking the ones which should have been picked tomorrow. We had as many narcissi as you wanted so we sent you four thousand instead of the two thousand you ordered. We wish you had not put us down as petty-minded by mocking and writing that you had sent the money, when we regard your children as if they were ours and we respect and love them more than our own, and therefore we are as pleased as you and their father ... Sarapas will confirm what I said about the roses, that I have done all I could to send you the number you wanted, but couldn't find them.
>
> (*POxy*. 3313; translation Muir 2009: 41–2)

John Muir (2009: 41) suggests that the letter was written in late January, before the month of February when festivals named "Rosebearing" were held. The numbers of flowers given in the letter—a thousand roses and four thousand narcissi—are astounding, and even allowing for exaggeration, they raise questions as to the social status of the family marrying their son.

Fresh cut flowers were highly sought in the ancient world. Dried ones were also in demand, especially for use in medicinal preparation. We now turn to the trade of medicinal dried herbs.

CRETAN AND LIBYAN HERBS: SELLING PHARMACOLOGICAL PLANTS IN ANTIQUITY

In his treatise *On Antidotes*, Galen wrote that many medicinal plants from some regions of the Empire were only brought to Rome once a year:

> Some drugs are not brought in every day [to Rome], for instance, those that come from Sicily or Libya and Crete every year in the summer. In Crete, herbalists are maintained by the Emperor (not just by the Emperor himself, but also by the entire city of Rome); they send out vessels called *plekta*, which are plaited from agnus castus. Many drugs are brought from Crete to many nations, so that nothing from this island should be missed: no plant, no fruit, no seed, no root, and no juice.
>
> (Gal., *De antid.* 1.2, Kühn 14.9)

Libya and Crete had had a reputation for their powerful pharmacological plants for many centuries. Libya was the country of silphium (see below), and Crete produced numerous herbs, several of which are mentioned in the recipes of the Hippocratic Corpus, the collection of medical treatises written in the fifth and fourth centuries BCE and attributed to Hippocrates (see Totelin 2009: 166–7; 2016). By the time of Galen, Crete's reputation had not only increased, but was sanctified by imperial patronage, for the emperor now employed Cretan men called *botanikoi*, literally plant people who grew and gathered the plants (for these people and other involved in the drug trade, see Boudon-Millot 2003).

After giving details on the vessels that contain the Cretan drugs, Galen went on to discuss the issue of adulteration, one that plagued the ancient drug trade:

> Most [of these Cretan products] are pure. Some of the juices are adulterated, but this occurs relatively rarely. Because of the abundance of plants on the island, its plant men

hardly try to adulterate drugs, which is unlike the great manufacture [of forged drugs] occurring elsewhere. However, instead of [selling] the pure juice of wormwood, they mix it with horehound, or they adulterate it slightly, so that the perfume sellers in Rome who, every year, buy the *plekta* vessels filled with drugs, must learn to recognise whether something has been altered.

(Gal., *De antid.* 1.2, Kühn 14.10)

Galen, then, presented adulteration as a given in the drug trade, even that which occurred under the patronage of the emperor. It was, in his opinion, the duty of the perfume sellers in Rome to detect fraud. These people, beside selling ready-made unguents, also traded in medicinal plants in the ancient world. However, these perfume sellers fell short of the physician's expectations. Indeed, while they were familiar with the plants imported from Crete in their dried state, they failed to recognize the same plants in their fresh state growing near Rome (*De antid.* 1.5, Kühn 14.30–1; see Hardy and Totelin 2016: 38). To Galen, anyone involved in drug preparation should be able to recognize plants at every stage of their growth, in order not to be defrauded. Elsewhere, Galen conceded that even people who knew drugs well could be deceived since forgers were very skilful. His solution was either to travel to the place where drugs grew (or occurred, in the case of minerals) or to procure them by the intermediary of a trusted friend (*De antid.* 1.2, Kühn 14.7).

The pharmacologist Dioscorides (first century CE) provides us with much information on drug adulteration in antiquity. Thus, he wrote that the bark of frankincense was adulterated by mixing it with the bark of several varieties of pine (1.68.5); the seed of lovage with other seeds that resemble it (3.51.3); and that the juice of silphium could be adulterated with *sagapenon* (perhaps asafoetida) or bean-meal before drying, adding that "this can be detected by taste, smell, appearance, or by moistening" (3.80.3).[5] It is worth pausing for a while on this last plant, silphium, as it was one of the most important pharmacological plants of antiquity, but its identification remains a mystery to scholars, although it is possible to determine that it probably belonged to the genus *Ferula*, which includes asafoetida (for summaries of the question of identification, see Amigues 2004; Totelin 2015).

Silphium came from Cyrene, which had been a Greek colony in Libya from the seventh century BCE. According to the Therans (the Greeks from Thera who had established the colony in Cyrene), the plant suddenly sprang up seven years before the foundation of Cyrene (in 638 BCE) after heavy rains (Theophr., *Hist. pl.* 6.3.3 and 3.1.6; *Caus. pl.* 1.5.1). These stories had as their purpose to present silphium as a Greek plant—one that appeared roughly at the same time as the Greeks started to explore Libya for colonizing purposes. In reality, the plant may have been traded by the indigenous populations of Libya for many centuries before the arrival of the Greeks (see Evans 1921: 284–5 for the suggestion of silphium Bronze Age trade). Nevertheless, the Greeks of Cyrene fully appropriated silphium: they made it the symbol of their city-state. A representation of the plant figures on most Cyrenaic coins produced from the seventh to the first centuries BCE (Koerper and Kolls 1999). A Laconian cup dating to the first half of the sixth century BCE and found at Vulci (in Etruria) also bears scenes that scholars interpret as activities linked to the silphium trade (see Figure 3.3). On the top scene, a seated character supervises four others while they are weighing a product; he is represented as taller and larger than all the others, indicating his high hierarchical status. An inscription designates him as Arcesilaus, which was the name of four kings of Cyrene: here we have Arcesilaus II, who ruled from 560 to 550 BCE. The smaller characters manipulate the large scale

FIGURE 3.3 Weighing and loading of silphium in Cyrene. Wash drawing of a Laconian cup, seventh century BCE. Bibliothèque Nationale, Paris. Courtesy of Wellcome images.

and carry net bags, which may be the bags in which silphium was transported. On the bottom scene, two men scurry around carrying net bags, under the watchful eye of a man wrapped in his cloak.

The Greeks and Romans used silphium as a condiment in cooking (see Dalby 2003: 303–4) and as a medicine, in particular for gynaecological ailments (Riddle 1991, 1992, 1997). Because importing the plant was so expensive (see e.g. Ar. *Pl.* 923–5), the Greeks repeatedly attempted to transplant it to the Peloponnese and Ionia, but without success. The author of the Hippocratic treatise *Diseases* IV stated that only Libyan soil had the "humour" that enabled silphium to grow spontaneously (*Morb.* 4.3, edition Potter 2012: 104; see Lonie 1981: 273 for commentary). Silphium was a wild plant that could only thrive in a small geographical area.

By the first century CE, however, something had happened to Cyrenaic silphium. Pliny reported that the plant had not been seen in Cyrenaica for many years:

> After these [other plants], will be discussed the very famous *laserpicium*, which the Greeks call *silphium*, discovered in the province of Cyrenaica, whose juice is called *laser*: it is very important in general use and in medicines, and it is worth its weight in silver denarii. For many years now, it has not been seen in that region, because the tax farmers who lease pasture land, projecting higher profit, lay waste to the land by grazing sheep on it. The only stem found in living memory was sent to the emperor Nero.
>
> (*HN* 19.38–9)

Pliny's account does, however, raise more questions than it answers. Since silphium was expensive, it is very unlikely that tax-farmers would have made more profits from grazing the silphium-bearing land than from gathering the herb. It is more plausible that overexploitation of the herb led to its near-total extinction. While Cyrenaic silphium might have become extremely rare in the first century CE, in the early fifth century CE, the bishop of Cyrene Synesius mentioned silphium in two of his letters, showing that the plant was not entirely extinct (*Ep.* 106 and 134; see Parejko 2003; Roques 1984). Whatever might have happened to silphium, by the time of Pliny, alternatives had been found to Cyrenaic silphium: plants growing in Media, Persia, and Armenia that grew in abundance but were inferior in quality (Plin., *HN* 19.40; Dsc. 3.80).

While it is not possible to determine whether silphium truly became extinct or not (the debate continues), it is clear that overexploitation of natural resources for economic purposes could cause major issues in antiquity. We turn to another case: the cedars of Lebanon, and the timber trade in antiquity.

HADRIAN'S CEDAR: TIMBER TRADE AND EXPLORATION IN ANTIQUITY

In the 1860s, while working in Syria, the French archaeologist Ernest Renan discovered numerous highly abbreviated monumental Latin inscriptions on rocks in the northern part of Mount Lebanon (Renan 1864). The longest inscriptions read as "IMP(eratoris) HAD(riani) AUG(usti) D(e)F(initio) S(ilvarum) A(rborum) G(enera) IV C(etera) P(rivata)," that is, "Boundary of the forests of the Emperor Hadrian Augustus; four species of trees; the remaining species are private [unclaimed]" (translation in Meiggs 1982: 85–6). The species that Hadrian thus reserved for himself were probably cedar of Lebanon (*Cedrus libani*), Cilician fir (*Abies cilicia*), juniper (*Juniperus excels*), and oak (*Quercus* spp.) (Abdul-Nour 2001; Breton 1970).

Scholars sometimes interpret Hadrian's attempt at delineating the forest of Mount Lebanan as one of conservation, aimed at tackling deforestation and depletion of resources in precious wood, especially in cedar wood (Bru 2011: 61–2; on deforestation more generally see Hughes and Thirgood 1982). While Hadrian might have attempted to preserve the cedar forest, this was certainly not entirely disinterested. There were economic gains to be made from the sale of cedar timber. The wood of this tall tree was used throughout antiquity to build ships and ceilings of large buildings (Meiggs 1982: 55). Thus, Theophrastus remarked that silver fir is best for making triremes, but that in Syria and Phoenicia people use cedar because they do not have access to fir; and Vitruvius noted that the ceiling of the temple of Artemis at Ephesus was made of cedar, as was that of many famous temples (*De arch.* 2.9.13). Its beautiful grain and scent, as well as its durability, also made it an excellent choice for the production of precious objects, such as statues. Pliny informs us that a statue of Apollo Sosianus found in a shrine in Rome, but originally from Seleucia, was made of cedar wood, for example (*HN* 13.53). By marking off the cedar forest, Hadrian could make sure the revenue from cedar trade reached the imperial coffers.

Kings before Hadrian had marked the limited geographical areas where certain plants grew as royal property. Thus, the kings of Judea were the owners of the only two gardens in which the balm tree, a precious tree the sap and bark of which were used in medicines and perfumes, grew (Plin., *HN* 12.111). It is impossible to know whether the balm tree naturally had such a limited distribution, or whether the kings had artificially reduced its distribution in order to create a lucrative monopoly. Whatever the case might be, when the Roman emperors Vespasian and Titus defeated the Jews in 71 CE, they chose to exhibit the balm tree in triumph (Plin., *HN* 12.112). Parading the tree in triumph served the dual purpose of entertaining the Roman crowds with marvellous natural products and affirming Roman power over the conquered land: appropriating trees suggested that the land was now fully in Roman hands, down to its roots (on trees in triumphs, see Östenberg 2009: 184–8; for links between botany and empire in the Greco-Roman world, see Flemming 2005; MacAuley 2008; Pollard 2009; Totelin 2012; for earlier examples, see Carroll 2003: 43).

In his discussion of the balm tree Pliny noted that trees had been shown in triumphs since the time of Pompey the Great (106–48 BCE). Elsewhere, he stated that ebony was one of the trees (or perhaps even the only tree) shown by Pompey the Great in his triumph over Mithradates VI and the pirates, on 29 September, 61 BCE (*HN* 12.20; see Kuttner 1999a). Pompey may have chosen to include ebony in his triumph both because it was so rare and in emulation of Hellenistic rulers who had also displayed its wood. Indeed, we hear that two thousand ebony logs were shown at the Ptolemaeia, the festival held in honor of Ptolemy II, king of Ptolemaic Egypt from 283 to 246 BCE (Callixeinus as preserved in Ath. 201a; on the Ptolemaeia, see e.g. Rice 1983; Wikander 1992). In any case, ebony did not grow in Pontus (Mithradates' kingdom) or any of the other territories conquered by Pompey, but rather in Africa and India, as ancient authors themselves knew (Dsc. 1.98; Plin., *HN.* 12.17; Solinus 52.52). The exact provenance of ebony in Africa, however, remained a mystery to the Romans. Indeed, Pliny noted that an exploration of the geography of Ethiopia (the land south of Egypt), which had been sent under the Emperor Nero, had shown that "the tree [ebony] is rare from Syene, border of the Empire, down to Meroe, over a space of 1996 miles, and that there are no other species of trees apart from the palm tree" (*HN* 12.19; see Ulrich 2007: 251–2). Unlike in the cases of the balm tree and cedar, the Romans could never monopolize the trade in ebony; they always remained dependent on intermediaries for their supply of the precious wood.

The fact that Pompey chose his victory over Mithradates to lead trees in triumph for the first time is significant. Mithradates was known for his interest in pharmacology and botany. He had allegedly invented an antidote that protected him against the threat of poison (see Totelin 2004 for references). Pliny noted that the king had collected information on plants from all his subjects and written down their properties in treatises that Pompey ordered his freedman Lenaeus to translate into Latin; the conquest benefitted both the Roman empire and human life (*NH* 25.7). By using trees in his triumph over Mithradates, Pompey used a symbolic language that would have been meaningful to the defeated king (who had by then committed suicide). In 55 BCE, Pompey went on to dedicate to Venus Victrix a portico and garden, emulating the rulers of the East who were famed for their beautiful *paradises* (see Kuttner 1999a; von Stackelberg 2009: 76–8).

To sum up, ancient rulers, kings, queens, and generals showed much interest in plants. They had the power to conquer land, transplant trees, plant beautiful gardens, and appropriate for themselves certain plant species. Plants were symbolically significant to rulers for they had roots that linked them to their territories of origins. More prosaically, however, plants were economically important to rulers: they could become a source of great revenue, especially when a monopoly could be created. That said, much of the actual

FIGURE 3.4 Papyrus letter with inventory, PSI 15.1558. Florence, Istituto Papirologico G. Vitelli. Courtesy of the Istituto Papirologico G. Vitelli.

work involved in growing and trading plants was carried out at much lower echelons of society. We end this chapter with a source that reveals intimate details of the life of these traders.

A FAMILY ARGUMENT: TRADING PLANTS IN ANTIQUITY

Ancient documents that name traders are rare and therefore deserve more attention than historians generally give them. Among such documents is a third-century CE letter (see Figure 3.4), perhaps found at Oxyrhynchus, which inventories several products that, with one exception (asphalt), are all vegetal:

> Calleas to the most esteemed Flavius, greetings. You will do well to take delivery of the specimens from the seaman Perouan, so that you might hand them over to Spartas' wife [or 'to the lady Sparta']. When I took delivery, at the Coptos tariff, I also supplied them to her, not as I would to strangers, but rather to family members. As she did not trust this, she did not accept them. All the specimens were left in a large basket (creel), with the exception of the saffron. It was left outside, lest it crumbled. Here is the register:

> Mastic, 1 mina, and 160 drachmas
> Storax, 2 minas and 4 drachmas; of x ...
> Hepatic aloes (*Aloe perryi*?), 2 minas ...
> Myrrh, 2 minas ...
> Unguent [literally: something that is poured out], 2 minas ...
> [Pine] resin, 2 minas ...
> Syrian asphalt, 1 mina
> Saffron ...
> Excellent [product], 4 ounces
> Excellent amomum, 2 ounces.
>
> (*PSI* 15.1558; translation in Totelin 2020; see also Bonati 2016: 280–8)

The situation described here is complex. Calleas attempted to sell products to Spartas' wife, as if they were "family." However, she did not trust the deal and left the products. Calleas is therefore writing to Flavius to ask him to act as an intermediary in the transaction. A shipman named Perouan is also mentioned. Calleas specifies that he took delivery of the products "at the Coptos tariff." Coptos was an important trading station on the Nile. It was there that goods that had traveled on the Red Sea to ports such as Berenike, then through the desert, were loaded onto river boats (see Young 2001: 47–51, 209). The reference to a "Coptos tariff" seems to indicate that prices were fixed at Coptos, where taxes might have also been levied. An example of an ancient tariff has survived: the so-called "Alexandrian tariff," a rescript of the emperors Marcus Aurelius and Commodus, preserved in a legal compilation called the *Digest* (39.4.16.7). It lists various products subject to the *vectigal* tax. Among these are several plants that we have encountered in this chapter and/or that are listed in Calleas' register: cinnamon, long pepper, white pepper, cassia, myrrh, amomum, and *laser* (silphium).

The products that Calleas lists came from all over the known world and beyond: myrrh came from Arabia (Dalby 2000: 117–20); styrax (the scented gum of *Styrax officinalis*) came from Syria (Dalby 2000: 137–8); the best mastic came from the island of Chios (Dalby 2000: 136–7); the best saffron originated in Cilicia (Dalby 2000: 138–9); aloes might have come from the Arabian island known today as Socotra (Dalby 2003: 6–7); and amomum might have grown on the slopes of the Himalayas (Dalby 2000: 102–3). All these vegetable products would have found use in perfumery, medicine, and mummification; others (mastic and saffron) would also have been consumed as foods. The amounts listed are a good indication of the relative values of the products: while Calleas inventories 2 minas and 4 drachmas of storax (a mina contains 100 drachmas; a drachma weighs a little less than 4.5 grams), he only lists 2 ounces of amomum.

THE LIMITS OF EXPLORATION

The agricultural treatise of Varro, *On Agriculture*, is written in the form of a dialogue between various interlocutors. One of these is Scrofa, who reported the following:

> When I led the army in the interior of Transalpine Gaul, near the Rhine, I came to several territories where neither vine nor olive tree nor fruit trees grew, where they fertilized the fields with a white clay which they dug.
>
> (*Rust.* 1.7.8)

By the time of Varro, Transalpine Gaul was a well-established province of the Roman empire; yet its flora remained surprising and mysterious to one of its generals. "How could the indigenous population survive without wine, olives, and fruits?" we can almost hear Scrofa say. Scrofa, like many generals before and after him, turned himself into an ethnologist and scientist, observing with curiosity the customs of peoples and the flora and fauna of the regions he visited. Exploration in antiquity rarely occurred without political expansion. Rulers (kings, queens, emperors, generals, and other leaders) did not explore new lands for the sake of scientific knowledge: they explored lands that they had already conquered, or lands that they intended to conquer. They did so with a view to exploit the natural resources of the land. Some did so wisely, with something resembling an understanding of the risks of overexploitation; others did not. However, there was always a limit to what rulers could conquer and explore in antiquity. Greek and Roman rulers never managed to conquer or explore beyond the Indus Valley for instance. Yet plant products from eastern Asia, such as cinnamon, did reach the Mediterranean world. Greek and Latin authors did not really know where the ebony tree grew, yet ebony wood reached Athens, Alexandria, and Rome. That is, trade went further than exploration and political conquest.

CONCLUDING REMARKS

Plant products passed through the hands of many traders and retailers in the ancient world. These are often shadowy figures of whom we know very little. Still, we can piece some information together. We hear of both male and female traders and retailers, and indeed of couples involved in trade; family bonds and networks were central to the ancient plant trade. Some lived a hand-to-mouth existence, while others became quite prosperous. We know that traders argued and haggled, and that some of their practices were considered dishonest by onlookers. They used tales to promote their wares, and perhaps also to create a sense of community. They took deep care of their products, carefully choosing containers for their transport. They live on in portraits drawn by authors such as Galen, fragmentary papyri from Egypt, and provincial funerary steles.

CHAPTER FOUR

Plant Technology and Science

PATRICK HUNT

Plant and wood use and the technologies thereof have a long human history—and indeed an even longer prehistory—but while antiquity has not always preserved documentation for human interaction with plants, the material record in archaeology complements some of the fragmentary documentary textual record of literate antiquity. Because the observable technology of *Homo sapiens* can be easily traced back far beyond the last 10,000 years of the Holocene alone, and in what can be termed Neolithic use up to the last 5,000 years in some Old World contexts, there must be some arbitrarily selective chronological bracketing used in this discussion, as well as geographical constraints, in order to focus the specific human activity. This chapter will at times extend back to the Copper Age of around 5,000 years ago, but its main focus will be the Early Bronze Age to the Roman Era (3000 BCE to 500 CE), covering approximately three and a half millennia. While some of the recorded global material summarized here will extend to the Far East, the chapter will concentrate on the Old World—mainly Mesopotamia, Ancient Egypt, and adjacent regions in the Ancient Near East together with the Mediterranean world of pre-Europe and central Asia.

Because organic materials such as plants and wood products are subject to active decomposition over time in most environments, comprehensive discussion is limited by a lack of preserved materials (Hurcombe 2014: 2). On the other hand, deserts, bogs, and alpine environments often provide better conditions for the preservation of plants and woods; even carbonization and volcanic episodes can seal vital information for better understanding the range of plants used and technologies employed. For example, while the narrow Nile River basin with its tropical riparian conditions accelerates the decomposition of Ancient Egyptian and earlier plants and wood products, the adjacent deserts on both sides of the Nile have preserved plant and wood materials well for nearly five thousand years. Similarly, the high Ötztal Alps (now between Austria and Italy) preserved some of the most diverse and best-documented examples of wood and plant uses of all time at the 10,500 foot elevation: frozen materials associated with the glacial find of "Ötzi the Iceman," who belonged to a preliterate culture of some 5,300 years ago. Extremely low year-round temperatures inhibited decomposition in the high-altitude Alpine context. Sporadic findings of plant materials in European bogs—including a few rare textiles and even consumed food—can also be mentioned in this context. There the presence of anaerobic conditions due to lack of oxygen and humic acid in the bog, along with certain phenomena such as the pH levels and winterish cold water at the time of burial, inhibited decomposition. Bog findings include those associated with the Lindow Man, Tollund Man, Yde Girl, and Grauballe Man, mostly from the late Celtic or early Roman periods several millennia ago.[1]

The sciences associated with plant and wood use in antiquity, particularly in their infancy, are generally not well recorded, but many practical, empirically derived observations that nonetheless classify as scientific in any era would have passed on from generation to generation, sometimes by oral tradition only. Botanical practices, agriculture, and horticulture are in no way modern developments, as we have ample evidence of treatises on these subjects from the ancient world: Egyptian, Mesopotamian, Phoenician, Greek, and Roman. Much material has been lost, since papyri and other organic materials that recorded knowledge of plant and wood science in antiquity have not usually survived, but even when the texts themselves did not survive, references to these texts by other, later authors provide useful insights into botanical knowledge and plant uses in antiquity. Ironically, some ceramic texts of ancient plant science were preserved due to firing of inscribed clay tablets when the cities in which they were housed were destroyed by fires, as in the case of Nineveh in 612 BCE. Another fortuitous survival was the work of the Greek scientist Theophrastus, often called the Father of Botany (*c.* 371–287 BCE). An example of systematic plant science in antiquity, his *Enquiry into Plants* is an important primary source for recorded Greek plant knowledge, affirming the study of plants (*phuton theoria*) to pursue dual investigations in dual fields:

> The first investigation deals with plants that grow of their own accord, and here the starting point belongs to their nature; whereas the other starting point is that which proceeds from human ingenuity (*epinoia*) and contrivance (*paraskeue*), which we assert helps their nature to achieve its goal.
>
> (*Hist. pl.* 3.1.1)

Epinoia has alternatively been translated as "thought," and *paraskeuē* as "preparation," but both ideas clearly reference a scientific approach based on the systematic study of plants that Theophrastus carefully delineates. As for ancient technology, its definition will not be limited to usages by ancient authors of the Greek word *technē*, which usually implies skill derived over time, practice, or craft, but also has a connotative domain covering the systematization and propagation of knowledge about plants, trees, and their products. Other words for taxonomy or technology from various ancient cultures will also enter the discussion, and all suggest similar thoughtful approaches rather than accidental use.

The following sections cover selected examples of ancient plant science and technology related to the creation of plant-derived textiles, wood products, perfumes, paper, and plant-sap rubber use. While not intended in any way to be comprehensive, this mainly Old World summary of the range of human use of plants, trees, and their products is more or less representative of human ingenuity for however long such practices have been features of human enterprise.

PLANT TEXTILES AND TECHNOLOGY IN ANTIQUITY

Historic and prehistoric uses of fibers in textiles are not limited to clothing; rather, textile garments are part of the usage that includes plant accessories to be carried by humans or service animals as portage; thread to be sewn for a variety of purposes other than garb;

cordage tied to other materials; netting for fishing; food production; and other plant fiber uses. Technologies for textile production range from tying, looping, and interwoven splicing, to spinning, looming, netting, and other methods of working fibers after they have been separated from their original plant material. In textile production, threads could be tied, plaited, or looped in a variety of ways, ranging from simple—including knotless netting known as *nålebinding* mostly from early use in Coptic Egypt, Scandinavia, and the Balkans—to more complicated patterns for additonal integrity, with many knots forming those most intricate.[2] Textiles could be woven by hand, or they could be loom-woven; the oldest warp-weighted looms that have survived even fragmentarily may be from the Neolithic Starčevo culture in the Balkans about 7,000 years ago or earlier (at least sixth millennium BCE), and painted and carved imagery of looms from historic cultures is abundant (Nielsen 2005: 129–35). As in the production of other basic, multi-patterned fiber textiles, the fibers to be interwoven optimally should be coming from opposing directions. Thus, in looms the vertical warp threads were often under tension, aided by gravity in order to introduce the horizontal weft threads, even in plain weaving; loom weights at the bottom of the warp thread make that possible. In the case of looming, the support structures needed for interweaving a template with diametrically opposing warp and weft are also often plant-sourced, constructed of wood. As regards fiber sources, usable fiber plants come from a broad spectrum of environments, including natural forests and riparian stream beds or adjacent areas, as well as planted fields and groves of trees (Barber 1994). A few of the plants (or trees) used for fiber in antiquity were reed, rush, cotton, linen, silk (a special case secondarily derived from silkworms eating mulberry leaves), hemp, palm, tree bast, and nettle.

Reeds and **rushes** have been used as plant fibers for many millennia predating history, basket weaving having been an important historic and prehistoric use. Lacustrine reed plants like the lakeshore bulrush (*Schoenoplectus lacustris*) are common throughout Europe, and while basket weaving is a known craft of Late Neolithic and Copper Age peoples in the mid-fourth millennium, for example, as in the case of the Arbon Bleiche 3 dwellings on Lake Constance (bordering what are now Germany, Austria, and Switzerland) (Jacomet 2009: 47–9), it remains unknown exactly which reeds were used based on the very fragmentary organic evidence. Speculation that the roofs of Copper Age Swiss Lake dwellings were at times thatched as textiles with such reeds remains unproven. On the other hand, it is well established that rafts and boats made of bundles of reeds lashed together with rope saw early and widespread use for purposes of fishing, irrigation, and trade (see Figure 4.1). Egyptian Old Kingdom reed boats were made from papyrus (*Cyperus papyrus*), and even earlier (*c.* 5000 BCE), woven reeds appear in the archaeological remains of a reed boat sealed with tar from Subiya, Failaka Island, Kuwait, near the Euphrates-Tigris delta. Reed boats in the Tigris-Euphrates were certainly known since the Sumerians (4500–1900 BCE), and the Old Babylonian (*c.* 1900–1700 BCE) account of Atrahasis (protagonist of an eighteenth-century BCE Akkadian epic) relates the making of his reed-rope and bitumen-sealed, giant round boat (a modern *gufa*) of around 70 meters (223 ft) in diameter for the Mesopotamian "Noah" Flood stories (Finkel 2014: 122–3), continuing an already long tradition of reed boats in that region for river crossing and transport (Lawler 2002: 1791–2). The Hebrew text of *Exodus* 2 where Moses is hidden in a floating "bulrush" (*gome* in Hebrew) or wicker basket of reed also provides literary evidence of woven reed boats, possibly referencing papyrus use. As for other types of reed textiles, these include sleeping mats or blankets that are known, among other places, from eastern China, where water reed (*Phragmites australis*) appears as early as 4,700 years ago based on plant phytoliths discovered at the Tianluoshan site (Zhan et al. 2016). What are likely **hemp** (*Cannabis sativa*) cord impressions are found on pottery as early as in China's Yangshao culture *c.* 4800 BCE (Barber 1992: 17). Classical Greek and Roman texts make frequent reference to hempen rope, and hemp cordage has been recovered from wells in Roman forts at Saalburg and Zugmantel on the northern frontier of the province Germania, possibly sourced from Asia Minor, which produced high quality hemp fiber (Plin., *HN* 19.174; Wild 1970: 17). Other plant fibers, however, were also twisted into ropes. For example, the Roman fort at Bar Hill on the Antonine Wall in Scotland yielded rope of elder.[3] Rope textile of an unknown plant source can be

FIGURE 4.1 A long, narrow reed boat, at the stern of which stands an Assyrian soldier conducting captives across the water, *c.* 668–627 BCE, the reign of Ashurbanipal. Gypsum alabaster relief fragment from the Assyrian palace at Nineveh. New York, The Metropolitan Museum of Art. © Gift of John D. Rockefeller Jr., 1932. Courtesy of The Met Museum.

seen on a first-centry CE Roman terracotta from the British Museum where a satyr presses round coils of rope in wicker-work mats used for a winepress because the rope can be easily compressed.[4]

For clothing and related textile types, cotton, linen, hemp, nettle, and even moss have considerable ancient use. **Cotton** (*Gossypia arboreum*) grows in a boll (seed pod) on the seeds of the cotton shrub (Gleba 2012) and is harvested by picking the boll when it springs open and pulling out the fibers for spinning. As a vegetal textile, cotton artifacts can be found as early as the Early Bronze Age (2500 BCE) Indus Valley Culture of Mohenjo-Daro, with evidence of cultivated cotton in archaeological sites, including pieces of cloth and lengths of yarn in both 12 ply and 24 ply (Broudy 1993: 20). This Indus Valley cotton predates—by millennia—the account of the Greek historian Herodotus, who praises it above sheep's wool for clothing and who elsewhere in his *Histories* describes the cotton garb of Indians under the Persian officer Pharnazathres of Xerxes' army (3.106; 7.65). Theophrastus derived accounts about organized plantings of cotton growing in rows "like vines" in India and Bahrein from Alexander the Great's sojourn in those regions (*Hist. pl.* 4.4.8), and Arrian also describes cotton textiles worn by Indians (*Anab.* 7.16). It is known that in the sixth century CE cotton was exported from Gupta India to China, and at an earlier date Indian cotton had made its way to the Roman West as recorded in the *Periplus Maris Erythraea* of the first century CE (Casson 1989: 17; Schlinghoff 1974: 81). In India the old Sanskrit-derived word *karpâsa* refers to cotton, and our modern word derives from the earlier Arabic *qutun* (Betts et al. 1997: 489–99).

Linen is another of the most common ancient textiles, made from the annual flax (*Linum usitatissimum*) plant. After its seed was sown, flax was harvested in about ninety to one hundred days depending on climate, usually when up to a meter in height, with dried bast fibers found on the inside of the stalk, which at that stage had yellowed. The next phase was retting, which involved soaking the stalks in stagnant water while they swelled. Brushing and combing down the stalks released the long flax fibers, which were polished in the process. The stages of spinning flax into linen that followed have not differed much since antiquity. It is important to note that linen was more expensive than cotton because the production has almost always been more labor and time intensive. The earliest usage of dyed flax fiber possibly goes back to the Paleolithic Caucasus (Balter 2009: 1329; Kvavadze et al. 2009: 1359), and Sumeria certainly produced it, for the Sumerian *Inanna and Dumuzi Hymns* precisely describe linen production from flax, beginning with growing and harvesting the flax to combing, spinning, braiding, warping, weaving, and bleaching (Wolkstein and Kramer 1983: 30–1). The world's oldest complete garment is the so-called Tarkhan Dress, a v-neck shirt of *c.* 3482–3102 BCE from the First Dynasty Tomb of Tarkhan, that survives almost intact and is now housed in the Petrie Museum of Archaeology, University College London (*UCL News* 2016) (see Figure 4.2). Linen was produced in great quantity in Ancient Egypt, from the Predynastic to the Old Kingdom onward—likely in the hundreds of tons per year at many points, based on what has survived alone (Vogelsang-Eastwood 1992)—as the Nile River Valley provided ideal growing contexts and therefore high yields. Egyptian linen production was so extensive—*byssus* is only one Egyptian word for fine linen—that it is shown on almost all depictions of garments in tomb paintings, and linen is also the primary mummification wrap textile. Though Greek biographer Plutarch (*c.* 40–120 CE) references Egyptian priests as wearing fine linen for reasons of purity and sanitation, its lightness was an advantage in the desert heat and Nile humidity (*De Is. et Os.* 4). Due to the region's climate and linen's ability to absorb and conduct moisture, the dominance

FIGURE 4.2 Pleated dress, thought to be the earliest extant garment in the world, Egypt. Ancient Egyptian. First Dynasty, *c.* 3100–2890 BCE. © Werner Forman. Courtesy of Getty Images.

of linen in Egyptian clothing, and likewise the extent of its export to the Mediterranean world beyond, should not be underestimated. The Phoenicians not only traded Egyptian linen but also likely took flax around the Mediterranean and beyond. Although there are at least five different words for linen in Classical Hebrew (the most common being *shesh,* while *pishet* is flax), biblical allusions to fine linen usually reference Egyptian linen, as with the Pharaonic gift in the Joseph narratives of Genesis 41:42; however, it is likely

Mesopotamian linen that is meant in the later Iron Age *Esther* narrative of Susa, where the Hebrew *bus* (derived from Egyptian) is used for fine linen (Hurvitz 1967: 117–21). Cultivation of flax and linen production also had a long history in Greece, some of the earliest evidence coming from Mycenaean Linear B tablets at Pylos that name linen as *li-no* in texts about linen production and the female linen workers as *li-ne-ya* (Robkin 1979: 469–74; Gillis and Nosch 2007), whence came the Classical Greek *linon*, which the Romans borrowed as *linum*. Writing from a Roman perspective, Pliny the Elder—who describes linen production in detail—records that the most lustrous and finest quality European linen was imported from Spanish Saetabis (now Xàtiva), with Italian linen from Faventia (very white) and Retovium (very fine) vying for second place in estimation, and that from Alia ranking third (*HN* 19.1–2).

Hemp (*Cannabis sativa*) textile for clothing, which is more robust than many other textiles, has been sufficiently attested for millennia but apparently was never used in the same quantity as linen or other vegetal material. Herodotus, describing the western Scythians in Thrace—barbarians in the eyes of the Greeks—mentions their hemp cloth that resembles linen but is coarser (*Hist.* 4.74). In the Early Bronze Age, Italy apparently used hemp textile for cloth at times in burial, as noted in burial contexts at Gricignano d'Aversa, Caserta, and at Himera, Sicily from the fifth century BCE (Gleba 2008). This was the case, too, in ancient China, where corpses were shrouded in hemp cloth before interment. Hemp corpse covers were recovered from Western Han Dynasty (206 BCE to 24 CE) tombs in Gansu province: there, hemp-cloth outer shrouds, secured with ropes of hemp, covered bodies clad in silk garments (Li 1974).

Silk textile falls within the unique hybrid category of being vegetal in source through the mulberry leaves eaten and spun by insect larvae, specifically the silk moth (*Bombyx mori*). While sericulture as an industry reasonably can be dated back to Bronze Age China, *c.* 1600–1050 BCE in the Shang Dynasty (Guo and Qiu 2013: 41–9), Aegean wild silkworm (*Pachypasa otum*) use is also now suggested from a cocoon found in Theran volcanic contexts from Akrotiri in the Bronze Age Aegean, mid-second millennium BCE, although clearly not a *Bombyx* silkworm from China or Asia (Panagiotakopulu et al. 1997: 420–9), possibly confirming Aristotle's later notes about a long tradition of "wild silk" (*Hist. an.* 5.19.6). New studies suggest silk and sericuture in the Indus Valley in Early Bronze Age Harappan and Chanhu-Daro sites *c.* 2450–2000 BCE (Good et al. 2009: 457–66). One of the first known appearances of *Bombyx mori* in the Eastern Mediterranean world consists of silk strand traces found in mummy hair from Twenty-first Dynasty Egypt at the middle of the eleventh century BCE in a Deir el-Medina royal worker's burial (Lubec et al. 1993). Although speculative, the first documentation of silk textile might appear in biblical literature: Joshua 7:21, in which the Early Iron Age narrative states Achan purloined a "Babylonish garment"—possibly silk?—from Jericho that was at least as valuable as a gold bar of fifty-shekel weight and two hundred silver shekels. As a gauge of relative value, the gold and silver are listed after the garment (*'aderet shinar* in Hebrew where *'aderet* references garment and *shinar* references Babylon, possibly the easternmost locus known to the "author"). By the Exilic period (sixth century BCE) in or around Babylon, the Hebrew prophet Ezekiel may also reference silk, as Ezekiel 16:10 notes with the singular Hebrew word *meshî*. By the end of the first millennium BCE, Han Dynasty silk certainly made its way to the Roman world. Silk was highly prized in wealthy Roman courts, delivered via both overland and sea trades (Thorley 1971: 71–80). In an effort to bypass expensive middlemen en route, sericulture was eventually brought directly into the Byzantine world from the peripheral Chinese silk industry and the Silk Road around

563 CE by the emperor Justinian's clerical "spies," who were given potted mulberry trees as gifts, likely in Khotan, and intentionally purloined silkworm larvae hidden in bamboo tubes in order to begin an imperial early Byzantine silk industry in the West (Hunt 2004: 31–2).

Palm (date palm, *Phoenix dactylifera*) is not a plant readily available in most environments as a clothing textile, its bast fiber use limited to mostly arid and desertified regions or Mediterranean coasts, but it is attested, especially in the Saudi Arabian Peninsula even in antiquity and in Hellenistic Egyptian Serapis contexts (Retsö 2003). **Nettle** textile for clothing, on the other hand, is well attested in northern Europe from the Bronze Age onward (Hald 1942: 26–49), as burials often show considerable nettle fabric uses even when flax is available, for example at the Lusehøj woven nettle textile from Voldtofte, Denmark, *c.* 800 BCE, still the Bronze Age in Scandinavia (Bergfjord et al. 2012). **Moss** (*Polytrichum commune*, only one attested moss using scanning electron microscopy), is another plant used for textile since the Bronze Age, not just in padding or stuffing but also in woven form for what have been identified as hats and shrouds in Glasgow, Galloway, and South Lanarkshire, Scotland, and from Roman forts like Vindolanda (northern England) along Hadrian's Wall *c.* the second and third centuries CE (Harris and Gleba 2015). Other textile fibers used for clothing certainly existed, including generic **grass** used by "Ötzi the Iceman," discussed more fully in following material.

WOOD USE AND TECHNOLOGY IN ANTIQUITY

The Old World human landscape has been greatly modified for at least the past 6,000 years. Geographer Pomponius Mela, encyclopedist Pliny the Elder, historian Tacitus, and others in the Roman era describe the vast and dense, "trackless," once almost continuous Hercynian Forest (*Silvia Hercynia*) stretching across Europe from Germany eastward to the steppes, and they write, too, of the Ardennes forest and its neighbors, including the *Silvia Carbonaria*, to the west (Pompon. 3.29; *HN* 4.25; *Germ.* 28, 30; *Ann.* 2.45). Pliny makes particular note of the Hercynian Forest's giant oaks, whose veneration by the Celts led in part to Druidism, *druos* being the word for oak in Greek (*HN* 16.2). The oak as a focus of these forests may have given rise to the name Hercynian as well, because *erkunia*, derived from the older Proto-Celtic *perk-u* (later dropping the prothetic *p-*), is a likely Celtic name for oak that is probably linguistically connected to *quercus*, as the genus is now named from Latin (Pokorny 1959).

These and other ancient forests have been considerably reduced not only by clearing to enable intensive farming but also by harvesting their diverse timber resources. Exploited for wood by every ancient culture that lived within access, forests of both hardwood and softwood ranged from deciduous lowlands to high altitude conifers, although the latitude-altitude geographical correlate—the higher the latitude, the closer to high altitude a region appears climatically and in vegetation zones—means that in Scandinavian culture contests, the tree line is much lower and more conifers dominate than deciduous the higher the latitude (Meiggs 1982). Some of the more commonly utilized tree varieties include oak, holm-oak (ilex), beech, pine, fir, cedar, cypress, elder, ash, maple, spruce, larch, and elm. The types of trees used by humans in antiquity fluctuated based on purpose, geology, and soil conditions favoring the growth, and thus availability, of certain tree species. As for harvesting and working these various timbers, this required not just the use of axes and adzes but also the invention of the saw, which occurred in an unknown time frame. In Greek mythology, Daedalus is often claimed to be the inventor of the saw, and in a dark myth variant his nephew Perdix invented the saw, and his uncle Daedalus killed him out of jealousy (Hyg. *Fab.* 274; Ov. *Met.* 8.246). In ancient Egypt saws can be verified from at least the New Kingdom, as exemplified by a bronze example in the British Museum.[5] Many-toothed saws—(*priōn* in Greek; *serra* and *serrula* in Latin) ranging from frame saws, to wire saws, buck saws, and keyhole saws—of bronze and iron have been found in Greek and Roman archaeological contexts and are described in various Classical texts (Verg. *G.* 1.143; Sen. *Ep.* 57, 90; Plin. *HN* 16.43). These saws varied in use from cutting planks to cutting fine veneers. Other woodworking tools known in the Roman world are specified as axes, adzes, planes, chisels, gouges, rasps, files, and augers (Ulrich 2007: 13–6). For technology and engineering applications in mass production of timber, Roman sawmills for wood and stone—whether hand-driven, animal-driven, or dependent on some other power source like water—are well-attested, more than a few with crank and connecting rods and even toothed gears, also utilizing early examples of metal crankshafts from the second century CE or earlier. Though sawmills for stone from antiquity have survived in greater numbers than those for wood (Hierapolis, Asia Minor; Ephesus, Asia Minor; Gerasa, Jordan; Augusta Raurica, Switzerland), the same technology would have been used for cutting carefully measured, uniform timber planks (Wikander 2000: 401–12; Wikander 2008: 136–57; Wilson 2002: 1–32). Watermills with gear trains are depicted on Roman sarcophagi, as on that of Marcus Aurelius Ammianus, a miller from Hierapolis in Asia Minor (Ritti et al. 2007: 138–63).

Architecture, ranging from the creation of shelter to the construction of palaces, temples, and other civic buildings on a grand scale, constitutes the most basic and important use of timber in the ancient Mediterranean world. As walls were typically built of mud brick and, when available (or budget permitting), rubble or masonry, timber was used for wall, window, and roof-framing as well as for columns and doors. Because their construction was well documented, and because such buildings were more likely to be built of stone, we have considerably more information about palaces and temples than private dwellings. In the case of Mesopotamia and Egypt, the regents' campaign records and palace inscriptions are invaluable. Egyptian king Seti I, for example, recorded Lebanese subjects felling cedars for him on a relief at the Amon temple at Karnak (Meiggs 1982: 67). Mesopotamian structures, especially larger palaces in the Neo-Assyrian period after the tenth century BCE, required roof beams to span their many rooms, but timber was a rare commodity to the south in Babylon and even earlier Sumer, hence wood-length limitations on the size of rooms built mostly of mud brick. Thus, a recurrent theme in surviving royal records is tree-felling, and it was cedars from Lebanon that were particularly sought-after, as these trees grew to great heights, supplied timber of great lengths, resisted rot and insect infestation, and were aromatic, straight, and easy to work. Assyrian king Tiglath Pilesar III recorded of his palace at Nimrud that he roofed it with "long cedar beams, whose fragrance is as good as that of the cypress tree, products of Amanus, Lebanon, and Ammannama," and the prophet Isaiah indicated the intensity of the Babylonian Nebuchadnezzar's exploitation of Lebanese cedar: "The pines themselves and the cedars of Lebanon exult over you. Since you have been laid low, they say, and no man comes up to fell us" (Meiggs 1982: 78, 82). Biblical texts including I Kings 7 describing Solomon's Palace in Jerusalem make Solomon's use of aromatic cedar seem lavish as it was even named the "Palace of the Forest of Lebanon," indicating the use of many cedar columns (at least forty-five in four rows), great cedar beams, and, in the case of the Jerusalem Temple, cedar paneling. Cedar, of course, was not the only wood employed. In Assyrian cities like Nineveh in northern Mesopotamia along the Tigris River, the wooden roof beams were often smaller conifers from forests of the Mosul Mountains northward and the Zagros Mountains eastward, as for example, in Sennacherib's Palace of the early seventh century BCE (friezes now in the British Museum). And the Assyrian king Assurnasirpal II's palace (completed 879 BCE) contained paneling not only of cedar but also of cypress, juniper, boxwood, mulberry, pistachio, and tamarisk, much of this wood likely being locally sourced.

As for Greece, where royal inscriptions are absent, contemporary representations of architectural forms in artistic media are invaluable as Bronze Age sources: of particular note for use of timber—as columns, doors, windows, horizontal beams supporting additional storeys or ceilings, and half-timbering (structural frame of load-bearing timber is left exposed on the exterior of the building)—are the terracotta house model from Acharnes, the so-called "Town Mosaic" (faience plaques), and the Grandstand Fresco, the last both from Knossos (Dinsmoor 1975: 1–15; Lawrence 1983: 19–109; Meiggs 1982: 88–115) (see Figure 4.3). For the period following the Iron or Dark Age, public construction records and literary works complementing scant physical remains shed light on architectural timber use and species favored, especially for temples. These were initially constructed of wood and mud brick, with stone gradually supplanting mud brick from the eighth century BCE, a generalization to which lore surrounding the eighth-century BCE temple of Apollo at Delphi may add a level of nuance. The Greek lyric poet Pindar (in the fragmentary eighth *Paean*) and the historian/geographer Pausanias (10.5.9–13) attest a first temple of Apollo made of laurel, with a second of wax and feathers and a third of bronze. While the laurel temple has largely been discounted as mythological, it has nonetheless been suggested that, given the importance of laurel in Apolline cult, it is reasonable to entertain the possibility of a temple with walls not of mud brick but of woven laurel branches (Sourvinou-Inwood 1979). We rest on firmer ground with accounts of historical temples. In conjunction with the construction of the Parthenon, for example, biographer and essayist Plutarch (*Per.* 12–13) records the use of cypress and ebony. At a minimum these precious timbers would have been used for doors and special features, with fir and pine—which philosopher and scientist Theophrastus describes as primary construction timbers—for roof timbers, but widespread use of cypress for the temple is possible, as is the use of elm, for which there is inscriptional evidence (Oleson 2008: 250).

Pliny the Elder opined that trees and forests were Earth's "ultimate gifts to humankind" (*HN* 12.1), and architect Vitruvius, a vital source for construction methods in the Roman world, would likely have agreed. For Roman architectural members and carpentry, Vitruvius often provides ample discussion of wood based on function. He remarks on tree quality and highlights certain trees based on usefulness, starting with oak, elm, poplar, cypress, and fir; he finds conifers, like fir, somewhat less useful for building due to combustible sap but still strong, and oak, while harder and more difficult to work, more durable (*De arch.* 2.9.5). Vitruvius also provides ample discussion about felling different trees and drying their woods by draining sap before use (*De arch.* 2.9.3–4). Distinguishing hardwoods from softwoods, he notes that different woods are appropriate for different types of structures; for example, temporary theater seats should be made of lighter wood and longer-term seating of harder, and different types of plays—tragic, comedic, satiric—require entirely different types of timber sets (Mayor 1943: 198–203). He suggests larch for many purposes because it is somewhat fire resistant—very important for urban building as the Great Fire of Rome in 64 CE attests, termite resistant, as well as straight-grained (*De arch.* 2.9.15–16). Needless to say, there were considerable timber industries in the Roman

FIGURE 4.3 Faience inlay/plaque depicting a two-story house façade with dormer, Middle Minoan IIIA period, *c.* 1800–*c.* 1750 BCE. From the so-called Knossos "Town Mosaic." © Heritage Images. Photo by Ashmolean Museum/Heritage Images/Getty Images.

world, with many diversified wood sources and uses employing a wide range of people across many industries (Ulrich 2007: 1–2). Urban and rural structures consumed much timber, whether for entire buildings or only parts. In Roman colonial contexts in particular it is more likely that wood made up a considerable part of structures (Maxwell 1976: 33–8; Milne: 1982), and Roman exploitation of the vast deciduous lowland German forests (*Hercynian* and *Carbonarian*) on both sides of the Rhine is manifest in timber palisades of Roman forts like Saalburg, extending from the river to the frontier hinterlands covering hundreds of miles both north and south along the river as well as ultimately east and west from the river inland much later (Manning 1975: 105–29; Manning and Scott 1979: 19–62). Timber use was so extensive in Italy—growing as it did with expanding populations and an increase in scale of public building projects over time—that in the extended Roman world deforestation ultimately became disastrous in certain marginal regions, including the Atlas Mountains of North Africa (Cheddadi et al. 2015: 242–4; Hughes and Thirgood 1982: 60–75). And where local resources were depleted, woods were actively imported by sea from increasingly distant sources.

Shipbuilding is an activity for which wood was almost exclusively used. Among ancient Mediterranean peoples, the most famous for maritime navigation and shipbuilding were the Phoenicians, who were likely the first to survey the Mediterranean Sea and venture beyond it, sailing past the Strait of Gibraltar into the Atlantic Ocean. Their ships were designed and built with the best techniques available (Linder 1981: 31–42). Assyrian friezes depict Phoenician ships co-opted by Assyrian military: for example, a planked bireme boat with pointed bow ram at Nineveh *c.* 700–692 BCE.[6] Such carvings at Nineveh and Khorsabad, as well as texts like the book of Ezekiel in the Bible, reveal that Phoenician ship types included warships with a convex stern, single mast, and two banks of oars; heavier transport vessels with hulls large enough to accommodate cargo; and smaller trade vessels with a single bank of oars. Records about the Cothon "Secret" or double Harbor at the Phoenician colony Carthage in turn suggest that colony's widespread maritime experience. That their ships were mass-produced in an assembly line is indicated by the Roman discovery, in the course of the Punic Wars, of a foundered Carthaginian ship every timber of which was numbered, including maple keel and oak ribs; such numbering allowed the Romans to copy Carthaginian ships, as seen in the recovered Punta Scario ship off Marsala, Sicily in 1969 (Frost 1982: 42–50; Frost et al. 1973: 33–49).

The importance of ships and seafaring in Greece, too, cannot be overestimated. For the Greek city-states, timber, especially for ships, was a vital necessity, with the best timber coming from Macedon (Borza 1987: 32–52). Though mythological, the tale of 1,186 ships in the excerpted Homeric Catalogue of Ships (*neōn catalogos*) heading for Troy in *Iliad* 2.494–759, thought to be the oldest section of the poem (possibly dating to the late Bronze Age), suggests the capability of shipbuilding on a large scale and the huge timber resources needed even at an early period (Andersen 1998: 181–91; Huxley 1966: 313–18);[7] meanwhile, details about techniques, tools, and timbers employed for shipbuilding emerge from Homer's *Odyssey*, when Calypso leads Odysseus to tall fir, alder, and poplar trees for rebuilding his raft on her island (5.235–61). In 480 BCE at the Battle of Salamis, the Greeks alone gathered at least 371–8 galleys to the Persians' 300–600, according to Herodotus (*Hist.* 98.48; Lazenby 1993: 93–4, 174). By the fourth century BCE, even long after the dissolution of the Delian League—the Greek defensive alliance against future Persian threat to which members, at first, contributed ships—Athenian fleets are known in comparable numbers, with over four hundred triremes recorded in the Athenian Naval Yard (Hale 2009: Intro.).

Roman ships that have survived to yield timbers for identification show that while the woods utilized sometimes vary greatly, the most common woods deriving from Apennine forests include oak, pine, fir, and sometimes cypress, although not necessarily in that percentage order (Thommen 2012: 37–41, 85–9). One of the best hardwoods, dense oak has greater water resistance than many other woods, especially conifer softwoods that are more porous and can easily rot; oak is vital for the keel and hull planks that take more buffeting from agitated water. In ancient ships, the coniferous woods are most often used above the water line. Since they grow tall and straight, pine and fir, however, were used not only for masts but also for oars.

Some of the most famous Roman ships in history can be reconstructed, or at least imagined, on the basis of literary accounts. In his *Ploion* (2), Lucian of Samosata has the sophist Samippus describe the noted grain ship *Isis* (Casson 1950: 43–56; Sarton 1965: 126) *c.* 150 CE at Athens' Piraeus harbor as 36 meters (180 ft) in length and 13 meters (44 ft) in depth. Even if the sophist's statement is hyperbolic, saying that it carried enough grain to feed everyone in Attica for a year, *Isis* must have been fairly typical for Alexandrian grain ships. The emperor Caligula had several famous boats, one a huge barge of 800 tons (the so-called "Giant Ship") just north and outside Ostia that had six decks and measured 104 × 20.3 meters (341 × 66 ft). It was filled with concrete and deliberately sunk to serve as the base of his Pharos-like lighthouse after it had apparently transported an obelisk from Egypt—the one now in St Peter's Square in Rome, but originally in the Circus of Gaius and Nero nearby—and Pliny refers to this vessel as the "most wonderful construction on earth" (*HN* 36.14.16–77; Keay 2001). Caligula's other famous wooden ships or barges were extravagant and luxurious party boats—floating palaces—sunk at Lake Nemi immediately after his death. These boats were filled with high technology for the day, including piston and crank pumps with plumbing for hot and cold baths, were covered with mosaic floors, and outfitted with gilded brass, gems, marble, and other lavish items. The main ship, 73 × 24 meters (240 × 79 ft), was recovered in stages between 1895 and 1927 and again in 1932, long after its discovery in the Renaissance (in 1446) by famous artist and architect Leon Battista Alberti, who had read Classical texts like the historian Suetonius. After the lake was drained several times, the palace boat remains were housed in a museum that was later destroyed by fire during the war in 1944 (Carlson 2002: 26). Remains of another five ships, barges, or boats from the Roman Port of Claudius can be seen at the Museum of the Roman Ships adjacent to Rome's coastal Fiumicino Airport. The Napoli C shipwreck and other recent finds have been well documented and their wood taxa provenanced from oak (*Quercus* sp.) to cypress (*Cupressus* sp.), elm (*Ulmus* sp.), pine (*Pinus* sp.), fir (*Abies* sp.), poplar (*Populus* sp.), and even plane (*Platanus* sp.), among others (Allevato et al. 2009: 33–42). Roman ship construction and a wide range of ship types can be extrapolated from archaeological remains in such ancient ports and harbors, including the tidal Thames of London (Roman *Londinium*) where ships had to anchor mid-stream and where both long- and short-range craft have been accumulating for thousands of years, providing enormous study resources not just for nautical archaeologists but historians in general, even when remains are fragmentary (Milne 1985) (see Figure 4.4).

Tools and weapons are usually metal, yet wooden hafts and shafts were necessary for spears and arrows, as well as adzes, picks, hoes, shovels, and other tools, with the types of wood utilized dependent on local sourcing. Balkan ash (*Fraxinus* sp.) and dense cornel (*Cornus mas*) are recurring woods named for shafts of spears and tools in Homer's *Iliad* (e.g. 4.540–620) and other texts, since both woods are heavy and often grow straight for at least 1.5 to 2 meters (5 to 6 ft) in length. Cornel wood was considered superior for weaponry to almost any other woods by craftsmen (Verg. *G.* 2.447); cornel even became synonymous in Hellenistic poetry for "spear" (Markle 1977: 323–9). In Austria the early Celtic Hallstatt B (from *c.* 1000–800 BCE onward) salt mines have preserved some wooden tools and implements, including shovel hafts and even wooden stairs, *c.* 1344 BCE, presumably made from local Dachstein Salzkammergut conifer or deciduous trees: larch (*Larix decidua*), black pine (*Pinus nigra*), beech (*Fagus sylvatica*), and the mountain ash (*Sorbus aucuparia*) also known as rowan. Ever observant, Vergil noted that strong but resilient yew wood was "bent" to be used for bows (*G.* 448). Wooden *ballistae* and other catapults were common to Roman siege warfare, and the Greek inventor Ctesibius (fl. 285–222 BCE) made large wooden weapons like the *gastraphetes*, essentially a large crossbow cocked and held at the waist by a yoke.

Sculpture in wood has not survived well in temperate continental climates, belying its universal popularity for the making of both small and large statuary. While information

FIGURE 4.4 Roman ship discovered at Arles, France, in 2011. The boat is 31 m long and was accompanied by 450 artifacts reflecting trade and commerce in the city of Arles during Roman times. © Patrick Aventurier. Courtesy of Getty Images.

regarding Roman statuary of wood is limited, quite a bit can be learned from Greek literary sources where archaeology is relatively silent. From Pausanias, who documents sanctuaries and cult statues of the Greek world in detail, and also Theophrastus, we learn that the Greeks, at a minimum, used ebony, cedar, cypress, oak, yew, lotus (likely nettle tree, *Celtis australis*), boxwood, olive, wild pear, and fig for the making of statuary, with ebony and cedar likely being the only imports (Meiggs 1982: 300–24). The most ancient and hallowed cult statues were wooden *xoana*, and they were credited with remarkable longevity, lasting in some cases thousands of years. Among the most memorable wooden statues that appear in literary sources is the statue of Artemis that Iphigeneia and Orestes steal and convey from Tauris (southern coast of the Crimea peninsula) to Brauron—by sea and in the face of death—in Euripides' *Iphigeneia at Tauris*. The theft by Odysseus and Diomedes of Athena's ancient wooden statue, the Palladium, from the Trojan citadel, was critical to the Greeks' capture of that city (Apollod., *Bibl.* 3.12.3). The Palladium, like the Taurian Artemis statue, had reputedly fallen from the heavens, as had the olive wood statue of Athena Polias housed in the Erechtheion on the Athenian acropolis (Paus. 1.26.6). Meanwhile, the wooden Silenus statuettes to which Alcibiades likens Socrates in Plato's *Symposium* had an unambiguously earthly origin (215b). These were on display in statuaries' shops and could be pulled open to reveal a smaller statuette in their hollow core. In addition to being a primary material for Greek statue carving, wood saw important use in forming the core of sphreylaton (beaten bronze) and chryselephantine (ivory and gold plate) statuary, of which the bronze Apollo from Dreros, Crete (seventh century BCE), and Pheidias—the famed, lost cult statue of Athena (fifth century BCE) from the Parthenon—are examples.

Egypt, with its arid climate, remains our best source of ancient wood sculpture; there, unlike in Greece, it was not eclipsed by sculpture in stone and has been found side by side in tombs with objects made of precious metals and stones. The numerous examples include Early Dynastic cedar reliefs of the high official Hesi-Re, who is depicted repeatedly and at various ages (Dynasty 3, *c.* 2650 BCE, at Saqqara, now in the Cairo Museum); the strikingly animated sycamore statue of the priest Ka-Aper, Fourth Dynasty, *c.* 2500 BCE (Cairo Museum, 1 m [3.67 ft] high); the acacia wood statue of the royal carpenter and builder Kaipuneset, Fourth Dynasty, *c.* 2528 BCE (Metropolitan Museum, New York); and the wood statue of the courier official Ihy, Dynasty 6–8, *c.* 2200 BCE, 91 cm (36 in) high (Metropolitan Museum, New York, see Figure 4.5). Considerable examples from Middle Kingdom Egypt include wooden statues like that of the High Priestess Imertnebes, *c.* 1991–1783 BCE, Twelfth Dynasty (Leiden Museum) and of Wepwawetemhat, once covered in gesso and painted, from Asyut, *c.* 2134 BCE, Dynasty 11 or soon thereafter (Museum of Fine Arts, Boston, 112 cm [44 in]). Many of these slender wood sculptures are carved from single branches of sycamore, *Ficus sycomorus*, a tree growing up to 20 meters (65 ft) in height; the narrow natural size of the sycamore limits its width and accentuates its relative height. In Dynasties 3–6, ancient Egypt had its own modest African wood resources, since the sycamore fig was planted in quantity from around the third millennium BCE in the Nile Valley, and some residual modern relict copses still exist (Zohary and Hopf 2000: 164–5). New Kingdom wooden statuary is also extant, like the shea wood (*Vitellaria paradoxa*) statuette of Queen Ahmose-Nefertari, Dynasty 18–19 (transitional) from Deir el-Medina, now at the Louvre (N 470).

Household goods, furniture, and decorative items of various kinds have survived where preservation conditions have allowed. The Vesuvian events of 79 CE that destroyed Pompeii, Herculaneum, and neighboring towns, sealed many materials under volcanic ash and carbonized individual pieces, especially Roman everyday household items that can be identified with respect to timber and sometimes date. Carbonized wood items from Pompeii, Boscoreale, Oplontis, and especially Herculaneum include roof beams, folding doors, and partitions as well as benches, beds, tables, couches, cupboards, lararia, and a baby's cradle (Hatcher 2002: 217–24; Ulrich 2007: 214).

Egyptian carved cosmetic spoons survive in quantity, especially from the New Kingdom in Dynasty 18 and 19, *c.* 1500–1100 BCE, with many beautiful examples (e.g. representations of lute players) at the Louvre with broad fenestrations and open space,

FIGURE 4.5 Painted wood statue of an official Ihy, *c*. 2200–2100 BCE. Egypt, Old Kingdom, Dynasty 6–8. Memphite Region, Saqqara, Djoser Pyramid precinct, near, Ptolemaic tomb, cache of statues of Ihy, SAE Excavations. New York, The Metropolitan Museum of Art. © Gift of Edward S. Harkness, 1927. Courtesy of The Met Museum.

likely carved with small keyhole and wire saws. Egypt is also the source of a variety of well-preserved Roman toys ranging from figurines to toy daggers, game sets, and pull-toys such as horses on wheels. While desert Egypt is poor in natural wood with forests being virtually nonexistent, items of indigenous sycamore, persea wood (*Persea* sp.), and acacia (*Vachellia* sp.) are common along with imported cedar. For New Kingdom sarcophagi, imported Cedar of Lebanon (*Cedrus libani*) is the most represented timber after the major Egyptian sycamore sources were depleted. Although Egyptian temples often had accompanying groves, these were not so much for timber as for practical shade, humidifying and cooling the air in addition to their aesthetic value. The same is true for Egyptian gardens shown in tomb paintings and papyri of the *Book of the Dead*.

Vehicles and wheels. Celtic wagon burials usually preserve bronze or copper fittings rather than wood, but frequent movements and migrations with such wagons required ample wood for the wagon box, axles, wheels, yoke, post, and other parts. Examples are the Celtic grave wagons, buried at Hochdorf-Eberdingen (Hallstatt D period, *c.* 530 BCE. Wooden box 1.71 meters [5.6 ft] long) in Germany and at Vix, France (also Hallstatt D period, *c.* sixth–fifth centuries BCE), where wood boxes and wooden wheel parts are almost gone. The same is true for Etruscan wagons buried and preserved in part in Etruscan tombs, as at the Cerveteri Regolini-Galassai Tomb, seventh century BCE. The earliest wheels depicted in surviving Early Bronze Age art may well be on the Sumerian War Standard of Ur (*c.* 2500 BCE), where what presumably are wooden war chariots appear to have wheels of solid wooden, unspoked rounds with stout cotter dowels on their axles, making them heavier than later spoked wheels.[8]

PLANT PERFUME USE AND TECHNOLOGY IN ANTIQUITY

Roman historian Quintus Curtius Rufus (first century CE)—who called Arabia famous for its perfumes (*Alex.* 5.11)—and Plutarch relate famous anecdotes about the volume and range of Persian perfumes encountered by Alexander the Great upon capturing Gaza in 332 BCE and, in the following year, taking possession of the baggage trains of Darius III after the Battle of Issus. On those occasions a kingly fortune in perfumes fell into Alexander's hands. Normally being ascetic, he was apparently unused to the luxuries associated with a private, perfumed hot bath as was found in the abandoned pavilion of Darius, a tent "marvelously fragrant with spices and unguents" (Plut. *Alex.* 20.13; Hunt 2016: 59). Plutarch notes that on conquering Gaza, Alexander sent five hundred talents (about 28,500 lbs) of frankincense and one hundred talents (about 5,700 lbs) of myrrh—all from the civic stores—back to his tutor Leonidas in Greece, a high volume difficult to verify but suggesting enormous trade wealth in these commodities (Plut., *Alex.* 25.4).[9] Persian royal parks and gardens, as at Pasargadae—especially adjacent to Palace P and nearby pavilions—were well known and highly regarded for their trees (Arr., *Anab.* 6.29) as well as the associated floral production and "sweet scents" (Xen., *Oec.* 1.4.18).[10] It should be no surprise that species like saffron and rose, prized and favored in antiquity, are still essences in Persian sherbet today. The Latin name *Arabia Felix* ("fortunate Arabia"; in Greek, *Eudaimon Arabia*), describing essentially the entire Arabian Peninsula, alluded to its wealth and profusion of perfumes. Many desert perfumes and spices would have derived from what is now Saudi Arabia and Oman—location of the city of Ubar, once thought to be mythical but now known to have existed as the center of the 5,000-year-old trade in frankincense—as well as what is now Yemen and the "Holy Island" of Socotra off the south tip of the Arabian peninsula (Edgell 2003: 105–20).

The Late Hebrew poetic collection of the *Song of Songs*, certainly redacted in the Persian and Hellenistic periods, refers to many perfumes from plants in Ancient Near Eastern use, including henna (*kopher*), spikenard (*nered*), saffron (*karkom*), calamus (*qaneh*), cinnamon (*kinnamon*), frankincense (*libonah*), myrrh (*mr*), aloes (*'ahal*), balsam (*besem*), and generic spice[s] (*bosem*), all described as growing in one particular garden

(Song of Songs 4:13–16). Even if the garden is largely a literary construct, a real garden being unlikely to support so many different species, it is nonetheless significant that perfumes, spices, and related words appear at least forty-two times in this singular, poetic, biblical text (Hunt 1996: 188–94; Hunt 2008: 103–9, 115–30).

Perfume technology in antiquity was not limited to collecting the essences from wild plants, although likely the best sources; sometimes essences were collected from plants deliberately grown and cultivated on earmarked estates. As most of these perfume and spice plants were from desert contexts, the juices were already concentrated at high viscosity due to low ambient moisture, except those of a few plants like rose and galbanum. These natural perfumes were generally not diluted when merchandised. Storage and transport required that none of the perfume essence was lost along the way. In the case of nard, for example, it was likely carried in sealed containers, and sometimes the gummy essence was pressed into balls. It may even have been preserved in olive oil to avoid desiccation and then sold in sealed glass or similar containers as described in John 12:3 ff. as pure nard. In addition to olive oil, other oils used in antiquity for suspending perfume included sesame oil and almond oil, the latter aromatic in its own right (Theophr., *De odor.* 15). In Egypt, Dynasty 18 tomb scenes like that of Rekhmire (*c.* 1430 BCE) show perfumers at work crushing plants, heating the mass in a cauldron, then mixing it with other ingredients including oils for saturation and suspension (Brun 2000: 277–308). Following the tradition of his other works on plants, Theophrastus (*c.* 371–287 BCE) wrote a treatise on fragrant plants entitled *De odoribus,* a compendium of contemporary knowledge on perfumes and their sources, that included the plants in the following discussion and many others not highlighted here. Dioscorides' *De materia medica* (first century CE) also describes various perfumes and related medical unguents in the first-century CE Roman world.

Nard (*Nardostachys jatamansi*) takes its name from the Greek word *nardos* and is often referenced as spikenard. Whether from the Asian nards of the Himalayas or others, it was extremely fragrant and prohibitively expensive if sourced from great distances. Plant essences and extracts like nard, whether sap-based or from juices, would often need to be even more concentrated or adulterated with other perfume bases or oils, which sometimes required a high volume of plants in boiled or heated reductions as the Egyptian tomb scene attests. Other times the nard was pure; the New Testament story of Magdalene tells of a nard vessel whose perfume alone was worth three hundred denarii, about an average annual wage. The vessel in the story was an *alabastron*, a typical glass or ceramic container that was permanently sealed by heat or other means to reduce evaporation. In order to access the *nard*, the sealed *alabastron* vessel was broken, so it was the perfume that was priceless, not the vessel. In Hebrew the same word is *nered* for spikenard in Song of Songs 1:12 and 4:14 (Munro 1995).

Attar of Rose (*Rosa gallica* and others) is the essential oil extracted from the petals of various types of rose, the most common rose of antiquity likely being the red *Rosa gallica*, which yielded a concentrated rose perfume when oil was harvested from many hundreds or more of roses (Osmun 1975: 110–12). *Ittar* or *attar* are later Urdu or Turkic words for perfume oil. Theophrastus and Pliny knew several kinds of roses, as they describe different roses attested for medical and other purposes. Theophrastus, for example, distinguishes the *kynorodon* (dog rose, *Rosa canina*) from the *kynosbotos* (white rose, *Rosa sempervirens*); the *hekatonphyllon* (a predecessor of the famed modern cabbage rose, *Rosa* x *centifolia*); and the *rhodon agrion* (wild rose, *Rosa* x *dumetorum*) (*Hist. pl.* 6.6.4–6). Pliny suggests that different soils and pruning heighten rose fragrances (*HN* 21.68–9). Theophrastus and Pliny also rank scented roses, with the Rose of Cyrene as "most fragrant" (*odoratissimus* in Pliny); in antiquity wild roses were not known as much for strong scents and, according to Pliny, only Egypt grew more roses than Campania (*HN* 13.6). Roses were associated with religious contexts, too, being sacred to Aphrodite and, as attested by Apuleius' *Metamorphoses,* to Isis (Bohm 1973: 228–31). Diocletian's *Price Edict* of 301 CE gives a value for the highest quality rose oil as eighty denarii per pound (327 grams) relative to the higher four hundred to six hundred denarii per pound of myrrh—with an agricultural laborer's daily wage being about twenty-five denarii (Brun 2000: 299–300).

Balsam is a generic word for several aromatic plant perfumes with medicinal properties about the botanical sources of which there has been confusion and debate since antiquity. Theophrastus names *balsamon* from Syria and remarks that its fragrance is "exceeding great and rich"; its gum was collected from incisions made on the shrubs' trunks, as was the case also with frankincense and myrrh (*Hist. pl.* 9.1.2, 9.1.7, 9.4.1, 9.6.1–4). Pliny, Diodorus Siculus, and Tacitus refer to balsam as a thorny prized plant indigenous to Palestine and its deserts, although Diodorus states that it grew in Arabia as well, hence the overlapping source region (Diod. Sic. 2:48, 3.46, 19:98; Plin., *HN* 12.54; Tac., *Hist.* 5.6). In his discussion of balsam, Dioscorides notes that the *balsamon* tree, which yields a fragrant sap called *opobalsamon* (Latin *opobalsamum*), grows only in Judea and Egypt. Arab tradition suggests that balsam was brought by the Queen of Sheba to Solomon, who planted it in his gardens at Jericho, as referenced also in I Kings 10:10. Balm is known generically in Arabic as *beshem* (Feliks 2007), and as *boshem* in *Song of Songs* biblical texts (Song of Songs 5:1, 5:13, 6:2; Josephus, *Antiquities of the Jews* 8.6.6). The biblical exilic prophet Jeremiah queried, "Is there no balm in Gilead?" (Jer. 8:22), Gilead being situated in modern-day Jordan, yet his contemporary colleague Ezekiel mentions it as a commodity in Tyre (Ezekiel 27:17; Filipczak 1993: 171–85). All of these balsam plants, whether from Arabia, the Judean desert, or elsewhere are now recognized as *Commiphora gileadensis*.

As for the uses of balsam, the word "balm" is derived from balsam, which was clearly used for "embalming." Balsam's gum (*opobalsamum*) from Judean desert plantations at the En-Gedi and En-Boqeq oases was a rich source of revenue for Rome (Taylor 2012: 261). Given its fragrance and utility, the extremely high expense of Judean *opobalsamum*—one thousand denarii per *sextarius* (approximately 546 ml or 17.5 fluid oz.) according to Pliny—must have been justified for wealthy Romans (*HN* 12.112–23; Brun 2000: 299).

Myrrh (*Commiphora myrrha*) was an astringent perfume used in unguents and medicines, partly for its preservative powers and ability to mask foul odors, hence its application as a burial perfume. An aromatic resin produced in Arabia and the eastern coast of Africa (now Somalia), myrrh was a profitable import for the Romans (Beek 1958: 141–52) and was one of the important products of the Arabian trade mentioned in the *Periplus Maris Erythraea*, an anonymous work from around the middle of the first century CE written by a Greek-speaking Egyptian merchant (Casson 2012: 122). One of the Eastern Magi in the Matthew 2 gospel narrative brings the precious gift of myrrh as proleptic for the ultimate burial of the Infant Jesus as the Son of Man, acknowledging in its expensiveness a rare gift for a royal personage, a threat to Herod; myrrh was later used in Jesus' "embalming" (Tucker 1986: 425–33).

Frankincense (*Boswellia sacra*), *libonah* (Hebrew), *libanos* (Greek), *olibanum* (Latin), is an aromatic resinous sap that has been utilized in religious contexts as ritual incense since antiquity, but is sufficiently fragrant to have been used in perfume and was even chewed as a spice gum. Pliny the Elder states that frankincense was known to no country other than Arabia, but there are various Boswellia species that yield frankincense, including *Boswellia carteri*, *B. glabra*, and *B. frereana*, some of which varieties grew in what is now Somalia (*HN* 12.30; Casson 2012: 122). Because frankincense was in high demand from Europe to Asia, the kingdoms of southern Arabia became an integral part of global economy with shipping connections to India, the Mediterranean, and the Silk Road. Phoenicia played an important part in the conveyance of frankincense, and Diodorus Siculus mentions that Minaea in South Arabia was a vital source that brought frankincense to the desert Nabateans around Petra, who profited as perfume and spice trade middlemen (3.42). There are many biblical references to frankincense (as many as fifty-four), and Classical authors also reference frankincense frequently, including Herodotus, Theophrastus, Strabo, Vergil, Pliny the Elder, Arrian, and Suetonius (Hdt. 3.107; Theophr., *De Odor.* 9.11.10–11; Strab. 16.4.18; Verg., *G.* 2.117; Plin. *HN* 12.32; Arr., *Anab.* 41; Suet., *Aug.* 35.3; McLaughlin 2010: 194n11). Pliny documents different kinds of frankincense and how it was harvested: at the hottest time of year by slitting the small desert trunks and collecting the milky sap that hardens into resin (*HN* 12.32). Nero was said to have burned an entire year's Arabian harvest of frankincense, the entirety of Rome's supply, after his (likely) murder of his wife Poppaea Sabina (*HN* 12.41). Apart from use in perfume and as incense, frankincense found application as a repellant, driving away insects like mosquitoes, and also as a medicine to treat ulcers and other ailments.

Balanos (*Balanites aegyptiaca*), still abundant in Sudan, is called "Egyptian" by Theophrastus, who describes its oil and gum, comparing the plant to myrtle in appearance and with perfume derived from the caper-size fruit—but only from bruising the skin thereof, not the fruit itself (*Hist. pl.* 4.2.1, 6). It was excellent as a base for other perfumes, being prized more for longevity than strength of scent; Theophrastus notes that, of perfumes, the longest lasting are "Egyptian" or *balanos* (*Balanites aegyptiaca*), spikenard or *nardos* (*Nardostachys jatamansi*), and most of all myrrh (*Balsamodendron myrrha*) (*De odor.* 38).

Henna (*Lawsonia inermis*) was a perfume from the fragrant flower clusters mentioned in the Song of Songs 1:14, "a cluster of henna is my beloved to me in the vineyards of En-Gedi." Traded throughout the Ancient Near East, it was sourced from desert oases of Palestine like En-Gedi and was mentioned in various perfumery contexts in the Greco-Roman world (Brun 2000: 299; Conan 2007: 27). Though its modern use is primarily as a colorant, it was used in antiquity as a fragrant dye.

Galbanum (*Ferula gummosa*) was an aromatic plant resin originating from Persia, possibly from the Elbruz Range near the Caspian Sea or further west on the Zagros Mountains. It possessed a musky scent and a bitter taste, and it was often mixed with other spices or aromatic plants. It was apparently named *sammim* in Hebrew and used in the *ketoret* incense mixture in the Hebrew Tabernacle (Exod. 30:34), although there said to be a "sweet spice." Hippocratic scoliasts claim it had medicinal use, as does Pliny, who also mentions that it kills vipers (*HN* 24.13). *Galbanum* was also noted on the clay record tablets at Mari in Mesopotamia from the time of Zimri-Lim (1780–1759 BCE) (Brun 2000: 278). *Ferula gummosa* is still used today in modern perfume as an intense "green" fragrance base.

Lotus (*Nymphaea* spp.) is the name for a generic flower in Egypt and elsewhere, although it often specifically references a water lily, which is the source of persistent confusion. In Egypt *Nymphaea caerulea* or blue lotus (blue water lily)—known in Egyptian as *seshen*, the most sacred flower—was grown mostly in pools and associated with religious themes like eternity based on its own longevity as a flower. Egyptian tomb paintings show both men and women holding the lotus, but they are also represented seated or standing with a perfumed wax cone on their wigged heads, which the heat melts, distributing the perfume. Many surviving Egyptian linen dresses have residual stains from the perfumed wax. Achaemenid Persian kings were also often depicted holding a lotus in their left hands, possibly a symbol of the ephemerality of power as well as beauty. Theophrastus describes the marsh yellow water lily (possibly *Nuphar luteum*), *nymphaia* in Greek, as "sweet-smelling," and Pliny describes a similar water lily, although it is important to remember the term "lotus" can be generic for several related flowers and was used for a variety of purposes (*De Odor*. 9.13.1; *HN* 25.75).

Cinnamon (*Cinnamomum verum*) and **Cassia** (*Cinnamomum aromaticum*), fragrant wood bark from shrub trees of South and Southeast Asian origin, were both often referred to as "cinnamon" in antiquity, though stemming from different species of the same tree genus. Cinnamon is mentioned in many ancient texts—including Herodotus, Theophrastus, Dioscorides, Pliny the Elder, and the *Periplus Maris Erythraea* (Mclaughlin 2010: 143)—and was so expensive that it was more often used in medicinal applications and perfume than for culinary purposes. The poetic Hebrew Song of Songs 4:14 refers to it as *qinnamon* (Hunt 2008: 127–8), more spicy to the taste than fragrant, but when mixed with other fragrances is noted for its pungent odor. Theophrastus, noting that their valuable bark has a strongly pungent fragrance used for perfume mixing, associates cassia with cinnamon but also draws a distinction between them, calling one *kinamon* and the other *kasias* (*Hist. pl.* 9.5.1, 3). Theophrastus relied on accounts of other Greeks returning from Alexander's prolonged entourage in Persia and India; his awareness of the distance Asian plants had to travel justified their great cost—up to three hundred denarii per pound in the Roman era)—in the West (Brun 2000: 299. In the Roman world cinnamon (or cassia) was sometimes named as *malabathrum* in Latin from the Greek *malabathron* (Miller 1969: 157–9) and was mentioned in the *Periplus Maris Erythraea* as well as by Pliny and Dioscorides (*HN* 12.26; *De mat. med.* 1.76).

Sandalwood (*Santalum album*) was one of the generic aromatic woods said to have been imported by trade-keen King Solomon (I Kings 10:11–12) from either Ophir, on the Red Sea, or India via King Hiram of Tyre's Phoenician ships, although trade in sandalwood is not easily deducible until the mid-first millennium BCE (Villiers 2001: 24–43). Sandalwood was named *'almug* in Hebrew, as in the above-cited biblical text. Its oil and roots are also aromatic, and although there are nineteen known species in the *Santalum* genus, it is unknown exactly which (if any) sandalwood was exploited and traded as far west as Kings purports. Its woody, spiced fragrance, still used in modern perfumes, is long lasting and was a fixative for other more ephemeral floral scents.

Cedar (*Cedrus libani*) or Cedar of Lebanon has already been noted as the aromatic wood in Solomon's Palace of the Forests of Lebanon (I Kings 7), and it is mentioned in Song of Songs 1:17, "the beams of our house are cedars," again implying Solomonic use. Cedar wood oil was also an aromatic in antiquity, known to possess insecticidal properties, and it is used in modern woody perfume bases as well. While many of the perfume products of antiquity were based on either sap or oil—these being reduced to either liquid, gum, or powder—wood or bark, as in the case of cinnamon and cedar, also remained important ingredients that could be suspended in oils and, in turn, be mixed with others.

Other plant perfumes saw use in antiquity but the above appear to be among the most common and widespread in production and trade.

PAPER USE AND TECHNOLOGY IN ANTIQUITY

Papyrus (*Cyperus papyrus*) from Nilotic Egypt is the most common writing material not only in Pharaonic and earlier Egypt but also for the Mediterranean as a whole. Our modern word "paper" derives from the Greek word *papyros*, a word used at least since the sixth century BCE. Papyrus can be traced back to Old Kingdom use (First Dynasty *c.* 3000 BCE), although more common in the Middle Kingdom, and was abundant in the New Kingdom and Ptolemaic periods. Great volumes of papyrus have been found in tombs, and whole trash dumps, as in the Greco-Roman site of Oxyrhynchus, have preserved considerable numbers of papyrus documents, especially demotic (Hunt 1988: 159–61). Nilotic papyrus preparation was similar to that of other bast plant materials. The stages necessary for using papyrus as a writing or imaging medium included collecting the wedge-shaped plants from aquatic contexts, stripping the outer cambium, and removing coarser fibers from the fresh strips. The strips were generally less than 30 cm (12 in) in length, trimmed from the internal middle part of the reeds since papyrus can grow up to 2 meters (6 ft) high, but only the middle is optimal for this application. Individual papyrus sheets were formed by laying the strips side by side, first in one vertical layer, and then in a second horizontal layer. These two layers were placed one upon the other and hammered until the sap suffused the whole, the natural cellulose liquids acting as glue. After drying, the papyrus "piece" was sanded with abrasives like pumice until fairly fine for writing or painting, then finally trimmed to size. Papyrus pieces were glued together to make longer writing media as in the New Kingdom *Books of the Dead*, funerary scrolls containing incantations, spells, and details of the afterlife.[11] Other materials, however, also certainly found use in antiquity as bases for writing. Paper experimentation was a historical state enterprise in China, and the Han Dynasty (*c.* 200 BCE–200 CE) seems to have determined that hemp paper was sturdier with hemp's 80 percent cellulose compared to 40–50 percent cellulose in paper made from wood pulp (Huang 1990; Malakowska et al. 2015: 134–5). It is known from the Etruscans that they sometimes wrote on linen, e.g. *liber linteus* (Gleba 2008: 52–3, 67–8), and other paper sources, including palm leaves, are also known for ancient writing media, although infrequent once papyrus dominated as a writing medium.

"ÖTZI THE ICEMAN": WOOD SPECIALTIES IN THE COPPER AGE (*c.* 300 BCE)

Ötzi, a Copper Age (*c.* 5,300 years ago) ice mummy, was discovered in 1991 in the Ötzal Alps at the 3,200 m (10,500 ft) elevation of the Similaun Glacier between what is now Austria and Italy (see Figure 4.6). This remarkable case deserves discussion here because of the volume of worked wood material that has survived. Ötzi's considerable belongings were spread out over a 10-square-meter (30 sq. ft) radius in a glacial gully on the summit, nearly all with traces of his DNA preserved in the frozen context. Among many other objects were products made of eighteen different types of wood, including yew, ash, larch, viburnum, birch, maple, cornel, and hazel. His bow was yew, his arrow shafts were of viburnum and cornel, his dagger handle was ash, his backpack was larch and hazel, one of his "bucket" containers was birch bark, and he carried his hot embers wrapped in slow-burning green maple leaves. Plants and seeds, too, were among the belongings that he carried. One of his garments was woven of grass, and he even had grass and moss tucked inside his leather shoes.

FIGURE 4.6 Display recreation of iceman Ötzi with a bearskin hat, goatskin coat and leggings, and leather sandals filled with straw, on October 1, 1997. He carried a flint knife in his belt, and is pictured with his bronze axe making a bow out of yew. Ötzi is the name given to the frozen mummy of a man from around 3300 BCE, found by two German tourists in 1991 in the Schnalstal Glacier in the Öztal Alps. © Patrick Landmann. Courtesy of Getty Images.

What is perhaps most significant about the diversity of the wood products specifically is that Ötzi seems to have understood the best properties of each wood: there may not be a better, more resilient wood than yew for bows under enormous tension, just as Vergil notes for bows in the Roman period (*G.* 2.448)—and as attested by later English longbows of yew from Crecy and Azincourt in the medieval period. Likewise an axe handle of yew, which he also had in his possession, could withstand enormous compression. The tensility of his dagger's ash handle could also take the stress of penetrating cuts. The larch and hazel for his backpack are lighter softwoods for easier portage; Vitruvius would later note larch as a wood resistant to both fire and termites. The straight shafts of cornel and viburnum are easily adapted for as arrows, and they would serve as the ideal weapon shafts for the Greeks and Romans. In other words, Ötzi's wood equipment suggests that he had a sophisticated understanding of the wood resources in his environment, their practical applications reinforced millennia later. Perhaps even more impressive is the fact that that the plant medicines in his possession—including the birch bracket fungus *Piptoporus betulinus* that Ötzi purposefully carried tied on thongs—were natural antidotes to some of his ailments, like Lyme disease (*Borreliosis*) as well as whipworm, being antiviral and antibacterial. For Ötzi the modern distinction between plants and fungi was nonexistent. Whether or not he was representative of his culture remains to be seen, since there are no other comparable organic materials surviving from his era, but the range of ideal wood and plant uses that Ötzi apparently knew well suggests a strong working relationship with his world of organic materials (Fowler 2001; Hunt 2007: 26–7; Price et al. 2013: ch. 10).

CHAPTER FIVE

Plants and Medicine

ALAIN TOUWAIDE

The plant kingdom provided abundant resources for both nutrition and therapy in the ancient world, particularly in the eastern Mediterranean, which was characterized by an exceptionally abundant biodiversity. It is therefore no surprise that these rich resources were exploited by humans from the beginning of civilization for both the preservation of health and the treatment of diseases. In the Mediterranean region, the collection of botanical knowledge for medicinal purposes is known at least as early as Mesopotamia—Sumer, Assyria, and Babylon—as well as Egypt, but it was ultimately the Greeks who generated an extensive body of knowledge that became the basis for further developments in Rome, Byzantium, and the Middle Ages.

Other parts of the world developed similar knowledge of medicinal plants and collections of data during the same period. In India, the most ancient medical text of Ayurveda, the so-called *Sushruta Samhita*, dealt in great detail with the uses of plants for the treatment of disease. Its title presents it as a compendium (*Samhita*) by Sushruta, a physician said to have received his knowledge from the Hindu god of medicine who taught at an undetermined time between 1200 and 600 BCE according to modern scholarly research. It is more probable, however, that the text resulted from successive accretions, the most ancient layers of which might date as far back as the early Indian world, with the whole going through a gradual and cumulative process of edition possibly concluded before the fourth or fifth century CE. After it was fixed in a canonical, immutable form, this text became the foundational book of Ayurvedic medicine together with the *Caraka-Samhita*, the *Bhela-Samhita*, and parts of the so-called *Bower Manuscript* (Wujastyk 2003: 149–60).

Further East, early Chinese botanists and physicians collected medicinal plants, described them, and studied their effects on the human body, thereby formulating a complex theoretical system which remains a source for present-day practice in China. Although these scientists probably wrote down the results of their experience and knowledge about the therapeutic uses of plants, no such work has been fully transmitted because of the fragility of ancient books, the continuous creation of new compendia of medical knowledge that made previous ones obsolete, natural catastrophes, and also upheavals of history. Many such early works are known only through citations in later encyclopedias. The most ancient currently known is the *Shennong bencao jing* (Divine Husbandman's Materia Medica) dating to the late first or early second century CE. The most comprehensive such later compendium is the *Bencao gangmu* (Compendium of materia medica) compiled by Li Shizhen (1518–93) and published posthumously in 1596 after first being released by its author in 1577.

Because the knowledge of medicinal plants that they possessed was extensive, well documented, and influential, this chapter focuses on the Greeks and the transmission of *materia medica* (the body of knowledge about substances used for therapeutic purposes and their properties) to Rome, Byzantium, and the West in the Mediterranean region. This body of knowledge generated a tradition that expanded both east and west in subsequent centuries. In the east, it was transmitted to the Arabic world and beyond from the ninth century on. In the west, it reached what is now Italy, Spain, France, England, and Germany, as well as central and northern Europe. This western transmission proceeded in several phases, first through translations into Latin from the fourth to sixth centuries, then via translations into Latin of the Arabic versions of the Greek legacy from the late eleventh century to the early fourteenth century, and finally with the massive transfer of Greek botany, medicine, and science from the east after the Fall of Constantinople and the end of the Byzantine Empire in May 1453, which accelerated the emigration of Greeks to the west. The sum of material generated through these several processes of transmission became the foundation of the medicine now identified as Western.

REFRAMING THE KNOWLEDGE OF MEDICINAL PLANTS IN THE ANCIENT GREEK WORLD

The beginning of the Greek tradition has usually been identified in scholarly literature with the work of the physician Hippocrates (460–between 375 and 351 BCE), from the Aegean island of Kos, and of those physicians who practiced medicine after his manner, the so-called "Kos School" of medicine. In this view, Hippocratic medicine is characterized by observation of patients, critical analysis of their cases, and deductive reasoning resulting in the formulation of general recommendations. This philosophy of medicine is best represented by the Hippocratic *Aphorisms*, which express fundamental methodological rules—including the therapeutic uses of plants—in brief sentences. However, a closer look at the documentation indicates that generalizations allegedly based on the observation of multiple cases going back to Hippocrates are already attested earlier in the Greek world, though in a different form. For example, Greek myths accounted for the creation of many parts of the natural world and connected plants with specific gods. These tales cannot be dated with any precision; they derived at least in part from the Near East and may have had their origins in the Bronze Age (3000–1050 BCE). Take, for example, Artemis, a daughter of the Father of the Gods, Zeus, and twin of the god Apollo. A complex figure with a still unexplained name and various, often seemingly conflicting attributes, Artemis might have been venerated among Greek populations as early as the first millennium BCE. Whereas the goddess, generally speaking, was the complete personification of pure, intangible femininity, in Ephesus, she was instead the *Great Mother Goddess*, as her statue with multiple breasts—possibly dating as far back as the seventh century BCE—vividly expressed (see Figure 5.1).

As early as the most ancient written record of Greek medicine (the *Hippocratic Collection*), the very name of the goddess also was that of a plant (*artemisia*, mugwort, *Artemisia* spp.), which was—and still is—used to treat feminine conditions (see Figure 5.2). Giving the plant the name of this goddess suggests not only the clear perception of the mugwort's efficacy in the treatment of gynecological conditions, but also the formulation of the specific gynecological use of the plant through its assimilation with the goddess best personifying femininity. Later this notion of the specificity in the use of

FIGURE 5.1 Roman marble copy of the colossal statue of many-breasted Artemis (125–175 CE). From Turkey, Ephesus, city hall "Prytaneion." © DEA PICTURE LIBRARY. Courtesy of Getty Images.

plants expressed in a concrete, figurative way was reflected in technical terms through the abstract concept of *therapeutic property*. The most important point is that, already at the time of the development of mythology or even before, a notion of specificity in the use of a given plant existed, most probably developed on the observation and comparison of multiple cases, the identification of a common element (the action of the plant), and its expression in a language (here, the "language" of mythology) that would be understood by all the members of the Greek community.

After a period traditionally identified as the Dark Ages (*c.* 1100–800 BCE), to which we shall return, medicine emerged in the Greek world as both a rational science and a practical art. One of its most salient characteristics was the rejection of super-natural phenomenona and, instead, the introduction of perceptible and explicable causes in the

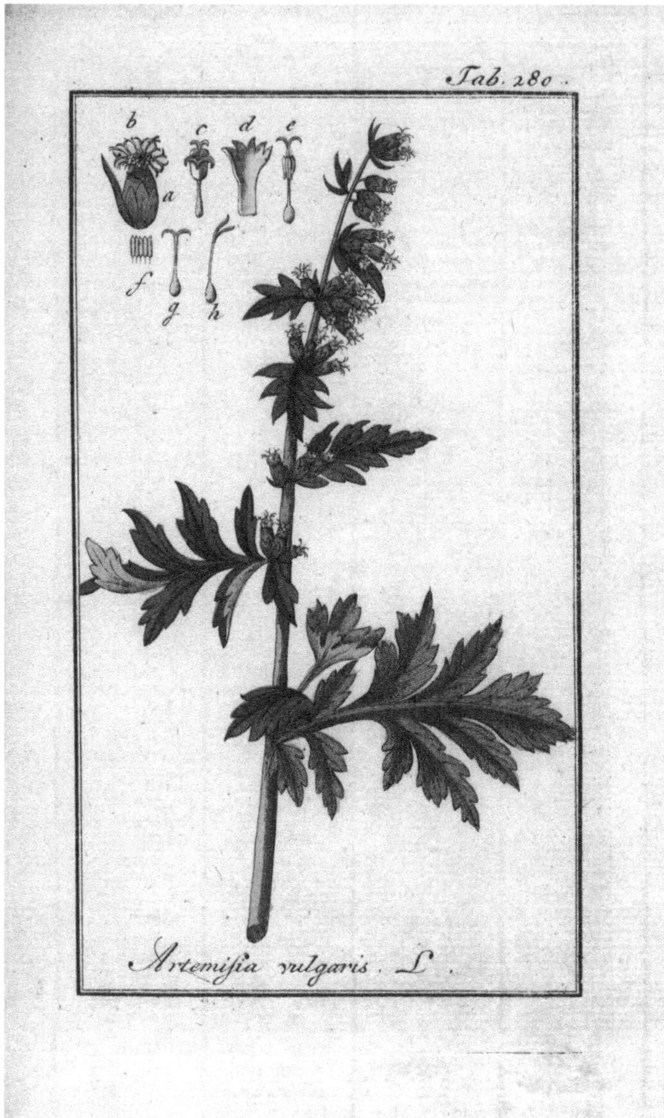

FIGURE 5.2 Mugwort, *Artemisia vulgaris*. Hand-colored copperplate botanical engraving from Johannes Zorn's *Afbeelding der Artseny-Gewassen*, Jan Christiaan Sepp, Amsterdam, 1796. Photo by Florilegius. Courtesy of Alamy Stock Photo.

processes of both disease and therapy. As in the case of the Hippocratic treatise *On Sacred Disease* (late fifth century BCE), which contested the interpretation of epilepsy as a disease provoked by a supra-natural cause, dismissal often took the form of denouncing the *Magoi* (*Magi*), Persian priests purportedly acting as magic healers. In spite of its claim to be an indigenous creation, Greek medicine was indebted to the knowledge and practice of previous civilizations, including Persia and Egypt. The tales relating the return of the Greek warriors from Troy, which were not limited to the *Odyssey* but included other complementary and sometimes diverging narratives, clearly hint at this debt of Greek medicine to Egypt: Helen, Menelaos of Sparta's wife—who, according to legend, was the cause of the Greek expedition against Troy—sailed back to Greece via Egypt. As she was deeply depressed over the loss of life she had caused, she received a potion to alleviate her suffering from the Egyptian queen. Although no information is available about the substance used to prepare this beverage, it is believed that it was poppy, which might be native to Egypt and was used there at least as early as 2500 BCE, as the so-called Edwin Smith Papyrus dating to the seventeenth century BCE but reproducing an earlier text of *c.* 2500 BCE indicates. Poppy was known and used in Classical Greece for its sedative action.

The debt of Greece to earlier botanical and medicinal knowledge and practices resulted, in part, from the introduction to the Mediterranean of plants such as pomegranate, which is native to the Iranian environment and was domesticated in Mesopotamia, between the Euphrates and the Tigris, as early as the third millennium BCE. Several plant species used in the Mediterranean world and in Greek medicine came from even farther afield (India and China), traveling through Iran and Mesopotamia. Pepper was one of them. Native to the mountain range of the Western Ghats in Western India, it was introduced to the Greek world by the Persians and the Medes in the fifth century BCE, if not earlier, along an itinerary identified in contemporary historiography as the Silk Road. Interestingly enough, its therapeutic uses were modified along this journey, though an original application of Ayurvedic medicine was preserved: the treatment of ophthalmic pathologies. Some plant derivatives also entered the Mediterranean world from further afield, coming from Egypt, the southern part of the Arabian Peninsula, and even further south. This, for example, is the case with myrrh, an exudate of several species of *Commiphora* tree species growing in present-day Yemen; aloe originally native to the island of Socotra off the Yemeni coast; and cinnamon, the bark of different species trees of the *Cinnamomum* species that might have come from the island on the east coast of India once identified as Ceylon and now as Sri Lanka.

DOCUMENTING MEDICINAL PLANT HISTORY IN THE GREEK WORLD

The first written sources about the medicinal use of plants that have come to us from the Greek world is the vast set of over sixty treatises attributed to the so-called Father of Medicine, Hippocrates of Kos. This collection currently identified as the *Hippocratic Collection* was the product of a gradual assemblage of several texts dating from the late fifth century BCE to the second century CE that share a view of medicine as a rational art based on the careful observation of the signs of disease presented by patients as well as on the use of a broad range of plants for the treatment and prevention of disease. None of these treatises was specifically devoted to medicinal plants, but all of those dealing with therapy mentioned several plants as ingredients for medicines. Most of these therapeutic treatises were about gynecology and dated to the late fifth or early fourth century BCE, and they contained material of an earlier time collected by members of Hippocrates' circle or by others in his time. This body of knowledge certainly resulted from a tradition that gradually developed, including by assimilating contributions of earlier civilizations as explained above.

In contrast to the Hippocratic method of recording what were already traditional practices, members of Aristotle's (384–322 BCE) philosophico-scientific school in fourth-century BCE Athens analyzed the action of medicinal plants in an investigative way. In the notebooks of the school currently known as the *Problems*, they recorded the questions that they were interested in, including the processes through which plants acted on the human body. Several of Aristotle's students and colleagues noted their answer in the *Problems*. Through collective reflection, they tested several hypotheses and came up with different ideas and answers to the questions they were studying, giving to present-day readers of the *Problems* a sense of their debates and exchanges.

Aristotle's life was also the time of the expedition led by Alexander the Great (356–323 BCE) to India. Educated by Aristotle, Alexander invited scientists and chroniclers to join his troops during his military odyssey, as later sources such as the historian Plutarch (*c.* 45–before 125 CE) and the biographer Diogenes Laertius (mid-third century CE) indicate. While the names of all those who accompanied the expedition are not known, they included Callisthenes, who wrote a now-lost history of the expedition, and Nearchos, the admiral of Alexander's fleet, who described their voyage at sea. The scholarly cohort was entrusted with the task of exploring and describing the many countries that they were expected to cross, together with their populations, customs, natural resources, and other products. Besides describing their sailing the Red Sea and the fauna and flora of the regions they discovered as far as India, they brought back plant species from these regions to Alexandria in Egypt. Although their writings have not been preserved, data about these plants can be found in the botanical literature of the subsequent centuries, starting with the work of Theophrastus, Aristotle's pupil and his immediate successor as the head of his school.

A native of the Aegean island of Lesbos, Theophrastus (*c.* 371–*c.* 287 BCE) specialized in botany. It might be significant that Lesbos is characterized by a great diversity of environments, from desertic to coastal, each of them with a typical flora. Theophrastus followed the Aristotelian method in science, which consisted of compiling all available information on a topic (in ancient Greek this was called *historia*, literally translated as "research") so as to have it conveniently collected for further study. The result is the *Historia plantarum*, aptly translated as Inquiry into Plants, in the ninth book of which Theophrastus summed up the knowledge of medicinal plants in his time.

Another Aristotelian of the subsequent period is Diokles (fourth/third century BCE), who was born in Carystus on the island of Euboea but lived in Athens. He pursued the lines of questioning opened in the *Problems* and made original contributions in developing the abstract concept of the action of natural substances on the human body. Unfortunately, his writings are not preserved, but they are known through citations in other, later treatises, which present a clear idea of his contribution. Diokles seems to have built a model through analysis of the deleterious agency of animal venom (particularly scorpions and snakes), which he may have transferred to the therapeutic action of plants.

Documentation about the study of the medicinal uses of plants in the Greek world is scant for the period from Diokles to the first century CE. Judging from available later texts, production of such documents was abundant, however. This largely took place in Alexandria and in the eastern Mediterranean world. Not much is known about Alexandria and the scientists of the Museum, who appear to have analyzed the abundant material brought back by the scientists who accompanied Alexander's troops in their expedition to India. The activity in the eastern Mediterranean—more specifically, in the circle of the sovereigns of the kingdoms that arose from the division of Alexander the Great's empire—is better documented, though only through later reports. In the kingdom of Pergamon in Asia Minor ruled by the Attalids, Attalos III (r. 138–133 BCE), who died without heirs and donated his kingdom to the Roman Republic, had a special interest in poisonous plants. If later accounts are reliable, he cultivated them in the royal gardens and studied their effects in ways that we shall see below. It might be in this context that Nicander, born in the nearby town of Colophon, wrote his two poems *On Venoms* and *On Poisons*. Imitating the epic style of the *Iliad*, he described the effects of animal venoms and poisonous plants on people, casting this as the fight of humans against an inescapable death. Besides describing the effects of venoms and poisons with great accuracy, he enumerated multiple remedies, most of which were made from plant sources. Slightly later, Mithridates "Eupator" (r. 120–63 BCE), who ruled the kingdom of Pontos on the southern shore of the Black Sea, was interested in counter-acting the effect of poisons. Interestingly, he experimented on himself, ingesting increased doses of lethal substances on a regular basis so as to create immunity. A fierce enemy of Rome, he was defeated in 63 BCE by the Roman general Pompey (106–48 BCE). As the story goes, when he was on the verge of capture by the Romans, he wanted to commit suicide and drank the poison he always carried with him in case he was caught off-guard or captured. Because of the immunity he had developed, however, he was unaffected by the drug and had no other recourse than to ask his own soldiers to kill him. Believed to have known many of the multiple languages spoken at that time in Asia Minor, he was credited with having recorded the results of his experiments in notebooks preserved in the capital of his kingdom, Amasya. Pompey is reported to have seized this precious documentation and to have brought it to Rome where he commissioned one of his slaves, Lenaeus, to translate it into Latin.

The Greek physician Kratevas (first/second century BCE) is thought to have collaborated with Mithridates and to have written works on the therapeutic uses of natural substances, including plants. As a result of the influential work of the German medical historian Max Wellmann (1863–1933), it is widely believed that Kratevas authored the first illustrated herbal, even though this view is not actually supported by available evidence. At any rate, no such plant representations have been preserved, with the exception of some botanical tables in an early sixth-century manuscript allegedly coming from Kratevas' original text. Whatever the exact nature of his contribution, Kratevas is typical of the period that saw transfer of both power and science from the Hellenistic Greek to the Roman world, into which, in turn, the eastern Mediterranean was gradually absorbed from the mid-second to the mid-first century BCE. Kratevas was not the only case. The city of Tarsus in Southeast Asia Minor, famous as the birthplace of St Paul (d. *c.* 62 CE), might have been the seat for both the capital of a province and a school specializing in the study of the plants and other natural resources used for medicinal purposes.

Dioscorides, the most accomplished pharmacognost of Classical antiquity, was a native of Southeast Asia Minor. Born in Anazarba, a town not far from Tarsus, he lived during the first century CE and authored the largest assemblage in antiquity on the products of the three natural kingdoms (vegetal, animal, and mineral) credited with medicinal properties. The result is the vast treatise *De materia medica*, according to an exact Latin translation of its Greek title. Divided into five books (which did not correspond to organic divisions of the whole field under study, but were in effect five rolls of papyrus), it analyzed over a thousand substances, three-quarters of which were plants or plant derivatives. Each such substance was dealt with in a chapter constructed on a typical template with both the most common and the other possible names of the substance, its physical description (this is particularly the case for the plants), and its therapeutic properties expressed in an abstract way and immediately followed by its medical applications. Interestingly enough, the chapters included methods of detecting adulterations when appropriate. In manuscripts dating from the fifth to the sixteenth century, the work contained colour representations of many plants, the origin of which is debated: did they reproduce some of Kratevas' illustrations, as the presence of his name alongside some representations of plants in manuscripts containing *De materia medica* has suggested, had they been commissioned by Dioscorides, or were they introduced into the work through the centuries at a yet unknown point in time?

Dioscorides and his work are representative of the transition of medicine and botany from the Greek to the Roman world. Not only did he mention Kratevas in his work (with all that this implies, as we have shown above), he also might have frequented the circle of Roman administrators in the province of Cilicia in Southeast Asia Minor. He appears also to have reproduced data on medicinal plants similar to those in an earlier compilation by the Latin scientist Sextius Niger (first century BCE/CE). At this time the Roman world was shifting from a traditional agricultural society—perhaps best represented by Marcus Porcius Cato (234–129 BCE), author of a manual on agriculture and rural economy embodying the values of the old Roman society, and Varro (116–27 BCE), a writer who compiled an encyclopedia on agriculture—to a type of "global" world, characterized both by the adoption of Greek culture and science and by the acclimation of plants native to the East when introduced into the Italian environment as a result of Rome's conquest of the eastern Mediterranean world. This shift reached a first culmination with the transformation of the Roman Republic into an Empire in 27 BCE that extended over all the known world of that time.

Another manifestation of this process of universalization is the *Naturalis Historia* (Natural History) by the Roman author, naturalist, and commander Pliny the Elder (23/24–79 CE), who died in the eruption of the Vesuvius. Pliny compiled all the knowledge of his time, from cosmogony to arts and crafts, including plants and their medicinal uses. He listed all his sources in a methodical way at the beginning of each of the books that constitute the *Naturalis Historia*. Interestingly enough, he borrowed from the same material as Dioscorides, Kratevas, and Sextius Niger, as a comparison of texts and cross-references make clear.

After an extraordinary efflorescence of scientific works in Latin in the first century CE, science in Rome reverted to Greek in the following centuries, beginning with Galen (129–after ?216 CE). A native of Pergamon in Asia Minor, a student of medicine in Alexandria, and a physician to the gladiators in his native city, Galen moved to Rome where he both practiced medicine in his capacity as physician to the imperial court and wrote a vast oeuvre touching upon all parts of medicine. He collated a vast quantity of data on medicinal plants and spent years and considerable financial resources in assembling a collection of *materia medica* (plants, minerals, and animal parts and products, together with their respective derivatives), as he repeatedly stated in his treatise on materia medica. Indeed, he traveled throughout the Mediterranean world for this purpose, stockpiling the collected materia medica in Rome, in what can be assumed to have been the ancestor of a medieval apothecary, and using these substances not only to treat patients but also to experiment, as we shall see. He summed all the knowledge of the medicinal plants that he gained in the immense treatise *De simplicium medicamentorum temperamentis et facultatibus* (On the Mixtures and Properties of Simple Medicines) in which natural substances are divided by natural kingdoms (vegetal, animal, and mineral) and listed in alphabetical order on the basis of their most common Greek name. Curiously enough, this work was not as influential in the following centuries as many other of Galen's works. Although it was preserved through time, it never became as extensively used as Dioscorides' *De materia medica*.

DISCOVERING THE PROPERTIES OF MEDICINAL PLANTS

Discovery of the medicinal plants native to the Mediterranean environment and knowledge of their therapeutic activity were first explained in the Greek world through mythology. A Latin manuscript dating to the thirteenth century containing a text probably dating back to the sixth century and apparently resulting from the translation of Greek material into Latin (manuscript 93 of the National Library of Austria in Vienna; see Figure 5.3a and b) and an Arabic manuscript completed in 1224 CE, probably in Baghdad (accession number

FIGURE 5.3a and b Asklepios finding the betony plant. *Herbarium Apuleii Plantonici*, Vienna Cod. 93, f. 5v and 6r. Sicily, *c.* 1220–49. Courtesy of Österreichische Nationalbibliothek.

1960.193, Arthur M. Sackler Museum, Harvard University; see Figure 5.4), present a variously interpreted illustration of this: Asklepios revealing the therapeutic applications of betony (*Betonica officinalis*) to humankind.

This tale is illuminating: Asklepios was a son of the god Apollo and the first mythical healer of the Greek world. He learned medicine and the medicinal uses of plants from Cheiron, who was a centaur (having the torso of a human and the body of a horse). Whereas all the other centaurs were violent and died in a battle, Cheiron was saved due to his wisdom. He was living on Mount Pelion in the east of continental Greece. It is probably no coincidence that he was considered wise and was credited with the knowledge of the therapeutic properties of plants, as Mount Pelion was and still is famous

FIGURE 5.4 The plant Kestron (betony) is flanked by two men—presumably a physician wearing a hooded cloak and his youthful companion shouldering a spear. The text advises on ways to prepare the plant for use as an emetic, purgative, or antidote (painting, recto; text, verso), folio from a manuscript of the *De materia medica of Dioscorides* (1224). Middle Eastern (Iraq?). Ink, opaque watercolor, and gold on paper. Harvard Art Museums/Arthur M. Sackler Museum. © Harvard Art Museums/Arthur M. Sackler Museum, Bequest of Abby Aldrich Rockefeller.

today for the richness of its medicinal flora. Returning to Asklepios, he appeared in the *Iliad*, where he was a hero, that is, a personage between the world of the gods and that of humans, whose sons Machaon and Podalirus were the healers of the Greek soldiers fighting Troy. In spite of this early attestation in the Greek world, the healing cult of Asklepios, originally centered at Epidauros in the Peloponnese, did not seem to have developed before the fifth century BCE.

Another relevant myth is story of the legendary healer Melampous, who knew the medicinal properties of plants. Asked to treat the daughters of the king Proetos, who wandered the land mooing believing themselves to be cows, he soaked hellebore (*Helleborus niger*) in the river where they were accustomed to drink water. The plant provoked a violent purgation that brought Proetos' daughters to reason, as insanity was believed in the ancient Greek world to result from a congestion of the brain. Again, the person at the center of this episode is not insignificant: Melampous—whose name (the *Black-footed*) gave him a special status—was blind. A seer, he knew the world from inside, without being lost in, and distracted by, the spectacle of the world. As such he had an innate perception of plants and their properties.

The *Iliad* and the *Odyssey*, which together recount the story of Achilles in the tenth year of the war and of Odysseus' ten-year homecoming, also recount uses of plants by individuals situated between the mythical and human worlds. One such creatures is the sorceress Circe. Not only did she seduce Odysseus, but she mixed a poison in the food that she gave to his companions, transforming them into pigs (*Od.* 10.234–43). The botanical agent, though not identified in the *Odyssey*, most probably was a psychotropic substance possibly of the Solanaceae type. A deft manipulator of all kinds of substances, Circe knew poisons well, as the story of her niece Medea attests. According to the Greek poet Apollonios of Rhodes (third century BCE) in the *Argonautika* (3.275–98), Medea instantly fell in love with Jason when he arrived at the palace of Aeetes, Medea's father, during his expedition to find the Golden Fleece. She was so deeply troubled by this sudden passion that she wanted to commit suicide and was already taking a vial containing a dark poison, when her sister caught her and prevented her from an instantaneous death. Later on, however, Medea engineered a dramatic death for Jason after they married and moved to Greece, when she discovered that he was cheating on her (Ap. Rhod., *Argon.* 3.802–9).

Psychotropic action might also explain the case of the famous Lotophages, the lotus eaters, mentioned in the *Odyssey*. Driven by adverse winds, Odysseus' ships landed on the island of the Lotophages. Those of his companions sent to explore the place received lotus from the inhabitants of the island. Once they ate it, they forgot about home and did not want to leave (*Od.* 9.82–104). Lotus (*Nelumbo nucifera*) is indeed known to have a sedative, hypnotic, and anxiolytic action that might account for the forgetfulness and indolence of Odysseus' companions. These and many other, similar tales captured the early knowledge of medicinal plants probably gained by trial and error over centuries, if not millennia, in vivid narratives. Only later, in the historical period, did research for new medical applications proceed in a methodically organized and rational way. This was principally to the credit of the so-called Empiric medical school developed in the third century BCE. Empiric physicians believed that the only knowledge that could be deemed reliable was that obtained by direct experience (*empeiria* in ancient Greek). Although they initially rejected theoretical speculation, they did ultimately accept some. Starting from the agency of a plant observed in a perceptible and measurable way, they expected that the same agency would reside either in another species of the same genus, or in a plant of a different genus with similar botanical characteristics.

Looking back in time, we have no explicit information on how the therapeutic properties of plants were discovered in the Greek world before the appearance of written documentation. However, on the basis of later data and comparative ethnobotanical and ethnopharmacological analysis, we can infer that multiple plants were used for therapeutic purposes early in human history, even before plant domestication and agriculture. This might be the case with the fruits of trees such as acorn, chestnut, and walnut, which were consumed in the time of hunter-gatherers. It is also possible that in a pre-agricultural phase, plants used for medicinal purposes might have been noticed because of remarkable botanical or organoleptic characteristics, be it their color, shape, flowering time, taste, or any other notable fact, most of which result from evolutionary adaptative mechanisms aimed at defending the plants from their predators. The red color of the poppy flower certainly drew attention, as did also the white lily, the dark-purple black hellebore, and the iris—which has all the primary colors of the rainbow—the source of its name (iris) in ancient Greek. The interesting shapes of the testicle-like double bulb of orchids, the stellar anise seed, and the deep flower—a trap to capture insects—of some species of birthwort were certainly noted. The flowering time of St John's wort corresponding to the summer solstice was most probably also noticed, as was the taste of plants: sweet, bitter, salty, or astringent. Other strange characteristics likewise probably provoked curiosity, like the projection of an irritating seed by the ripe "squirting cucumber" (*Ecballium elaterium*) when lightly touched or even closely approached. Some plants offered several of such remarkable characteristics together, like the red, round, and seed-filled pomegranate (*Punica granatum*), having juice similar in appearance to blood. According to the so-called "doctrine of signature" developed much later by the German philosopher Jakob Böhme (1575–1624) in his *Signatura Rerum* (The Signature of All Things) first published in 1621, these characteristics acted as indicators (*signatures*) of the therapeutic applications of the plants. If red species are good for blood and the cardiac system, yellow ones are indicated for the gall bladder and liver. Such plant lore appeared in the first written documentation in the Greek world of the therapeutic uses of plants, the *Hippocratic Collection*.

In historical times, no report on the discovery of the therapeutic properties of plants appears before Theophrastus. In the ninth book of the *Historia plantarum*, he mentioned several interesting cases, not only shepherds who observed the behavior of the cattle, but also the fourth-century BCE Thrasyas of Mantinea, who claimed to have discovered a poison from plants. It might be no coincidence that Thrasyas came from Arcadia, in the Peloponnese, the region where the seer Melampous treated Proetos' daughters. This might hint at a local tradition of herbalism also illustrated by the Pelion mountain and Cheiron. Returning to Theophrastus, he also referred to common knowledge among people, healers, and herb-providers (then identified as *root-cutters*), in addition to merchants at markets, some of whom might have been quacks boasting the efficacy of plants or selling inefficacious plants.

Theophrastus' compilation technique—the Aristotelian *historia* as we have noted—is echoed in Dioscorides' *De materia medica*. As he himself stated in the introduction of his treatise, he was interested in materia medica in his early years and he "*explored much land*." On the basis of the latter affirmation, he is traditionally credited with acting as a military physician accompanying the Roman troops, though this is not confirmed by any evidence. As he explicitly specified in the introduction, he collected most of the information contained in his work through careful personal observation (*autopsia* in Greek). For substances regarding which he could not have personal experience, he reviewed information that had become generally accepted as well as the opinions of local people. Whatever the origin of the material, he reported it with great accuracy and organized it in a vast, organically structured synthesis, as we shall see. As Dioscorides himself defined it, the vast compilation resulting from this search was a *historia*, as already was the case with Theophrastus' *Inquiry into Plants*.

Whereas Dioscorides' approach was largely descriptive, Galen was inquisitive. Even though he, too, collected information and substances from multiple sources, he did test the properties of the plants and even organized trials in a way that might have been inspired by the experiments of Attalus and Mithridates. Unlike Attalus, who experimented on humans, and Mithridates, who did so on himself, Galen experimented on animals, possibly establishing the prototype of modern animal testing. For example, in order to determine the lethal dose of a certain poison, he prepared small bread balls into which he mixed vegetal poison. He fed hens with these balls and observed the physiological effects of the poison on them, ranging from death, dizziness, and equilibrium troubles to survival, depending on the dosage of the poison.

GROWING AND COLLECTING MEDICINAL PLANTS

Since the range of foods and medicines attested in later written documentation closely correspond, one may surmise that after the domestication of plants and the spread of agriculture, plants grown for alimentary purposes were also used as therapeutic agents. We do not know what the species grown in the early phases of agriculture were. An early source in the historical period is the poem *Works and Days* by Hesiod, who seems to have lived toward the end of the eight century BCE. In the poem, Hesiod mentions only wheat and grape vine, probably reflecting his own agronomic misfortunes rather than the actual practice of his time, which was much more diversified. Indeed, the earliest Greek medical literature, the fifth-century BCE treatise *On Regimen* attributed to Hippocrates, lists a broad range of plants together with their nutritional value and medicinal properties. Although it might actually have been the case, it would be surprising if cultivation of this vast range of plants began only during the three centuries between Hesiod and *On Regimen*. The knowledge regarding techniques for growing, collecting at the best moment, and storing alimentary plants was slowly acquired through the centuries and transmitted across generations through oral tradition, being recorded in writing only much later. Since the range of alimentary and medicinal plants closely correspond, it can be expected that cultivation of grape vine, for example, as well as the production of wheat, developed similarly. If so, it might be speculated that medicinal plants were probably cultivated in small orchards adjacent to the houses in or outside the Greek cities, close to their walls. Given the distribution of tasks in Greek households, the care of such orchards, wherever they were located, probably was the responsibility of women who were also in charge of the transmission of knowledge across generations. Transmitting the knowledge of medicinal plants acquired over time in families might have been the purview of the elder women who might no longer have been able to engage in manual tasks outside and cared for the next generation inside.

In addition to collection of some species via cultivation, various others were gathered in the wild, as the documentation in historical periods makes clear. These were rarer species, mostly growing in the mountains as, for example, gentian (*Gentiana lutea*). Again we do not know who, in the communities, was in charge of collecting such plants, although we might suspect on the basis of the division of work typical of the social model of ancient Greece, that it was the responsibility of men. The existence of healers attested since the "age of legend," such as Melampous already mentioned, suggests that there might have been individuals who specialized in collecting useful plants in the wild. In that case, procurement of medicinal plants in early Greek communities might have been divided along a gender line, with the rare plants collected in the wild by men and the plants commonly used both as foodstuff and medicine by women. At a certain point in time—possibly as late as the fourth/third century BCE—a profession of drug providers existed, which is attested by mentions of them in Theophrastus' *Historia plantarum*. Interestingly, judging from Theophrastus, this profession was divided into at least two specialties: the so-called *root-cutters*, who appear actually to have collected medicinal plants in the wild without limiting themselves to *roots*, and the vendors of medicines, who might have sold manufactured products, ready to use, in addition to the raw products provided by the *root-cutters*. Granted, these may have been two different parts of the same person's activity. Though more difficult than harvesting of cultivated plants, collection in the wild was not made obsolete by cultivation because mountain species were believed to have greater efficacy, especially when compared with the species and individual plants growing close to water.

The case of the early Roman world was probably not different from that of Greece until the end of the second century BCE, although the range of plants might have been substantially smaller, possibly because of both environmental differences and explicit choice. Cato was famous for opining that the drug range of a household should be limited to broccoli. Nevertheless, in his treatise *On Agriculture*, he described and analyzed the techniques for growing fig trees, olive trees, grape vine, pulses (fava bean [*Vicia faba*], bitter vetch [*Vicia ervilia*], fenugreek [*Trigonella foenum-graecum*], and lentil [*Lens culinaris*]), besides wheat, barley, and unidentified fruit trees. Later on, however, as Rome became an overpopulated capital towards the end of the first century BCE and in the first century CE, production of plants, whatever their use (nutritional or medicinal), was transformed. Romans increasingly were no longer in direct contact with the natural environment, and an early form of intensive "commercial" agriculture was developed. Such production was transferred overseas (particularly to Egypt), where nurseries were established to meet the high demand of the Empire.

There were, of course, exceptions. In both Rome and Pompeii, gardens were created within the walls of the city. These were of different types: in broadest terms, in Rome, these were gardens allowing for ostentation of unlimited financial resources and extravagance, while in Pompeii they were smaller and were included in the structure of the houses as peristyle gardens. In his *Naturalis Historia*, Pliny lamented the creation of intra-mural gardens as the most recent extravagance of his time (*HN* 19.50). It might be the case that he did not fully understand the ongoing urban evolution and the avalanche of effects set into motion by the dramatic demographic increase that required the outsourcing of plant production, particularly from Egypt, where archeological traces of nurseries are attested together with the containers (amphorae) used to trade plants overseas. The gardens created in Rome and Pompeii in the first century CE might be a response to this new situation, as "productive" gardens reduced the distance introduced between plant cultivation and consumption, thereby eliminating the problems linked with distance, ranging from transportation to loss of freshness and quality.

In addition to this transfer of plant production, Rome witnessed another transformation, which might have started earlier: an expansion of the range of available plants through the introduction of Eastern species hitherto unknown in Italy. During their conquest of the East, Roman troops came into contact with natural environments different from the ones they were used to. In the process, they discovered plant species that had been domesticated and were cultivated in those regions, such as peach, apricot, and cherries. According to Pliny (*HN* 15.102), cherries were introduced to the Roman world in 73 BCE by the Roman general Lucius Licinius Lucullus (118–57/6 BCE), who discovered them while he was campaigning in the East. Interestingly, he also built sumptuous gardens in Rome.

ANALYZING AND EXPLAINING THE THERAPEUTIC PROPERTIES OF PLANTS

The various tales reported thus far, be they of Artemis, Odysseus, the Lotophages, Circe, or Melampous, attest to a practical knowledge of the agency of plants on humans. In no case do they translate such knowledge into abstract, theoretical concepts. The Hippocratic *Regimen* already mentioned did not proceed differently. It described effects on the human organism of a broad range of plants used as foodstuffs without offering any explanation or conceptualizing these effects through general inferences. More generally, none of the *c.* 250 different plants mentioned in the *Hippocratic Collection* have their agency described in general terms. All are simply named in the formulae for the many medicines prescribed in the *Corpus* for a broad range of diseases. Such knowledge, as stated above, likely derived from an earlier time, reflecting the uses of Greeks after the transition period from the Mycenaean world to the Early Archaic period in continental Greece.

For the plants introduced into the Mediterranean environment early in history, probably after their domestication, we do not know whether they entered the area with any knowledge about their nutritional uses and properties or their therapeutic agency. For the species introduced with the expedition of Alexander the Great to India written documentation does not allow us to establish whether they came with related knowledge. Similarly, we have no evidence that the plants of the Egyptian world generated new knowledge among the scientists doing research at the Museum in Alexandria. Later on, when several species of fruits (cherry, apricot, plum, and lemon) were introduced to Italy by the troops fighting on the eastern frontier of the Roman world, they seem to have been accompanied by some knowledge of their uses, as no debate regarding this question appears in contemporary literature. A special case is spurge (*Euphorbia* spp.) in the first century BCE/CE. As the brief mention in Pliny's *Natural History* (5.16) states, the strong cathartic effects of this plant native to the High Atlas were discovered by Euphorbus, the physician to the king of Mauritania (that is, northern Africa) Juba II (52/50 BCE–23 CE), after whom the plant was named by Juba himself.

The most ancient treatises in the *Hippocratic Collection* do not contain explicit theoretical analysis regarding the therapeutic action of plants; however, they implicitly rely on a broad range of notions defined by the nature of the medical conditions on which they were supposed to act. In Hippocratic medicine, health resulted from the equilibrium of the physiological components of the body identified as four humors (blood, phlegm, yellow bile, black bile). Disease was explained by the alteration, insufficiency, or excess of any of these elements, and treatment consisted of restoring the normal state or quantity of the defective humor. According to the principle *contraria contrariis curantur* (opposites are cured by opposites), which provides a theoretical basis for the allopathic treatment of disease, physicians prescribed remedies presumed to compensate for disequilibrium. The therapeutic action of plants can be extrapolated from the nature of the ailments they were used to treat. Hellebore, for example, mostly administered to reduce excesses of humor, can be identified as a cathartic.

Therapeutic substances were identified by their major action in Hippocratic medicine, and these major actions, in turn, led to the creation of therapeutic categories. For example, in the case of hellebore and other plants or natural products with a similar action, catharsis was identified as a therapeutic action. The therapeutic categories resulting from this descriptive method were clearly retained over time, as they appear in Dioscorides' *De materia medica*, and even until recently in the history of pharmacy—in the nineteenth century.

However accurately observed they might have been, the actions that generated these major therapeutic categories were not explained in the naming of them. Catharsis was an effect, not a mechanism. The first actual explanation of the therapeutic agency of plants appears in the *Problems* of the Aristotelian school. Positing that pathologies resulted from pathogenic substances in the body, the scientists of the Lyceum logically deduced that such substances needed to be eliminated from the body via catharsis proceeding from either purging or vomiting. Thinking in terms of physics, they concluded that purgation was provoked by the downward pressure of heavy therapeutic matter, which pushed down pathogenic matter, whereas vomiting resulted from the action of light therapeutic substances.

The first century BCE witnessed a new line of thinking developed by the Methodist school, famous for its claim that medicine could be easily learned and that its methods should be simple. Instead of considering that therapeutic products acted through the entirety of their "matter" or substance, as had been the case up to this point, Methodist physicians fragmented the substance and, consequently, also the action of therapeutic products. Building on atomism (that is, a theory according to which matter is made of indivisible particles that are somehow held together), they postulated that the human body was made of matter in constant movement. Excess or insufficiency of movement resulted in accumulation or rarefaction of bodily substance, both of which were pathological. Therapy was required to restore the normal quantity and movement of particles, doing so by introducing into the body substances with the right amount and type of particles. According to this view, the several materia medica used by Methodist physicians were defined by the structure of their matter, which relied on the shape and quantity of their particles.

In the second century, Galen synthesized many of the earlier theories that in a sense concluded antiquity. Taking Hippocratic allopathic thinking as a basis, he combined it with the atomistic fragmentation of the Methodists. This was not a simple undertaking, as the Hippocratic system of four humors was complex, in part because it linked the four physiological components with the four elements thought to constitute matter: earth, fire, air, and water. Galen specifically used this link to combine Hippocratism and Methodism. In his synthesis, therapeutic substances that were prescribed on the basis of the Hippocratic allopathic principle introduced into the body one or more of the four elements in the form of particles of different shape and weight. In this way Galen accounted for the interaction between remedies and the body that had remained unexplained until then.

Galen went even further by theorizing the variability in the intensity of medicinal substances, which was only vaguely defined in earlier pharmaco-therapeutic literature. Whereas neither the *Hippocratic Collection* nor Dioscorides provided explicit information on this, Dioscorides did group the materia medica credited with the same property and listed them, within each such group, according to the degree of intensity of this property, from the most to the least potent. Although he did introduce order into the listing of materia medica by applying this principle, Dioscorides did not explicitly address or theorize about it. It was Galen who is credited with this. A systematic theoretician, he defined the action of all materia medica in accordance with four qualities—hot, cold, dry, and humid—in a perfectly Hippocratic way, and he measured the intensity of these qualities in any materia medica on a scale of four degrees, from 1 to 4, from the least to the most intense. He paired these qualities in two groups of opposites—with hot and cold in one pair and dry and humid in the other—as did Hippocratic medicine. Consequently, this pairing implied that any materia medica did not contain only one quality, but both opposites: for example, dry and wet together. This explains the title of Galen's treatise on the topic: *De simplicium medicamentorum temperamentis et facultatibus* (On the Mixtures and Properties of Simple Medicines), in which the term "mixtures" does not refer to compound medicines, but to the fact that the substance of the materia medica studied in the treatise is not "single" (consisting of one property), but rather is *mixed*, as it always *combines* opposite qualities. As a result, it was necessary to determine more precisely the exact quality of a given materia medica, determining whether dryness was present to higher degree than wetness, or the opposite, for example. To avoid potential contradictions arising from pairing opposite qualities, Galen linked these qualities, measuring them both on a graduated scale from 1 to 4, so as to be able to define the respective intensity of paired, opposite qualities. Both qualities could not be present with the same intensity—that is, both having an intensity of 1—which would lead to contradiction or "nullification" of the qualities in question. The problem, then, was to determine the quality of a materia medica by measuring the intensity of each of the two opposite qualities that constituted it, and to determine which one was prevalent over the other. This process is what is expressed in the title of Galen's work through the association of the two terms "mixture" and "property."

However brilliant it might seem, Galen's complex theoretical system was more problematic than helpful. Its complexity might have contributed to the lack of interest in the treatise in following centuries and, conversely, to the greater interest in Dioscorides' work, which was certainly less conceptually worked out, but more easily accessible. The difficulty of Galen's system is further evidenced by the history of therapeutic substances in the Mediterranean, particularly in the Arabic world, where scientists designed new, exact methods of measuring the property of materia medica according to Galen's system, including mathematical formulas.

PRESCRIBING AND TAKING PLANT MEDICINES

No treatise on the prescription of medicinal plants has been preserved, and we do not even know if any ever existed. In the earliest works of the *Hippocratic Collection*, which also are the most ancient therapeutic documents currently known, the description of a broad range of clinical cases is followed by the report of the medications that were prescribed, together with the account of the evolution of the patients' health, including evaluation of the therapy. Some such cases are strictly individual but others are general, resulting in therapeutic recommendations potentially of universal use for identical or similar uses. Although this is not specified, the latter prescriptions probably resulted from the treatment and careful observation of multiple cases of the same pathology, the recording of the therapy, possible gradual fine-tuning, and codification in a manner that might not be so different from the formulae prescribed by modern pharmacopoeias. As explained above, the nature and action of the therapeutic agent was determined by the mechanism supposed to have caused the pathology to be treated, which was identified through diagnosis. Prescription was a deduction directly resulting from diagnosis.

Although it is to be expected that such strategy resulted in prescribing as many different natural products as possible, particularly medicinal plants, this was not the case. Of the broad range of botanical species mentioned in the *Hippocratic Collection*, not all were used with the same intensity. Strangely enough, the most frequently mentioned therapeutic substances were myrrh and frankincense—that is, exudates of trees typical of the south of the Arabic Peninsula, which were expensive. It has been speculated that the use of these rare and costly ingredients was intended to imbue the medicines into which they were mixed with greater "prestige" value and, consequently, higher "subjective" therapeutic efficacy. Even though a psychological component might have come into play, contributing to the efficacy of myrrh and incense, it is probably more realistic to hypothesize that both resins were used primarily for their real antiseptic and bactericidal properties—all the more so because they were abundantly administered in external use, in both gynecological and post-partum contexts, without doubt in the latter case to avoid the puerperal fever which killed so many parturients until quite recently. The second most frequently used plant in Hippocratic medicine was hellebore (*Helleborus niger*). However surprising this might seem at first glance because of the highly cathartic effect of the plant, its use might in fact have been justified by this very effect, which eliminated unwanted substances from patients' bodies—including possible intestinal worms and other parasites that affected populations until not so long ago—thus preparing for the subsequent administration of medicines and their enhanced absorption by the body.

Of the many other plants mentioned in the Hippocratic literature, which total over 250, a group of *c.* fifty emerges that yielded more than half of the uses of plants in the *Collection*. These are vegetables, pulses, and fruits, none of them being rare, expensive, or difficult to grow. Significantly, all were also used as foodstuffs, something that validates the principle attributed to, but not explicitly formulated by, Hippocrates according to which food is medicine and medicines are food. This reduced number might seem surprising at first glance. At the very least, it invalidates our expectation that a great variety of therapeutic agents could have been prescribed by Hippocratic physicians at the conclusion of diagnosis. Examination of the items that make up this short list of often-prescribed medicinal plants is telling. All of these fifty plants were foodstuffs, from onion and garlic to pomegranate and fig, including sage, oregano, and laurel, in addition to olive, fennel, and lentils, which fact vividly illustrates the adage about the identity of foodstuffs and medicine so typical of Hippocratic medicine. Furthermore, all these species are Mediterranean, and their twofold nutritional and medical use invites consideration that they had been consumed and administered for a long time before the compilation of the Hippocratic treatises in which we see them appear. Inversely, their presence in the *Hippocratic Collection* suggests that Hippocratic therapeutics was not a creation *ab nihilo*, but rather the inclusion in "scientifically based" medicine of practices that originated far back in time and had probably been orally transmitted for centuries. What's more, the very nature of the fifty most-used plants of Hippocratic medicine reveals certain other parameters taken into consideration by Hippocratic physicians when prescribing a therapy: all these fifty plants were common and most probably cultivated in orchards close to dwellings, possibly by women. Available all year round or capable of being easily preserved from one year to the next, they did not necessitate any additional investment or special work, and most of them did not even need to be specially stocked. They were consumed fresh, shortly after having been collected, assuring the best quality.

The other plants appearing in Hippocratic writings were rarely used. They were less common and needed to be collected in the wild, probably by men or by professionals, possibly the so-called "root-cutters" mentioned by Theophrastus. The infrequent use of these plants is no less revealing than the frequent use of the fifty plants that saw common use: their variety (over two hundred) correlates negatively with their respective number of uses. This hints at a certain awareness of the biodiversity typifying the eastern Mediterranean, together with avoidance of over-harvesting with its negative consequences on biodiversity. The case of silphium, over-harvested and supposedly extinct maybe as early as the first or second century CE, is indicative of the risks of excessive consumption. Returning to the Hippocratics, their uses of the two hundred plants referred to above were very specific and demonstrated a good knowledge of their therapeutic action, and this for well-defined pathologies. Again, such applications certainly attest to a long process of discovery resulting in a proper mastery of determined properties.

The Empiricists probably wished to break with such long-established traditions. In their effort to re-found remedial therapy on a solid basis, as we have seen, they formulated rules for the prescription of medicines. Apart from the substances whose agency they had already established by repeated personal observation, they did, however, accept some speculative reasoning. They thus believed that if a plant proved successful in treating a given pathology, other plants with similar botanical characteristics could be equally successful for the treatment of that pathology, thus expanding the range of remedies for it. Conversely, they speculated that if a pathology could be treated by a plant, then any other medical condition with identical or similar characteristics could be healed by the same plant, thus optimizing the use of available resources.

Hippocratic "traditionalism" and personal, autoptic observation in the manner of the Empiricists were merged by Dioscorides. This mix of approaches resulted in a complex system for the identification of the therapeutic agents to be administered to patients. In the introduction of *De materia medica*, Dioscorides explicitly stated that he classified therapeutic substances analyzed in the treatise according to a certain system—however, without explaining what this system was. Nonetheless, careful reading of the work allows for the identification of this system. Substances with the same action were grouped in coherent sets within which they were ordered according to two parameters, possibly in combination: their degree of possessing the property common to the group and their mechanism(s) of action. The resulting groups were then ordered according to the cultural values their action was credited with. This order proceeded from a group with positive cultural values to the opposite. Concretely, the first group of the whole collection was that of the plants with warming properties, used to eliminate any excess of humidity in the body, from swellings to abnormal gynecological discharges. The last group, for its part, was that of the matters considered to be cold and humid, mostly prescribed for the treatment of dermatological pathologies that supposedly resulted from a process of inflammation. In a way reminiscent of Galen's association of the opposite qualities of materia medica, these two groups were associated and became the two poles of an axis along which all other groups were ordered according to a gradual decrease of the positive qualities of the first group (the warming property) and a gradual increase of the negative values of the last group (the cooling and moistening property). This whole structure created a *scala naturae* (Ladder of Being) in which all natural elements are part of a single continuum. After he had made a diagnosis, a physician had first to locate on this *scala* the group of the matters that were required for the treatment of the pathology under consideration and, second, to identify the matter that offered the right degree of action together with the required mechanism in this group.

Further, this structure contained an implicit semantic system. Logically, the first group was that of scented and exotic plants, whereas the last one was that of minerals. Furthermore, the first plant of the first group was described as containing all colors in its flower, as it is indeed the case with the blue iris flower, in the heart of which all primary colors appear. Meanwhile, the last type of matter in the last group was soot, used for the preparation of ink. In the ancient Greek chromatic system black is not a color but the absence of any color. Through the opposition of iris with soot Dioscorides' whole system thus ranged from the full palette of colors to its opposite, the lack of any color.

On another semantic level, the presence of iris in the first chapter of the whole work introduced a further dimension of the construct. In ancient Greek the very name of *iris*, the plant, was also that of the *rainbow*. In Greek mythology the *rainbow* was personified as *Iris*, who was the messenger of the gods, that is, a connection between the ethereal universe of the Olympians and the sublunar world of humans, exactly as the rainbow links earth and sky. Delving deeper into cultural meanings, medications were defined as "the hands of the gods" by the Greek physician Herophilos (330/320–260/250 BCE) according to the historian Plutarch (*Mor.* 663B–B) and confirmed by Galen (*On the Composition of Medicines According to the Places* 6.8 = 12.965–6 Kühn). This sheds light on the full meaning not only to the position of iris at the beginning of *De materia medica*, but also—if not above all—to the entire field of materia medica and medicinal plants: they were a sign of the empathy of the divine world for the humans.

The great theoretician Galen abandoned Dioscorides' complex construction and its metaphysical dimension, and simply ordered the materia medica and medicinal plants alphabetically on the basis of their most commonly accepted Greek name. In his time, therapeutics was undergoing a deep transformation. The Roman world, which had become an empire in 27 BCE and covered most of the known world of that time, was experiencing a dramatic demographic explosion with ensuing expansion of cities, particularly Rome. These *megalopoleis* were cut off from the natural world, a fact that impacted therapeutics. It no longer was possible to have a direct access to nature and fresh plants as it had been in the time of the Hippocratics. To compensate for this, a new strategy for the making of medicines developed: compound medicines prepared in the manner of some modern products. Credited with having a long "shelf life," they offered the advantage of being standardized and stable over time, besides being broad-spectrum therapeutic agents thanks to the multiplicity of their components. The role of physicians was transformed accordingly and no longer consisted in identifying one or more natural substances to be administered as medicines, but only in selecting an existing compound medicine. The identification of the natural substances to be mixed in compound medicines was transferred to pharmacists who could use Galen's treatise and its alphabetical enumeration of substances with the enunciation of their properties and also the degree of such properties. Such medicines were compounded by synthesizing different therapeutic actions and balancing such actions on the basis of Galen's system of degrees. As already mentioned, Galen's system—which was characterized by the presence of opposite qualities in the same materia medica and required measurement of the degree of each of these opposite qualities in order to identify the exact property of a given materia medica—did not gain favor with professionals, who preferred Dioscorides' treatise and the implicit structure of the natural world it conveyed, however complex this construction might have been.

Galen's immense work was the last significant contribution to medicine in antiquity. While it was intended by its author to subsume all the previous knowledge on medicinal plants generated in antiquity—and possibly to make previous literature obsolete, since it offered a comprehensive encyclopedia solidly underpinned by a comprehensive theoretical system—this is not what actually happened, as it was too complex to be easily referenced and deployed in the daily practice of therapeutics. Forsaken by physicians and scientists in the subsequent epochs, it left a didactic void in the field of materia medica. As noted above, this void was occupied by Dioscorides' *De materia medica* and its simple, descriptive treatment of the natural resources used for the preparation of medicines mostly comprised of one major ingredient, often a single plant. Medicines made of several ingredients—the *compound medicines*—were nevertheless developed and used, even though therapeutics was equated for centuries with *simple medicines*, particularly medicinal plants.

CHAPTER SIX

Plants in Culture

Botanical Symbolism in Daily Life and Literature

ANNETTE GIESECKE AND MECHTHILD SIEDE

This chapter focuses neither on humanity's practical, physical dependence on plants to sustain life nor on scientific inquiry founded on botanical observation, but rather on that inherent spiritual connection between humans and plants arising from a host of the latter's notable characteristics, among them their burgeoning and growth, their life cycle from seed germination to maturity and dormancy or death, their annually recurring flowers and fruit, their diverse appearances, and their possession of a mysterious agency and vitality though rooted to a single place. The cultural perspectives offered here are those of ancient Greece and Rome, which civilizations have yielded a wealth of archaeological and textual material.

In Greek linguistic usage, the phylogenetic proximity of plants and humans is particularly clear; the Greek term "*phuton*," originally meaning "the thing that has grown, creature," narrowed over time in application so as to denote plants specifically. This, in turn, provided the basis for a broad spectrum of plant symbolism and botanical metaphors (cf. Pl., *Tim.* 77A–C; Eur., *Med.* 231). For example, one spoke of being in the prime or "bloom" of life (*anthein*), as is still the case today, and it was not without good reason that botanical excursus in the medical texts from the Hippocratic Corpus—in *De genitura* (On Generation), *De natura pueri* (On the Nature of the Child), *De morbis* (On Diseases) *IV*, probably dating to the last third of the fifth century BCE—exhibit a striking interest in analogies between plant and human physiology (*Genit.* 10:2; *Nat. puer.* 26ff.; *Morb.* 4.33–5). Similarly, the Greek philosopher and author Theophrastus (*c.* 370/371–288/286 BCE), who is credited with being the "father of botany," underscores the close relation between humans and plants by drawing numerous parallels between plants and humans in medical contexts (*Hist. pl.* 2.7.6; *Caus. pl.* 1.13.5ff., 4.16.3ff.; Plin., *HN* 17.37.218–19), and Roman encyclopedist Pliny the Elder (23–79 CE) evidences the greatest reverence for trees in his vast *Natural History* (12.1), noting that from earliest times they supplied "the most valuable benefits conferred by all of Nature upon humankind."

MEANING IN RELIGION AND MYTH: SACRED TREES

Long before humans constructed specific places of worship for their gods and created likenesses of them, deities were thought to reside in the natural environment. Special places in nature such as springs and caves as well as rocks and trees were believed to be filled with divine power, or *numen*, as the Romans called it; trees, according to Pliny the Elder (*HN* 12.2) were the original temples of the gods, and in his *Life of Numa* (8)—the legendary second king of Rome and purported founder of key Roman religious institutions—biographer and essayist Plutarch (*c.* 46–120 CE) notes that the sacred trees of Rome were of hallowed antiquity, antedating the city itself. In this early period, trees were held in particular reverence for their stately form and long-lived vitality; trees' palpable soulfulness and animation, indicated by the rustling of windblown leaves and branches outstretched like arms, were also presumably factors.

Tree worship is widely attested in the ancient world and can be traced to a very early period. For example, the Zoroastrian *Avesta* tells of an originary tree that bore the seeds of every plant (Erdmann 1954: 1–3). A similar belief is attested in Norse mythology: verses comprising the poetic *Edda* glorify Yggdrasil, the Cosmic Tree, whose branches touch the heavens and whose roots deeply penetrate the Earth. Commonly referred to as an ash tree, the epithet "evergreen" suggests identification of the tree as a yew. According to Egyptian belief, the Cosmic Tree was a sycamore fig that housed the gods in its canopy and was located at the easternmost horizon; in the Book of the Dead (Chapter 109) two sycamore figs serve as heaven's portal through which the sun god re-emerges in the morning (Kees 1956: 84) (see Figure 6.1).

In these various cultures the Cosmic Tree was clearly conceived of as connecting the earthly and divine realms, and certain trees were believed to be sacred to certain deities, generating myths that explained these specific associations. A prime example is the Greek cult of Zeus at Dodona, where invocation of the god was performed beneath a sacred oak. When petitions of the faithful were heard, rustling leaves indicated the god's attentive presence (Nicol 1958: 139). One of the most common and important deciduous trees in the Mediterranean area, the oak also had a prominent role in the belief systems of pagan cults beyond those of the Greeks and Romans (Herzhoff 1990: 257–72). Pliny the Elder notes that the Celts of Gaul had such reverence for the oak that the tree's Greek name, *drys*, gave Druidism its name (*HN* 16.95.249). Altars of the Near Eastern tree and fertility goddess Astarte were located either beneath large oaks and terebinth trees—distinct species but with similar outward appearances—or in sacred groves of these (Dtn. 12.2; I Reg. 14.23; Jer. 3.13). The ancient Oak of Mamre in Judea, located by a sacred spring (Gen. 18.1–8), was an ancient Canaanite cult site, as was the oracular oak of Sichem (Jdc. 9.37); at both, Jahweh the Lord made his presence known in a manner similar to Zeus at Dodona. Akin to Jewish belief, according to which the oak was present from the world's beginning (Josephus, *Bell. Jud.* 4.9.7, *Ant.* 1.10.4), Greco-Roman tradition held the oak to be the oldest species of tree, even going so far as to preserve a belief that the first humans sprang from oaks (*Il.* 11.86; Verg., *Aen.* 8.315; Juv. 6.12). This venerable tree, which yields honey and sweet acorns, epitomized the paradisiacal conditions said to have prevailed in the Golden Age, *aurea aetas* (Hes., *Op.* 232ff.; Verg., *Ecl.* 4.30; Ov., *Met.* 1.112). The importance of the oak to the Greco-Roman world was comparable to that of the sycamore fig in Egypt, both tree species at times signifying trees in general.

Other especially venerable trees associated closely with the early myth-history of Rome were three particular specimens of lotus (*Celtis australis*); according to Pliny the

FIGURE 6.1 The goddess Isis, appearing to Sennedjem and his wife in the sacred sycamore and providing them with water, bread, and lotus flowers. The couple wear perfumed wax cones on their wigged heads. Burial chamber ceiling, Tomb of Sennedjem, also known as TT1 tomb, dating back to the reign of Rameses II, Deir el-Medina, Theban Necropolis (Unesco World Heritage List, 1979). Egyptian Civilization, New Kingdom, Dynasty XIX. © DEA/G. DAGLI ORTI. Courtesy of Getty Images.

Elder (*HN* 16.86.236), Romulus himself was said to have planted a lotus—still visible in Pliny's day—by the Temple of Vulcan, which was of central importance to the Roman state. Another lotus of great antiquity—some 450 years old in Pliny's day—grew in the space in front of temple of Juno Lucina on the Esquiline, and a third, said to be older still, received its name "*capillata*" (covered in hair) from the fact that Vestal Virgins hung locks of their hair from its branches when they assumed their priestly duties.

In addition to their vigor and, in many cases, appearance of great antiquity, there are other qualities that induced the veneration of certain trees and underlay their dedication to particular divinities. The laurel, generally identified as sacred to Apollo, possessed cathartic and oracular as well as chthonic significance, derived in part from an earlier association with the earth goddess Gaia that persisted in the Apolline cult. Thus, Apollo was said to issue prophecies from his sacred laurel *(Hom. Hymn Apoll.* 396), and because of its strong, aromatic scent, the laurel lent itself to purificatory applications, thus naturally becoming linked to Apollo in his capacity of healing god. According to Greek mythology, Apollo purified himself—and crowned himself as victor—with laurel that grew in abundance near Delphi after slaying the serpent Python that guarded the oracle of its mother, Gaia, who issued prophecies from deep inside the Earth.

Another tree with links to the Underworld is the white poplar (*Populus alba*), whose leaves, dark on their upper sides, were thought to show deference to the chthonic spirits while their light upper sides, by contrast, were aligned with the world of the living. It was a wreath of white poplar that, in a symbolic gesture, the hero Heracles wore when he emerged alive from the Underworld after his victory over Hades' triple-headed canine guardian, Cerberus (Baumann 1993a: 50–72). The black poplar (*Populus nigra*), too, had associations with death and the Underworld in so far as petal-less flowers and small seeds gave it the appearance of sterility (Murr 1890: 19). In the *Odyssey*, Hades' queen Persephone's sacred grove in the Underworld is comprised of black poplar and willow (10.509–10). Appropriately, it was black poplars, weeping eternal tears of amber, into which the sisters of Phaethon were transformed; their source of grief was the death—at Zeus' hands—of their brother who, in an act of brazen hubris, threatened the safety both of heaven and earth when he recklessly took the reins of the Sun God's chariot (Arist., *Mir. ausc.* 836b; Ap. Rhod., *Argon.* 4.603–11; Diod. Sic. 5.23.4).

The attractive small tree (or shrub) myrtle, on the other hand, was sacred to Aphrodite, the goddess of love, fertility, and life, and enjoyed particular importance in marriage ceremonies (Siede 2013: 378–89). This link to Aphrodite may have had its origin in the Semitic cult of the fertility goddess Ishtar, from whom the Greek Aphrodite evolved, and was presumably based on the shrub's lush, evergreen aspect as well as its seasonal profusion of small, fragrant flowers (Steier 1933: 1179). The second-century CE Greek travel writer and geographer Pausanias (6.24.7) points to rose and myrtle as the plants most important to Aphrodite, as was the case with her Roman counterpart, Venus (Verg., *Ecl.* 7.62; Plin., *HN* 12.2.3, 15.36.119–21, and passim; Murr 1890: 85). Plantings of myrtle could be found in proximity to the goddess's temples, and the myrtle grove near the Temple of Apollo at Delphi was dedicated to her (Eur., *Ion* 120; Bötticher 1856: 447). As Pausanias also reports, myrtle wood was used to carve images of the goddess (5.13.7).

The mythology surrounding myrtle underscores this plant's relation to Aphrodite as well (Blech 1982: 250). After emerging naked from the sea's frothing swells, the goddess was said to have hidden behind a stand of myrtle until the Charites (Graces) and Horai (goddesses of the Hours) clothed her in finery, and in an alternate version of the tale, the goddess used branches of myrtle to shield her body from the gaze of shameless satyrs (Ov., *Fast.* 4.133–44). This tale, in turn, underlies Roman women's custom of bathing "*sub viridi myrto*," beneath green myrtle. As winner in the beauty contest judged by the Trojan prince Paris, Aphrodite was fabled to have been crowned with myrtle, whence the plant earned Hera's hatred (Schol. Nicand., *Alex.* 618a). For this reason, myrtle was disallowed in the Temple of Hera at Samos (Schol. Ar., *Ran.* 330), and it was hateful also to Artemis for hindering her beloved Diktynna's flight from an enamored Minos when he pursued her (Callim., *Hymn* 3. 197–203; Nicand., *Alex.* 681a; Blech 1982: 320). The mythical Athenian queen Phaidra's tale of suffering provides an *aition* (explanation) for the perforated appearance of myrtle leaves, in reality caused by the retention of an essential oil: while her stepson Hippolytus, the object of her unrequited love, engaged in sport at the stadium of Troezen, Phaidra watched him from the heights of the sanctuary of Aphrodite Kataskopia (The Observer). In anger at the goddess for not ensuring that her affections were returned, Phaidra assailed the goddess's sacred myrtle, puncturing its leaves with pins that secured her hair (Paus. 1.22.2, 2.32.3). Meanwhile, Servius (late fourth–early fifth century CE), the ancient commentator on Vergil, remarks that the lovely Myrene, a priestess of Venus, was transformed into a myrtle by the jealous goddess when she announced her resolve to marry (Servius on Verg., *Aen.* 3.23). A similar fate was said to have befallen an Athenian maiden named Myrsine, who was favored by Athena for her beauty and strength: when Myrsine was killed by angry youths whom she had bested in athletic competition, the goddess transformed her into a myrtle shrub bearing round, black berries—reminiscent of olives, Athena's hallowed fruit—so that she might thereby eternally enjoying divine favor (*Geopon.* 11.6). Different as they are, these stories all point to the link between myrtle, love, and marriage.

As already suggested in the case of oak and myrtle, the ability to bear fruit was crucial to the religious and cultic significance of trees and shrubs. A more striking example than myrtle is the pomegranate, having fruit filled with myriad seeds and yielding blood-red juice. Contrary to common belief, the pomegranate appears not to have been viewed solely as a symbol of fertility, however. In addition to being linked with Hera, goddess of marriage, and Aphrodite, goddess of fertility, the pomegranate was sacred to Persephone, thereby possessing a chthonic connection (Engemann 1983: 692–6); after her abduction by Hades, Persephone, the harvest goddess Demeter's daughter, was obligated to spend six months every year in the Underworld because, while there, she had tasted this fruit's seeds. As an aside, in the highly erotic context of the biblical Song of Solomon, the pomegranate unequivocally symbolizes the beauty of the bride, with the more complex Greco-Roman symbolism—allusion to both abundant life and death—playing no significant part, which is typical of early Christian literature (*Cant.* 4.3, 13; 6.7; 7.13).

A combination of chthonic importance and fertility symbolism may also be found in other plants that have many seeds, like the poppy, which was sacred to Demeter (Roman Ceres) and her priestesses (Callim., *Hymn* 6.44). Over time, the cult of Demeter spawned the lovely notion of the poppy seedpod as symbol and image of the earth, its surface covered with mountains and valleys and its core filled with countless seeds (Cornut., *Nat. deor.* 28). The Roman poet Ovid (43 BCE–17/18 CE), for his part, compares the extent of his suffering while in exile on the dreary Black Sea to the immeasurable number of seeds contained in a poppy's seedpod (*Tr.* 5.2.24).

Indicative of the importance of fruit trees in general and the esteem in which they were held is the mythical tradition regarding the competition between Poseidon and Athena at Athens to determine which of these deities would be awarded guardianship of that city. Athena produced an olive tree and Poseidon a saltwater spring, both on the Acropolis (see Figure 6.2). Deeming the olive more valuable, the judges, whether gods or citizens of Athens, awarded the city to Athena (Ov., *Met.* 6.5–145). Fruit trees possessed a connection with the divine not only in overtly religious contexts, as when growing in sanctuaries or forming sacred groves inhabited by gods, but also when growing in gardens that combined usefulness and aesthetics to paradisiacal effect, thus evidencing a divine aura. As a result, gardens lent themselves to idealization and symbolic characterization.

The Old Testament's Garden of Eden (Gen. 2.9) is a prime example of such a paradise. Fruit trees of every kind were said to grow therein, and the Tree of Life and the Tree of Knowledge resided at its center. The garden was the embodiment of divine abundance and perfection, which humanity lost of its own fault. At the same time, the garden is a source of consolation and hope: Jesus Christ is the embodiment of the Tree of Life, and the Holy Eucharist reverses the forbidden taste—taken in error and sin—from the Tree of Knowledge. The Tree of Life thus becomes a symbol of the eschatological paradise (Rev. 22.1–2).

In the cultural tradition of ancient Greece, the garden of the Hesperides, believed to be located at the westernmost margins of the world, was an archetype of the paradisiacal garden. The garden's trees, which bore golden apples (possibly quince) that were a source of eternal youth, had been a wedding gift from the earth goddess Gaia to Zeus' bride, Hera (Apollod., *Bibl.* 2.5.11; Hyg., *Fab.* 30). Another garden, in which atemporality and agelessness prevail, is the garden of King Alcinous on the *Odyssey*'s island of the Phaiakians. Therein all plants, whether fruits or vegetables, are well ordered and well tended, the result being an endless production of fruit, with no dependence on seasonality (*Od.* 7.112–32). The timeless beauty of this garden resides both in its eternal abundance but also in the functional, regular arrangement of cultivars. Both of these aspects can be found in other ideal gardens in the Classical tradition, among them the garden of Odysseus' father Laertes, also in the *Odyssey* (24.336–44), and the humble Roman farmer's garden of the *Moretum*, attributed (in all likelihood, falsely) to Vergil.

FIGURE 6.2 Reconstruction of the west pediment of the Parthenon according to a drawing by K. Schwerzek (1896), Acropolis museum, Athens. The contest between Athena and Poseidon during their competition for the honor of becoming the city's patron. Athena and Poseidon appear at the center of the composition, with the goddess holding the olive tree and the god of the sea raising his trident to strike the earth. © Photo by Universal History Archive/Universal Images Group/Getty Images.

MAGIC PLANTS

Magic potency was attributed to numerous plants, particularly to those with a strange or otherwise striking physical appearance, like the mandrake (or mandragora), whose root resembles a human body and found a range of medicinal applications (Diosc. 1.571–3, 4.75) (see Figure 6.3). Above all, the mandrake was one of the goddess Hecate's magic herbs and could be unearthed only with performance of certain rites, as also in Jewish tradition (Theophr., *Hist. pl.* 9.8.8; Plin., *HN* 25.94.147; Joseph., *Bell. Jud.* 7.6.3): King Solomon was said to have worn a piece of mandrake plant in a signet ring whereby to ward off evil spirits (syr. Herm. Trismeg. 2.14.606ff.).

Negative attributes, like sterility or thorniness, could also be a source of mysterious powers; in the case of trees with such attributes, these earned them a reputation as trees of ill fortune, *arbores infelices* (Plin., *HN* 16.45.108; Macrob., *Sat.* 3.20.2). The thorny growth of the wood pear was certainly a factor in Alcmena's being instructed to ignite pear branches at the witching hour in order to burn the bodies of the serpents Hera had sent to kill her baby Heracles (Theoc., *Id.* 24.88–92). In the context of this tale, as recounted by the poet Theocritus (*c.* 270 BCE), both the purifying aspect of *arbores infelices* when burned inside a polluted dwelling and—by way of sympathetic magic—the fact that snakes favored such shrubs as shelter must have been factors in their selection (Lembach 1970: 71–5ff.).

Strong scent was another quality that characterized plants held to possess magic potency. Apotropaic qualities were ascribed to myrtle for its fragrance, for example (*PGM* I.156). Those striving for honor and glory, meanwhile, could apply an ointment of felty

FIGURE 6.3 Greek physician, pharmacologist, and botanist Pedanius Dioscorides (40–90 BC) studying the mandrake root for his *De materia medica*. Engraving from *La-vie-des-savants-illustres* by Louis Figuier, 1866, private collection. © Photo by Leemage/Corbis/Getty Images.

germander (*Teucrium polium*), which happened also to be an antidote for poisons—for which reason, according to Pliny the Elder, it should be kept close at hand; this versatile plant could also drive off snakes as well as serve as a cure for ailments including cataracts, jaundice, and dropsy (*HN* 21.84.145–7). A similar range of medicinal uses were ascribed to sea onion or squill (*Urginia maritima*), which was hung over the door in order to guard against all manner of evils and misfortunes (Diosc., *De mat. med.* 2.202). It was presumably the fact that this plant's bulbs survived the arid conditions of summer, preserving sufficient potency to produce tall, narrow racemes of flowers (1.5–2 m [5–6.5 ft]) in fall, which underlay the diverse agency attributed to it. As already mentioned, the fragrant laurel, too, was seen to have apotropaic qualities—for example, warding off lightning strikes (Plin., *HN* 15.40.134)—and thus, like squill, was suspended over entryways. Another plant that can be mentioned in this context is what Greek sources called "*dodekatheōn*" (twelve gods' plant), possibly the lightly scented oxlip (*Primula elatior*), which Dryad nymphs (tree spirits) reputedly picked under a new moon and was regarded as a cure for melancholia, among other ailments.

A very different magic plant, spoken of only by the gods, is Homeric moly. Odysseus received this herb from the god Hermes as a means to ward off the evil spells of Circe, the witch at whose hands Odysseus' companions had been transformed into pigs. As Homer describes it, the plant "was black at its root, but with a milky flower. The gods call it moly. It is hard for mortal men to dig up, but the gods have power to do all things" (*Od.* 10.304–6). The Stoics and Neoplatonists later rationalized the herb and its characteristics, deeming it a symbol for the difficult path to knowledge or an image of the human struggle in the course of philosophical education, bitter (difficult to dig) but bearing white, sweet fruit (moly's white bloom) (Them., *Or.* 27.340a).

POLYVALENT PLANT SYMBOLISM AND IMAGERY: THE CASE OF MYRTLE

Ovid's *Metamorphoses*, which masterfully endows a host of mythological tales of transformation with cosmic significance, highlights the attributes of certain plants with particular thematic relevance to the poet's subject matter (Giesecke 2014). Plants retain the characteristics or emotions of those characters that metamorphose into them: grief, pain, joy, and close connections to certain deities. Because of the close relation between plants and the gods, mythology played a significant role in the symbolic usage of plants in antiquity. Plant symbolism, in turn, was firmly rooted both in private life, from birth to death, and in the political sphere, some plants having limited and others manifold symbolic resonances.

For example, myrtle had ties with multiple deities, among them Aphrodite (Roman Venus), Dionysus (Roman Bacchus), Hermes (Roman Mercury), and Athena (Roman Minerva). Consequently, this plant found an unusually broad spectrum of meanings in the human realm. Connection with the goddess of love resulted in its identification as suited to wedding contexts. The marriage god Hymenaios, son of Dionysus and Aphrodite, appropriately was represented wearing a crown of myrtle, and myrtle featured prominently in marriage hymns and erotic verses (Blech 1982: 293ff.). In his *De agri cultura* (8.2), Roman statesman and author Cato (234–149 BCE) notes several varieties of myrtle, among them "*myrtus conjugula*," whose nomenclature the elder Pliny links to marriage (*NH* 15.37.122). Myrtle's status as fertility symbol, deriving from its profusion of foliage and flowers, explains why phallic statues of the gods Hermes and Priapus were decked with myrtle (Paus. 1.27.1; Philostr. *VA* 5.15) and also why myrtle has a special resonance for Aphrodite/Venus as the goddess of fruitful spring (Artem. 1.77; Joh. Lyd., Mens. 4.65; Plut., *Num.* 19.2.72C). When, in a cultic context, images of the Bona Dea, Roman goddess of fruitfulness and women, were struck with branches of myrtle, the plant thereby acquired the status of fertility-bestowing "staff of life" (Plut., *Quaest. Rom.* 20.268DE). The Bona Dea's chastity, on the other hand, prefigured Christianity's later use of myrtle as a symbol of purity (Plut., *Caes.* 9.711e; Tert., *Ad Nat.* 2.9.22; Serv. Verg., *Aen.* 8.314; Macrob., *Sat.* 1.12.21–9; Fehrle 1910: 129ff.).

Meanwhile, myrtle's use in the making of wine (Ar., *Ran.* 328–30; Schol. Soph., *OC* 681) rendered it sacred to Dionysus, whose statues were often clad in myrtle branches (Paus. 1.27.1), and Greek author Athenaeus (second/third century CE) records the use of a fragrant wreath, presumably myrtle, in the Sikyonian cult of Dionysus called the "Iacha" (*Deip.* 15.678A; Murr 1890: 89). The tradition that Dionysus surrendered three of the plants sacred to him—the vine, ivy, and myrtle—to Persephone in order to secure the release of his mother Semele from the Underworld (Schol. Ar., *Ran.* 330) may be seen both as relating to the chthonic aspects of the plant and as linked with fertility and Persephone's half-year residence above ground—at which her mother Demeter rejoiced, allowing grain to sprout.

As a plant sacred to chthonic deities, it was important in Demeter's Eleusinian mysteries, and its use in cults of the dead can be attributed to its chthonic and apotropaic aspects (Istros 334F29, Apollod. 244F140). Mythological tradition tells of myrtle growing on the graves of Odysseus' fallen comrade Elpenor and of Polydorus, the Trojan prince slain by treachery in Thrace (Theophr., *Hist. pl.* 5.8.3; Verg., *Aen.* 3.22ff.). The custom of placing bodies of the deceased on beds of myrtle was widespread (Fehrle 1910: 241), as was that of crowning the dead and decorating their graves with wreaths of myrtle (Ath., *Deip.* 11.460B; Rohde 1925: 141n21, 189n40). Victors in funerary games were also crowned with myrtle in honor of the deceased (Verg., *Aen.* 5.72; Schol. Pind. 3.117), and in Vergil's description of the Underworld (*Aen.* 6.442–4), those who died of love-sickness wander, still consumed with passion, in a grove of myrtle that in this context bears both sepulchral and amatory significance. Myrtle's cultic, chthonic traits are further evidenced in the Hellotia, a Cretan relic cult of Europa Hellotis, the Phoenician princess pursued by Zeus in the guise of a white bull and later made a goddess; at this festival Europa's bones, couched in a large myrtle wreath, were carried in procession (Ath., *Deip.* 15.678AB; Schol. Pind., *Ol.* 13.56).

RITUAL USE IN DAILY LIFE: MARRIAGE, BIRTH, AND DEATH

As already intimated, certain plants' divine and mythological associations rendered them essential components of ritual. Indeed, plants played a particularly prominent role in rituals marking passage from one stage of life to another. One such transition, that to adulthood (particularly for women), was marriage, the primary goal of which was the production of legitimate children, and not surprisingly, agricultural metaphors for marriage and its anticipated consequences abounded. The point is vividly illustrated in a Greek declaration of betrothal by the father of the bride to her prospective husband preserved by the comic poet Menander (342–291 BCE): "I give you my daughter for the cultivation (lit. 'plowing') of [lawful] children" (*Pk.* 1013–14). While recovery of details about Greek and Roman marriage rituals has largely been incidental, gleaned from hints in the literary and artistic record, it appears that there was a significant overlap between Greek and Roman practices where plants are concerned.

In the course of prenuptial rites, as well as marriage rites themselves, brides, grooms, and guests donned crowns of flowers such as roses, violets, and lilies—all fragrant and associated with Aphrodite/Venus—or other vegetation such as sesame, a symbol of fertility owing to its prolific seed production, and mint, believed to be an aphrodisiac; of the Romans specifically, fourth-century historian Festus reports that "under her veil, the new bride wore a crown of flowers, herbs, and grasses picked by her own hand" (45); flowers were especially apposite since plucked blossoms symbolized virginity necessarily lost as a prelude to childbearing (Hersch 2010: 90–2). Marriage rites also involved making bloodless sacrifices of wine, a manifestation of Dionysus/Bacchus, god of burgeoning growth in plants, as well as of plant-based incense, which, like fragrant blossoms, was closely linked with Aphrodite/Venus without whose "blessing" a marriage could not hope to be productive. The houses of bride and groom alike were decorated with garlands and greenery: branches of beech, olive, or fragrant laurel and cypress (Blech 1982: 76; Bion, *Id.* 1.88; Lucian., *Dial. meret.* 2.3; Plut., *Amat.* 10, *Pomp.* 55; Catull. 64.292–3; Juv. 6.51–2, 227–8; Luc. 2.354; Stat., *Silv.* 1.2, 231; Apul., *Met.* 4.26).

In procession from the house of the bride to that of the groom, Greek couples were showered—via *katachysmata*—with various symbols of fertility: dates, figs, dried fruits, nuts, apples, myrtle leaves, crowns of roses, and garlands of violets are mentioned in our sources (Oakley and Sinos 1993: 26, 34). The throwing of nuts in order to encourage the couple's fertility appears to have been practiced by the Romans, too (Plin., *HN* 15.24.86). We hear that in Greece, and presumably also in Rome, brides carried objects signifying their ability to contribute to the household's economy; in Athens these were pans for roasting barley, a staple of the Greek diet, and sieves for sorting grains, all the purview of Demeter, goddess of grain and the harvest (Poll., *Onom.* 1.246, 3.37).

Before entering the nuptial chamber, the bride was said to have eaten an apple (*melon*, Greek; *malum*, Latin, which could have been either a quince or an apple) prior to entering the bridal chamber (Plut., *Mor.* 138d, *Quaest. Rom.* 279f., *Sol.* 20). This fleshy fruit, used as a love gift and one of the chief symbols of fertility, was sacred to both Hera/Juno, goddess of marriage, and to Aphrodite/Venus. The arrival, in time, of the couple's hoped-for progeny was, like the wedding itself, signaled by festive sprays decorating their house's portal: both olive and palm are attested (Eur., *Ion* 1433; Artem. 1.77).

Branches, leaves, and flowers were as much a part of rituals honoring the dead and securing their safe passage to the afterlife as they were of rites anchored in the desire to spawn new life—logically so, as birth and death marked critical junctures in a continuous cycle of "being." A wide variety of flowers and leaves were used for these purposes, and while seasonal availability must have played a part in their selection for use (Lucian, *De luct.* 11), parsley and celery (*selinon*)—perhaps because they grew in abundance and were fragrant, thus useful to cover the smell of death—were apparently favored by the Greeks for funerary crowns and to strew upon the corpse (Plut., *Tim.* 26; Suda. s. v.). Boughs of fragrant cypress and pine, symbols of death and mourning, were hung on the posts of doors of houses polluted, temporarily, by the presence of the dead (Toohey 2010: 364).

The use of flowers to bedeck the dead, biers, tombstones, burial chambers, funeral pyres, and cinerary urns is especially poignant, as blossoms, like the bloom of human life, are short-lived (Plin., *HN* 21.7.10; Dion. Hal., *Ant. Rom.* 11.39). Flowers' link to death is underscored by the story that Flora, goddess of flowers, first created flowers from the blood of the dead (Ov., *Fast.* 5.195–274). The inherent evanescence of vegetal funerary crowns and wreaths could, however, certainly be countered by substituting or supplementing them with wreaths of laurel, oak, rose, and other sacred plants in gold.

As for living plants that were sacred to the dead and marked sites of burial, the most famous of these is arguably Homer's asphodel. Fields of asphodel form a memorable part of Homer's relatively indistinct Underworld geography; it is there that Odysseus encounters his fallen comrade Ajax, and it is to this place that Hermes leads the souls of the men that Odysseus slew for threatening his Ithakan kingdom (Hom., *Od.* 11.539, 573; 24.13). A clumping perennial bearing white or pinkish flowers (*Asphodelus albus*, *fistulosus*, and *ramosus* are all contenders), asphodel is a typical Greek wildflower native to the Mediterranean. It was planted at gravesites, as its roots were thought to nourish the dead, and was held sacred to the chthonic deities. Even in antiquity, however, there was no consensus about the botanical identification of Homer's asphodel, and it is still suggested that Homer's drifts of asphodel were meadows populated by ash trees (Reece 2007). As already mentioned, poplar was a tree associated with death. According to Greek geographer and historian Strabo (63 BCE–23 CE), there was a fenced grove of poplar adjacent to the Mausoleum of Augustus (*Geogr.* 5.3.8). It is certainly the case that this grove had sepulchral significance, as the poplar's mythology suggests.

POETRY, PURITY, AND PROPHECY

A number of plants possessed a symbolism that placed them on a level more removed from day-to-day human life, like laurel for example. In its symbolism, laurel manifested Apollo's unapproachability or aloofness, a quality that the mantic dimension of Apollo's cult makes manifest, as the god did not prophesize directly to his petitioners but rather through the agency of his priestess, the Pythia (Lucr., *DRN* 5.112ff.; Lycoph., *Alex.* 3–7) (see Figure 6.4). Herein resided the association of laurel with poets as divinely inspired messengers of truth (Hes., *Theog.* 22–35; Kambylis 1965). Marking his consecration as a poet, Hesiod (fl. *c.* 700 BCE) receives a lush branch of laurel, which, serving as scepter and orator's staff at once, elevates him above the masses and enables him to discern the truth and disclose it in his verses. Indeed, laurel was inextricably bound to truth and purity in both a cultic and a medical sense (Eur., *IA* 759ff.; Ov., *Ars am.* 2.401ff.; Val. Flac. 3.417–58). In the work of Plato (*c.* 428/427–348/347 BCE), laurel is reinterpreted as a symbol of moral purity in so far as the Euthynoi (judges of accountability), who supervised public offices, were allowed to don wreaths of laurel as priests of Apollo and the sun god Helios (*Leg.* 12.947a 5–7). The honor of wearing laurel wreaths was also reputedly enjoyed by the mythical Hyperboreans, who lived in paradisiacal conditions beyond the source of the North Wind and who revered only Apollo among the twelve Olympian gods (Pind., *Pyth.* 10.40ff.).

Poetic consecration, as described above, is a recurrent motif in Classical literature, and, over time, various plants featured therein, being selected in accordance with their propriety for certain poetic endeavors as well as their political connotations. The Roman poet Horace (65–8 BCE), for example, intimates clear awareness of his poetic mission when stating that he writes under the aegis of laurel and myrtle, assuming the role of *vates* (prophet) of Apollo and as love poet, respectively (Braun 2010: 463; Hor.,

FIGURE 6.4 Red-figure volute-*krater* (mixing bowl), South Italian. Apollo, wearing a laurel wreath, holds a kithara, denoting his role as god of music, and his sister Artemis, the goddess of the hunt, is accompanied by her sacred deer. Their mother Leto stands at the right. On the left, the god Hermes leans on a pillar inscribed with his name. Attributed to the Palermo Painter, active *c.* 430–*c.* 400 BCE. © The J. Paul Getty Museum.

Carm. 3.4.18–20; and Philostr., *Imag.* 2.12.2). Myrtle gained particular importance as a poet's crown in Roman poet Ovid's work, composed during the reign of Augustus; in earlier sources, such wreaths tended to be made of ivy and laurel (Buchheit 1986: 263–6). Ovid's wreath of myrtle signals inspiration by Venus (*Ars am.* 3.53ff.; *Fast.* 4.15ff.) and is a direct reference to the proem of Vergil's *Georgics*, in which Octavian (called Augustus from

27 BCE), as restorer of justice, freedom, and morality, appears wreathed with a crown of myrtle, in itself a reminder of Octavian's—and all the Romans'—descent from Venus (Verg., *G.* 1.24–8; Ov., *Am.* 1.1.29, 1.2.23, 1.15.37ff.). The love poet Ovid's adoption of myrtle thus signals a new Rome under the reign of Venus, a reign marked by beauty and harmony, in a proclamation on par with that of Vergil. In the work of Tibullus (55?–19 BCE), too, the myrtle evokes an Elysian life dedicated to love and peace (1.10.27ff.).

PLANTS AND POLITICS

Cult-related functions of plants could spawn specific political symbolism. For example, myrtle's symbolic association with love translated to its connection with Aphrodite Pandemos (For All People), in whose honor the Athenian archons donned myrtle wreaths when conducting non-military versus military business. As sacred to Athena in the guise of protector of state order, myrtle also took on a conciliatory and peacemaking function (Preller 1860: 218–20). Similarly, in the resolution Rome's earliest conflict, the Romans and Sabines, whose daughters the former had abducted, were said to have purified themselves with branches from an ancient stand or grove of myrtle trees where they had sealed their alliance by intermarriage (Plin., *HN* 15.36.119; Eitrem 1915: 15ff.). The grove later became a cult site of Venus Cloacina ("the cleanser"). There is a tradition (Plin., *NH* 15.36.119–21) that myrtle was the first tree to have been planted in the public spaces of Rome, two of the oldest specimens having been located at the Temple of Quirinus, among Rome's most ancient temples. One was called the "Patrician" and the other "Plebeian"; their vigor increased or waned in accordance with the political fortunes of the group it represented.

Laurel, too, was a plant associated strongly with peace and political prosperity. As an embodiment of holiness and cultic purity, laurel lent itself ideally to the political program of the Roman emperor Augustus (63 BCE–14 CE), the entrance to whose house was adorned with laurel by order of the Senate—a testament to his courage, clemency, justice, and piety (Aug., *RG* 34). As a result, the emperor's residence became sacrosanct, underscoring Augustus' link with Apollo, whose temple was next door and whom Augustus claimed as special protector of his regime. This public gift of laurel, as well his status as laurel-crowned *triumphator* in the military conflicts through which peace for Rome was hard won, allowed Augustus to present himself as guarantor of peace and something of a god among mortals (*RG* 34.2). It is noteworthy that while Christianity eschewed pagan traditions, the laurel wreath was retained as a symbol emblematic of that faith's triumph. Thus, it was that a crown of laurels could replace Christ's crown of thorns, as seen in the Vatican's Passion-Sarcophagus (Dresken-Weiland 2018: 49).

WREATHS AS MARKS OF DISTINCTION

Wreaths made of a range of plant materials found significant use and symbolic importance in a variety of contexts, among them Greek symposia, aristocratic "drinking parties" that flourished in the Archaic period (from the early seventh to the late fifth century BCE) and were opportunities for attendees to engage in political and philosophical discourse as well as poetic endeavors. The so-called Naukratite wreath was a special wreath worn by symposiasts, and its use was traced to a feast organized by the Naukratite sailor Herostratos following his rescue at sea by Aphrodite and safe arrival at Naukratis (Ath., *Deip.* 15.675F–676C). According to rhetorician and grammarian Athenaeus

(170–223 CE), after sacrificing to Aphrodite and dedicating a statue to her, Herostratus "invited his relations and closest friends to a feast in the temple itself, giving to each a wreath made from the myrtle, which even at that time he called a Naukratite wreath."

Wreaths of myrtle, ivy, celery, mint, and rose were worn in symposia not only as a reflection of their festal nature but also as an antidote to drunkenness—and, more important in this context, to mark the transcendent, inspired status of those delivering speeches or composing poetry in these venues (Anacr. PMG 496, 151; Philox. PMG 836; Blech 1982: 71n43; Ath., *Deip.* 15.675E, 675C). Greek orators, too, donned wreaths made of myrtle or other types of vegetation, and sprigs of myrtle assumed a deictic function in the performance of drinking songs, *skolia*, being passed, like a speaker's staff, from one singer to the next upon completion of his song (Ar., *Nub.* 1364; Blech 1982: 292–4, 319).

As the elder Pliny writes, wreaths of leaves were first presented exclusively to divinities and later only to people who had achieved something close to divine status (*HN* 16.4). In Greece, the latter included athletes victorious at the various pan-Hellenic games. Wreaths of branches cut from an olive tree growing behind the Temple of Zeus were awarded to victors in the Olympic Games. At the Delphic Pythian Games, the victory crown was of laurel, culled from the Vale of Tempe; at the Nemean Games it was of celery (or parsley); and at the Isthmian Games, it was first of pine (until the fifth century BCE), later of celery (until the second century BCE), and thereafter either of pine or celery (Broneer 1962: 259–63).

As already mentioned, wreaths of laurel reflected the elevated status of victorious generals in Rome, but not in all cases; crowns of myrtle were awarded in cases of lesser triumph—formally called ovations—where the army was not put at significant risk and purification from bloodshed was not required (Dion. Hal., *Ant. Rom.* 5.47; Gell., *NA* 5.6.20; Fest. s.v. *ovalis corona*; Plin., *HN* 15.39; Plut., *Marc.* 22). Honorific crowns awarded for military achievements included the extremely rare and prestigious *corona obsidionalis* or *graminea* ("of the siege," "made of grass"), which was made of grass, or weeds and wild flowers growing at the location of the siege and presented by a besieged army to the general who liberated them (Plin., *HN* 22.4, 7; Fest. s.v. *obsidionalis*; Gell., *NA* 5.6.8; Liv. 7.34–7; Serv. ad Verg., *Aen.* 8.128). The *corona civica* (civic crown), which was also infrequently awarded—originally made of the ilex, afterwards of the aesculus, and finally of the quercus, three different sorts of oak all sacred to Jupiter, the king of the gods and guardian of the Roman state—was presented to soldiers who had preserved the life of a Roman citizen in battle (Gell., *NA* 5.6.11–12; Fest s.v. *civica*; Plin., *HN* 16.3, 5; Plut., *Quaest. Rom.* 92; Sen., *Clem.* 1.26.5).

BOTANICAL SYMBOLISM IN GREEK LITERATURE

The richness, versatility, and complexity of botanical symbolism provided authors with a welcome vehicle by which to enrich their texts and endow them with deeper meaning. This tendency is particularly evident in the work of the Hellenistic Greek pastoral poet Theocritus (fl. third century BCE), whose poetry is notable for its range of flora and evidences extensive botanical knowledge on the author's part. It is true that frequent reference to plants is wholly appropriate to the genre of pastoral, but closer inspection reveals that featured plants function as important interpretive guides as well as scenic backdrops.

For example, reeds, celadine, maiden hair fern, celery, and dog's tooth grass (*Cynodon dactylon*) are all moisture-loving plants and thus are suitable to the setting of the particular tale—the rape of Hylas—in which they appear (*Id.* 13.42): Hylas, Heracles' young page, was drawing water from a pool surrounded by these plants when its resident nymphs, taken with his beauty, pulled him beneath the surface. At first glance, these four relatively inconspicuous plants form a lush, green, tranquil backdrop for the unfolding tale: they offer a stark contrast to the showy, multi-colored, and deliberately beguiling flowers (rose, saffron, violet, iris, hyacinth, and narcissus) that lured Persephone and Europa to the meadows whence they were abducted—Persephone by Hades and Europa by Zeus (*Hom. Hymn. Dem.* 6ff.; Mosch., *Eur.* 63–71; Lembach 1970: 90–5). But for the observant reader, the hydrophilic nature of this plant community presages Hylas' water-bound fate.

Meanwhile in *Idyll* 11 (46), grapevine, laurel, cypress, and ivy form a different, special community of plants that surrounds the cave of the love-struck Cyclops Polyphemus. In conjunction with a cool spring, the plants form the components of a *locus amoenus* ("pleasant place"), an idealized place in nature characterized by physical beauty, shade, fragrant breezes, and cooling waters (Curtius 1953: 183–202). The place provides a range of sensory delights: taste (sweet grapes), sight (variously shaped foliage), and smell (scent of laurel and cypress, which are not only fragrant plants but also magical). It evokes a sense of tranquility and enchantment, its bewitching charm running contrary to the Cyclops' own hideous appearance, which serves as a reminder of his barbaric treatment of Odysseus and his companions in the *Odyssey*. Theocritus' Cyclops, however, is nothing like his Homeric counterpart, his character being aligned with the landscape that surrounds him.

Theocritean poetry, fueled by the Hellenistic age's intense interest in scientific inquiry and naturalistic artistic representations, in a sense marked the high point of botanical symbolism in Greek literature, but plants were employed figuratively much earlier too, and to great effect. In Homer's *Iliad*, which marks the beginning of the Western literary tradition, descriptions of the natural world are relegated largely to similes. Plants figure in some of the most memorable of these, and trees appear with much greater frequency than any other plant (Forster 1936), perhaps unsurprisingly given their cultural importance. In the thirteenth book of the *Iliad*, for example, the hero Asios, fatally wounded by Idomeneus' spear, falls to the earth like an oak, a silver poplar, or a pine felled by carpenters to make a ship's timber (*Il.* 13.389–93). Selection of these trees may have been motivated not only by their suitability for shipbuilding but also by their stateliness and resilience, both of which qualities echo those of the fallen hero.

Poplar appears also at the death of the young Trojan hero Simoeisios who, as he falls, is likened a young silver poplar, its bark still smooth, felled prior to maturation (*Il.* 4.473–89). Lying wilting on a stream's banks, the poplar will be used as the rims of chariot wheels, for which pliant wood was required. The simile can convey a range of meanings. Tragic as his passing is, Simoeisios, fighting as he was for Troy, will not have given his life entirely in vain, much as the young tree, too, will find a use. Alternatively, especially if the chariot will be used in war, Homer may have wished to emphasize the cost of war to human and nature alike. Simoeisios, son of Anthemion (translated as the "Flowery One") was, after all, a true child of nature, born of a shepherdess beside the river Simois.

In the *Odyssey*, meanwhile, plants play a more prominent role in the narrative proper. For example, productive, well-watered fruit trees planted in orderly rows on the island of Scheria by the mythical Phaiakians symbolize good governance, indicating that the utopian society on Scheria is a positive model for Odysseus' rebuilding of Ithaca, and an ancient olive, mainstay of the Greek economy, had been carved by Odysseus to form the base of the royal marital bed, a most apposite symbol of domestic stability (Giesecke 2007: 26–37).

After epic, the next literary "genre" to emerge in Archaic Greece was lyric, much of which, unlike epic, was composed in the first person and ostensibly proffers lived experiences. Apart from floral garlands mentioned in poems describing symposia, botanical imagery is scant except in the extant fragments of Sappho of Lesbos (*c.* 630–570 BCE), where it figures prominently. Fragment 96, for instance, likens a maiden to the moon, spreading its light over fields of rose, melilot (*Melilotus officinalis*) and chervil (*Anthriscus cerefolium*). These fragrant plants were used to make wreaths and appear as such in Sappho's other works where they are not only evocative of beauty and youth but also of festive occasions fondly remembered.

Meanwhile in Fragment 2, Aphrodite is summoned from Crete to a holy temple, her own temple, permeated by the scent of incense and surrounded by luscious groves of apple trees. There is a babbling brook, and a thick, soporific, cover of roses, as well as a meadow, swept by warm breezes, where horses graze and flowers grow. This is not merely a sentimental landscape conjured by a painterly and lovesick imagination. Rather, it is a *temenos*, a sacred space in Nature filled with divinity but inscribed by humankind for the purpose of interaction with the divine. This is not a wilderness but a garden that embodies the fertile, life-sustaining essence of Aphrodite, as evidenced in particular by the grove of apple trees and drifts of roses, plants sacred to the goddess (Giesecke 2007: 53; 2022).

Truly affective botanical imagery and symbolism reappear only much later, in the philosophical works of Plato, whose careful selection of featured plants evidences accurate botanical knowledge. A prime example is the dialogue *Phaedrus*, in which Socrates becomes enraptured by the natural setting in which he finds himself: Phaedrus and Socrates have strolled outside the walls of Athens and linger to converse at a particular spot along the Ilissos River. Phaedrus' description of the place is dispassionately utilitarian; as attributes to recommend it he mentions the shade of a plane tree, a gentle breeze, and grass on which to recline (229b). By contrast, in the words of Socrates, who waxes poetic, the place takes on an altogether different character. It becomes a most perfect *locus amoenus*. Here a spreading plane tree and a lofty chaste tree (*Vitex agnus-castus*), sacred to the goddess Hera and filling the air with the perfume of its flowers, provide welcome shade. The place's prevailing scent, beauty, coolness, comfort, and soothing sounds (chirruping of cicadas) engage all the senses, and Socrates, quite in spite of himself, is moved by his heightened sense of awareness from an embodiment of sobriety through a sort of Bacchic enthusiasm or frenzy to an embodiment of philosophy. The stages of his metamorphosis are, in fact, all prefigured in the landscape, for the chaste tree, as Hera's sacred tree, symbolizes sobriety, while the plane tree is sacred to Dionysus/Bacchus, who embodies the force promoting the luxurious growth of plants (Giesecke 2007: 87).

Botanical symbols do, of course, find a place in dramatic works as well, but not embedded in fully developed landscapes such as Theocritus, Sappho, or Plato conjure. In the context of tragic drama, theme-symbolic plant allusions quite reasonably take on an exalted nature. This is the case in Sophocles' *Oedipus at Colonus* (written 406 BCE), where an ancient, hollow, once-productive pear tree marks the location of a polluted Oedipus' passage from the earthly realm and consequent assumption of heroic status (*OC* 1581–97); the august, once-respected king will, after his death, bring prosperity to those who honor him.

As one might imagine, humble edible plants as well as those with noxious traits were useful especially in comedy to capture the personal quirks or character of protagonists in the most graphic manner possible. One humble plant that saw extensive culinary use and also possessed a potent defense mechanism—causing an acute burning sensation when touched and being intensely clingy—was the nettle, which appears repeatedly in comedic contexts. For example, Aristophanes (*c.* 446–386 BCE) avails himself of the nettle's metaphorical potential in the *Wasps* (884), when Bdelocleon prays that his father may cease to be so cantankerous, and again in *Lysistrata* (548), with reference to the tenacious aggression required by the women of Athens to steel themselves and seize the Acropolis in pursuit of peace. Meanwhile, unpleasant contact with this plant is what a character in a fragment of Pherekrates' *Automoloi* (fr. 29) wishes upon a performer; so painful is the audience's experience that the performer's song is said to warrant a crown not of laurel but of nettle.

Incidentally, weeds could bear meanings not unlike those that they would later convey in the Christian tradition: heathenism, heresy, and evil generally. Thus the orator Antisthenes avails himself of an agronomic metaphor to illuminate the deplorable state of affairs in government: it would be odd, he says, to separate out the ryegrass from grain and, in war, to remove the infirm from the ranks of the soldiery but not to dismiss incompetent politicians from the State (fr. 194.2). Antisthenes' intention is clear: the ryegrass and infirm soldiers are worthless to the point of being outright dangerous, the former compromising the nutritional value of the grain product and the latter posing a physical risk to the citizenry they ostensibly defend.

BOTANICAL SYMBOLISM IN LATIN LITERATURE

In literary use of botanical symbolism, the Romans took their cue from the Greeks, whom they emulated ingeniously, with all cases of borrowing—whether taking the form of allusion or outright "imitation"—involving innovation in order to convey entirely new meanings. In the case of Vergil (70–19 BCE), for example, who explicitly signals his debt to Homer's *Iliad* and *Odyssey* in the opening lines of his epic poem the *Aeneid*, abundant Homeric prototypes can be identified for botanical symbols. As in Homer, numerous similes involve likenesses with phenomena in the natural world: in the *Aeneid*, too, can be found rose-hued cheeks, the bloom of youth, fields blossoming with heroes, and warriors falling like felled trees (Keith 1933: 607). But there are also more intricately constructed and protracted allusions that involve multiple layers of reference to earlier texts. One example of many is Vergil's moving description of the brash but innocent young hero Euryalus' death. Fatally wounded by a sword, "his body turns in death. Gore runs across his lovely limbs, and his neck, sinking, comes to rest upon his shoulder—as when a purple flower, cut down by the plow, wilts in death; or as poppies, their necks fatigued, bow their heads under the weight of a sudden shower" (9.433–7). The double simile recalls the untimely wartime slaughter of another youth, a son of the Trojan king Priam, who when struck by an arrow "dropped his head to one side like a poppy, growing in a garden, weighed down by its seeds (*karpoi*) and the showers of spring" (*Il.* 8.306–8). Notably, Homer's poppy has already progressed from flower to fruit and seed, and it is merely bent, not broken, plucked, or shorn. There is no mention—and no apparent thought—of its premature removal from the cycle of life.

What Homer appears only to emphasize through this lovely image is the drooping shape of the slain warrior, but what Homer does not say, leaving the audience to infer it, speaks volumes. This poppy grows in a cultivated space and is ripe for harvest. Poppies had close links with the cultivation of grain, being planted in crop rotations to revitalize the soil. As such, they were sacred to Demeter, goddess of the harvest, and poppy seeds were said to be the first food eaten by Demeter after her long fast in grief over the loss of her daughter, Persephone, to the Underworld (Ov., *Fast.* 4.531–4). Thus, at a minimum, Homer intimates the cycle of life and death, as well as grief over passages from this life to the next, and the richness of the simile could not have been lost on Vergil when he appropriated it.

The differences between the Homeric prototype and Vergil's simile are revealing, however. Vergil dwells on the flower's blood-red color, a striking contrast with the youth's tender, snow-white skin, whereby he underscores the cost and violence of war as well as the innocence of its victims. Vergil further heightens the violence of the youth's death, bring in the notion of lost love as well, with an allusion to a simile in the work of Catullus (*c.* 84–*c.* 54 BCE), in which the latter likens the rejection of his love to "a flower on the meadow's edge, when it has been nicked by a passing plow" (11.22–4).

As in the case of Greek literature, plant symbolism appears not solely in epic poetry but in all literary genres, ranging from agronomic works and pastoral poetry—where plants naturally play an integral part—to philosophical mediations and satire. For example, plants have a particular prominence in the *Eclogues* of Vergil, the first Latin poet to imitate Theocritean pastoral. Indeed, the first lines of *Eclogue* 1, "You, Tityrus, lying under the cover of a spreading beech, play to the woodland Muse on a narrow reed" (1–2) are among the most famous in Latin literature. The beech and narrow reed clearly signal Vergil's generic leanings and constitute a debt both to Theocritus and to Roman poet Lucretius (*c.* 94–*c.* 51 BCE) with his portrait of rustic life (Breed 2000; Cairns 1999; Van Sickle 2004).

Meanwhile, Cicero opens his dialogue *De oratore* (On the Orator) with the image of a plane tree providing a pleasant, shady setting for a conversation between friends who have gathered at Crassus' villa at Tusculum (1.24–9). That the conversation will be philosophical in nature is clearly stated by one of the protagonists, who suggests that the group imitate Socrates in the *Phaedrus*; Socrates and Phaedrus had, as earlier mentioned, settled into conversation beneath a plane's canopy. Similarly, in Cicero's dialogue *De legibus* (On the Laws) a venerable oak serves as the venue and the starting point of discourse (1.1); again, the allusion is to Plato's plane, but with a contemporary Roman twist (Dyck 2004: 57).

Innovative, distinctly Roman treatments of botanical symbols included not only signifiers of genre and complex intertextual similes, like those illustrated in some detail above, but also, significantly, the appearance of such symbols in an expanding literary treatment of the garden and gardening efforts. In the Roman world gardens would become symbols of *Romanitas* (Roman-ness). Roman gardens took the form of a variety of cultivated spaces ranging from kitchen gardens (the *hortus* "proper") to expansive villa grounds and imperial parks (called *horti*). Gardens and their plantings could be viewed as repositories of nostalgic, humble, agrarian Roman values. At the same time, depending on context—as in the case of villa gardens and smaller scale ornamental domestic gardens in Pompeii and Herculaneum—the garden could be a powerful vehicle for display of wealth, good taste, erudition, exoticism, creativity, and capacity for philosophical repose. In consequence, plants and their symbolism naturally assumed a new prominence in physical and literary gardens alike.

As one may glean from Vergil's deeply political farming "manual" the *Georgics*, a typical, traditional Roman kitchen garden (*hortus*) would have contained plantings of endive, parsley, squash, narcissus, acanthus, ivy, and myrtle (4.119–24). It also might well have contained varieties of herbs; flowers like lily, poppy, hyacinth, rose, and verbena; productive plum and pear trees; and leafy lime and shady plane trees, as appear in the idealized description of the garden belonging to an old man—a model of self-sufficiency—living a quiet, apolitical life in Tarentum. Such plants were produced for subsistence, for use as food or garlands and votive offerings, but not for sale. For agricultural writer and statesman Cato the Elder, the kitchen garden, an essential part of the farm, is a repository of traditional, conservative values, but plants to be found there are in part "exotic," indicating just how far removed Republican Rome already was from its local, humble agrarian origins and presaging its imperial ambitions: "all types of vegetables, and all types of flowers for garlands—Megarian bulbs; 'conjugal' myrtle, as well as the white and black; Delphic, Cyprian, and wild laurel; smooth nuts, such as the Abellan, Praenestine, and Greek" (*Agr.* 8.2). There would always be a degree of tension between the nostalgic ideal of Romans as citizen-farmers and an increasingly "global" Roman world. Thus, later, in the garden of agricultural writer Columella (*c.* 4–70 CE) there appear specimens from around the world then under Roman rule: Spain, Gaul, Africa, Asia, and Greece. Plantings include plums from Armenia and Damascus; peaches from Persia, Gaul, and Asia; and figs from Caria, Chios, the Chelidonian islands, Libya, and Lydia (Pagán 2006: 29–30).

The most extensive descriptions of Roman villas and their pleasure gardens (*horti*) are those of Statius and Pliny the Younger. In both instances, gardens and their plantings are presented not as symbols of a humble, moral, agrarian life but as reflections of the author's or, in the case of Statius' patrons, erudition, social status, and ethical position (Myers 2005: 106). In both instances, too, the gardens are lavish displays in which Nature embraces the gardener/owner's artful manipulation of her (especially *Silv.* 1.3.10ff, 2.2.52–3). So conscious of Nature's gifts and so conscientious, for example, is Manilius Vopiscus that his Tiburtine villa preserves in its midst a tree, "rising through ceilings and doorways to emerge in the open, sure to [have] suffer[ed] the cruel ax under any other master" (*Silv.* 1.3.58–60). Atedius Melior's special plane tree, clearly the jewel of his estate, is ingeniously supplied with its own mythology and can thus be showcased as a true reflection of its owner's virtue (*Silv.* 2.3).

In Pliny the Younger's garden descriptions in particular, there is a heavy emphasis on the synaesthetic experiences provided by planted spaces—that is, aesthetic experiences engaging all of the senses—and, at the same time, there is a distinct lack of focus on their productivity. For instance, Pliny tells us that his seaside villa's park-like grounds contain, among other features, an exercise ground (*gestatio*) bordered by clipped hedges of evergreen box and, where the box will not grow, rosemary, an intensely aromatic shrub (see Figure 6.5). Inside the *gestatio*'s enclosed space there is a grapevine-clad pergola expending refreshing shade; the ground, presumably covered with grasses or creepers, provides a soft cushion even beneath bare feet (*Ep.* 2.17.15). Mulberries and figs grow here as well, orchard plants being an integral part of the decorative scheme. Elsewhere in the villa grounds there is a terrace, outlined with drifts of fragrant violet (*Ep.* 2.17.17).

Pliny's Tuscan villa, seated in what he describes as Nature's scenic amphitheater, a spreading meadow ringed by mountains, features a hippodrome-shaped garden whose beauty outshines that of the villa's gracious buildings (*Ep.* 5.6.7–8, 32–6). Here the garden's periphery is demarcated by a line of plane trees that are punctuated by box and

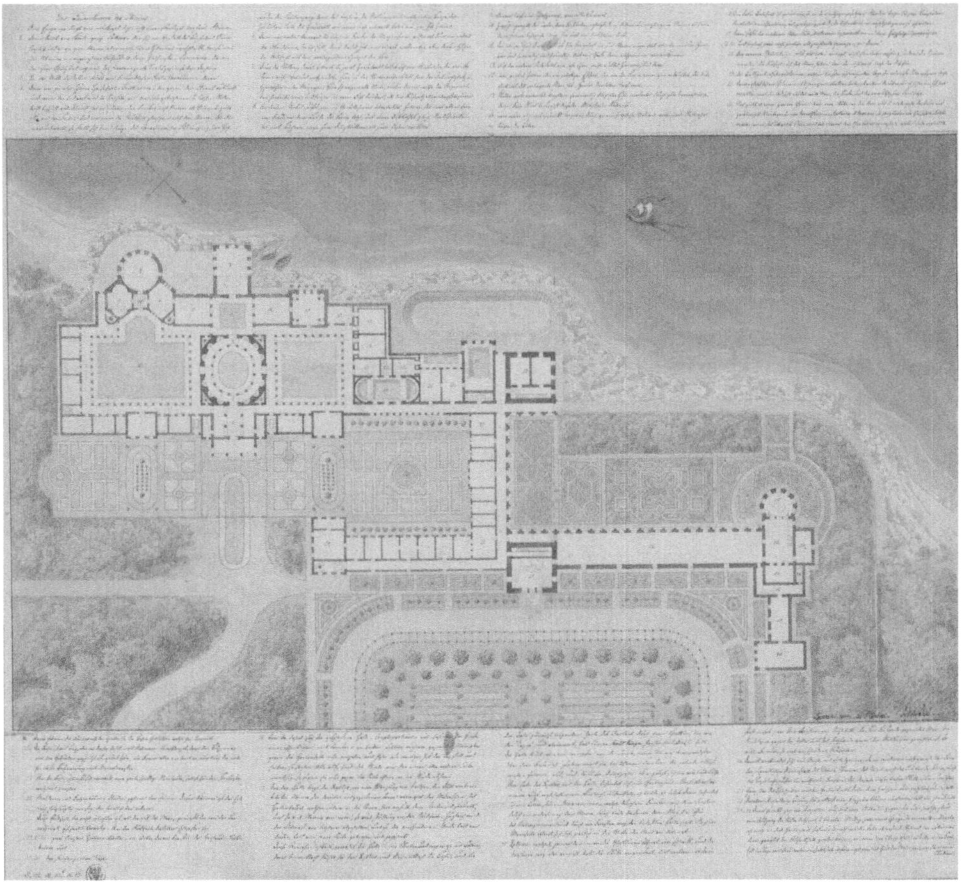

FIGURE 6.5 Pliny the Younger's seaside villa and its gardens, reconstruction. Rendering by Karl Friedrich Schinkel, 1835. Gray pen, graphite pencil, and watercolor. Kupferstichkabinett. Staatliche Museen zu Berlin. © Kupferstichkabinett, Staatliche Museen zu Berlin.

draped with ivy linking tree to tree. Behind the planes, a row of laurels adds their shade, and dense, dark cypresses constitute the mock-racecourse's "starting blocks." Circuits are defined by rows of box encasing drifts of roses and boxwood topiary in the shape of letters that spell out Pliny's and his gardener's names. Topiary obelisks alternate with fruit trees. And at the center of the whole is a planted space so natural in appearance that one would think it a slice of the countryside: there thick waves of acanthus are framed by low-canopied plane trees. It should be noted in this context that the link between Statius and Pliny's literary endeavors with the garden and garden plants—the memorialized gardens being a direct reflection of the authors' craft and literary persona—owes much to Vergil's humble, rustic garden and the image of its gardener; the farmer and his humble but bounteous plantings have been seen as representations of Vergil himself, wishing to ply his craft removed from the political arena that the *Georgics* represents.

BOTANICAL SYMBOLISM AND THE RISE OF CHRISTIANITY

Early Christian authors developed a very particular plant symbolism. Noteworthy is the fact that these symbols are aligned with Christ and not yet with Mary, Mother of God, around whose worship a rich plant symbolism would develop in the Middle Ages. Part and parcel of the rejection of pagan cults, magic, and superstition, Christian botanical symbolism eschewed meanings assumed through Greco-Roman religion and myth. However, while the Holy Scriptures served as its dominant textual point of reference, the new symbolism was necessarily grounded in a framework of Greek and Latin texts that formed the basis of education.

The Homeric epics persisted as canonical texts, and Christian authors mined these for plants whose pagan symbolism could be transmuted to suit their needs. One such plant was the lotus, which features prominently in the *Odyssey*. In the course of their ten-year, peril-ridden journey home from Troy, Odysseus and his companions come to the land of the Lotus Eaters, where they are offered the fruit of the lotus to eat (*Od.* 9.83–104). The Lotus Eaters mean the Greeks no harm; their offer is a genuine act of hospitality. But the honey-sweet fruit induces forgetfulness, and those partaking of it would forget their homes and their desire to return to them. In other words, it is a source of hidden danger. It is only with force that Odysseus is able to bring his companions back to the ship and so prevent them from tasting the fruit and abandoning their homecoming. The fruit that Homer speaks of is likely the sweet, date-like fruit of the tree *Zizyphus lotus* or *Zizyphus spina-christi* (see Figure 6.6). What fascinated the Christians about the fruit was its seductiveness. Thus, Gregory of Nyssa refers to the lotus metaphorically in an invitation to an otherwise unknown John: the latter should come to visit if he is not more charmed with the "lotus" (consuming allurements) of the city than affection for Gregory (*Ep.* 19.20). In the Christian tradition, the lotus assumes a positive connotation as well: Christ is compared to the sweet fruit of the lotus, of which one cannot have enough once tasted (PsPaulin. Nol. Carm. in App. 2.7ff.).

The early Christians also did not shy away from the Homeric herb moly, adopting it either directly from the Homeric poems or through the filter of later philosophical and epistemological reinterpretation. Even in antiquity, people were puzzled about the herb's botanical identification, which has preoccupied botanists through the ages. Today either *Peganum harmala* (wild or Syrian rue), or more recently, *Leucojum aestivum* (summer snowflake) are thought to be likely contenders. For Gregory of Nazianzus (fourth-century Archbishop of Constantinople) the fact that, according to Homer, only the gods, and not humans, had a word for moly provided a pretext for derision. He asserts that any given language is not solely the possession of its inventors, thus speech cannot consist of baseless signifiers (*Or.* 4.106). At the same time, the Neoplatonic interpretation assumed new life as, for example, when Prokopius of Gaza (*c.* 500 CE) urges the addressee of one of his letters to counter evil temptation with the moly of self-knowledge. Like the Homeric lotus, moly found its way into Christian thought directly from Classical texts in spite of the latter's pagan foundation, and while Christianity rejected the myths of Greco-Roman antiquity, the persistence of moly and lotus as botanical symbols certainly had to do with their unique and unusual character.

What ultimately spawned the rich assortment of plants and their manifold symbolism that feature in Christian thought was exegesis of the Holy Scriptures. The Bible mentions 110 plants by name, among them a great number of trees and also smaller members of the

A ZIZYPHUS SPINA-CHRISTI AT MAGDALA, SEA OF GALILEE.

FIGURE 6.6 Engraving of a *Zizyphus spina-christi* at Magdala, Sea of Galilee, 1889. Photo by Stephen Dorey—Bygone Images. Courtesy of Alamy Stock Photo.

plant world. Many of these appear in the Creation story, but the *Song of Solomon* is also replete with plants that gave rise to a burgeoning plant symbolism, as is the New Testament with its numerous parables drawn from agricultural life. The lily (*Lilium candidum*) is an example. For thousands of years, the lily, together with the rose, had assumed a place of prominence among flowers because of its beauty and its fragrance. Stories of the emperor Hadrian's adoptive son Lucius Aelius (130–169 CE) covering himself with a blanket of lilies (SHA *Ael.* 5.7–8) and of Elagabalus' (203–222 CE) spreading carpets of lily and other flowers on the floors of his palace (SHA *Heliogab.* 19.7–9), while likely anecdotal exaggerations, nonetheless have a core of truth. At a minimum, they attest to the value of this highly prized flower, whose white bloom symbolized beauty, purity, and grace even in pagan literature.

The lily did, however, also have negative connotations in Classical mythology. In particular it was said that Aphrodite disliked the lily intensely either because of its purity or because the flower sprang from her rival Hera's milk; her hatred induced the goddess to insert a yellow stamen in the shape of a donkey's phallus into the flower, thereby compromising its worth and honor (Ath., *Deip.* 15.683A). This tale underlies the lily's appellation as "Aphrodite's jest," and Aphrodite's dislike of this flower presumably contributed to the Christian notion of the lily as a symbol of chastity. Above all, however, the latter rests on the biblical verse from the Song of Solomon (2.1–2), "I am a rose of Sharon, a lily of the valleys. As the lily among brambles, so is my love among maidens."

Other points of reference for the lily as Christian symbol may be Jesus' words to his followers regarding the unimportance of material possessions: "Consider the lilies of the field, how they grow; they neither toil nor spin" (Mt. 6.28–30; Lk. 12.27). The context here is an exhortation to remain ever mindful of the claims of the Kingdom of Heaven without neglecting the necessities of life. Property and the work of one's own hands are permissible and necessary, but as so often in human life, correct estimation and management of these is required. The lily symbolism in Matthew is not strictly parallel to that in the Song of Solomon where, like the other meadow flowers, the lily of the valleys is a flower of the virtuous to be picked by the devout (Greg. Nyss. *Comm. in Cant.* 4; 5; 14). As a flower of the valleys and not of the mountains, which are the symbolic realm of the proud, the lily represents humility and, because of its white blossoms, chastity, as already mentioned (Hieron. *Comm. in Ioel* 3; Ambros. passim; Greg. M. *in Hes. Hom.* 1.6.4). Its perfume, on the other hand, is the fragrance of virtue (Hieron. Comm. in Zech. 3, 14, 20f.). On the basis of these qualities, the lily was dedicated initially to Christ, and only later, in the tenth century CE, to Mary; as theologian Ambrosius (before 212–*c.* 250 CE) remarks, it is like the immaculate deity, not choked out by thorns (*Lc.* 7.128). The lily of the valleys also symbolizes the resurrection of Christ, for, like the lily, he descended below the earth only to rise again (Hippol., *frg. arm. in Cant.* 2; Orig. *in Cant. Hom.* 2.6).

While the attributes of the lily were ideally suited to bear moralizing symbolic value, other plants could also embody Christ's suffering and death. The didactic Christian text *Physiologus*, for example, relates a notable allegory involving the fructification of the fig that takes into consideration every detail of the process: in the scoring of the fruit, the outflow of juice, the emergence of the pollinating insect, and the subsequent ripening of the fig the author sees a depiction of the death and resurrection of Christ (29). It is through Christ that humanity emerges into the light of Truth (= the pollinating insect) and receives nourishment (the fig's fruit).

The early Christian authors' resourcefulness and originality is evident in the fact that the same plant can assume a variety of different symbolic meanings. A particularly good example is the nut, the symbolism of which applies to the Holy Scriptures and their relevance to, or interpretation by, humanity as well as to the ontology of Christ (Christology) and eschatology (Siede 2013). The likeness of extracting true meaning from the Scriptures to cracking a nut's shell and removing the kernel appears vividly in the writings of theologian Origen (*c.* 184–*c.* 253 CE), who finds a correlation between the Old Testament's literal, moral, and mystical senses and the three parts of an almond: the bitter skin of the nut, the hard shell, and the kernel. Origen had a formative precursor for this symbolism in Clement of Alexandria's (*c.* 150–*c.* 215 CE) *Stromata* (patchwork), a miscellaneous collection of ancient wisdom through which he hoped to transfer Greek philosophical dogma to the Christian creed. "Much truth," Clement argues, "resides in the midst of Greek dogma, covered and concealed by the philosophers' writings, like an edible nut core in its shell" (1.18.1). In the works of St. Jerome (347–420 CE), Cassiadorus (*c.* 485–*c.* 585 CE), and Eusebius (263–339 CE), too, the physical character of nuts, namely hard shells surrounding a sweet kernel—or in the case of a chestnut, a spiny cupule or bur enclosing the edible seed—is compared to the multiple layers of the Holy Scripture and the difficulty of understanding their spiritual content (Hieron., *in Eccl.* 12.9f., Ep. 58, 9; Cassiod., *in Ps. 1*; Euseb., *in Cant.*). The four-fold "fruit" (edible seed) of the walnut, for its part, was said to reflect the four Gospels (Chromat. Auil., *Tract. in Math.* 2, 151; Fortunat. Aquil., *Comm. in Ev.* 1, 13). It should be noted that no distinction was made in these biblical texts between "true" or "botanical" nuts (fruit having an inedible hard shell containing an edible seed, as in the case of chestnut) and culinary nuts (seeds of drupes, as in the case of walnut and almond, which are sometimes called drupaceous nuts).

With respect to humanity, it is the Righteous—who have a bitter, caustic shield protecting them from temptation—that are likened, again by Origen, to nuts (Orig. *Schol. in Cant.* 6). Olympiodorus points to the constitution of humans as parallel to that of the almond: the human body, soul, and spirit correspond to the nut's green rind, woody shell, and kernel (Olympiod., *Comm. in Eccl.* 12.5). At resurrection, we shed the former two (body and soul), which are perishable (Caes. Arelat., *Serm.* 111.2). On the basis of its role in the Old Testament, the almond was found to be especially applicable to such discussions: its early bloom in *Ecclesiastes* 12.5, similarly to the Jewish interpretation of Philo of Alexandria (*c.* 20 BCE–*c.* 50 CE), represents youth, whereas the tree's bearing of fruit represents old age (PsJoh. Chrys., *in Eccl.* 12.5). Meanwhile, from a Christological perspective, the dry, infertile, almond staff of Aaron, which begins miraculously to bloom and bear fruit, is interpreted as signifying the virgin birth of Christ (Num. 17, 25; Cyr., *Catech. ad illuminandos* 12, 28; Chromat. Aquil., *Tract. in Math.* 2, 151). Aaron's staff of almond thus possesses the character (*typos*) of Christ and alludes also to the Root (or Branch) of Jesse, the family tree representing Christ's genealogy (Olympiod., *Comm. in Eccl.* 12.5). The garden of nut trees into which the bride descends in the Song of Solomon represents the arduous and trying depths of earthly life and, at the same time, indicates the bitter locale of Christ's crucifixion (6.11). The sweet flesh of the nut, so difficult to obtain, is Christ himself, according to exegesis of this passage. As for the blooming and fruit-bearing Staff of Aaron, together with the blooming almond tree prophesized in Ecclesiastes 12.5, it is symbolic, too, of resurrection: the flowers symbolize both the resurrection of Christ and the reawakening of the dead more generally in so far as the early-blooming almond epitomizes spring. Aaron's staff symbolizes the Resurrection also in that the mystery of Christ, that will in time be illuminated, resides within it and because, contrary to natural order, it bore fruit overnight, as quickly and fully as the course of the Resurrection.

CHAPTER SEVEN

Plants as Natural Ornaments

KAJA TALLY-SCHUMACHER

In May 2013, a couple in Miami Shores Village, Florida were ordered to remove their seventeen-year-old front-yard vegetable garden as the city had amended and clarified ordinance Division 17, Sec. 536[e] banning front-yard vegetable gardens. They would now only be allowed in backyards. The couple, unable to bear the $50/day fine for non-compliance, unearthed their garden. Reacting to the upheld ruling in August 2016, Hermione Ricketts said, "I am disappointed by today's ruling. My garden not only provided us with food, but *it was also beautiful* and added character to the community. I look forward to continuing this fight and ultimately winning so I can once again use my property *productively* instead of being forced to have a *useless* lawn" (Ovalle 2016). While the case may appear preposterous, similar cases have sprouted up across America and Canada in the last five years, including one in Michigan where the gardener was threatened with jail time (CBC News 2015; Fullbright 2012; Holt 2012; *Huffington Post* 2012/17; Kirpalani 2011; Sibilla 2013). At the heart of these legal battles is the concern over what constitutes an ornamental plant. Can a useful plant, such as an herb, a medicinal plant, a vegetable, or fruit, also serve an ornamental function? While chronologically disparate from the content of this chapter, these legal battles illustrate the contentious nature of ornamentation and the challenges one faces in identifying and defining plants which may serve an ornamental function. For these gardeners, ornamental is not synonymous with useless or unproductive, nor do they assume that a plant may only fill one function, a definition clearly not supported by the city ordinance writers. Furthermore, the limited confines of these particular discussions to North America highlights the cultural specificity of ornamentation and even the specificity of the kinds of discourse and disputes that occur in the process of defining and establishing what ornamentation is.

These disputes serve as a helpful entry point for discussions on ancient ornamental plants, in part because they are the product of Roman botanical thought. While the Miami Shores Village ordinance implies that useful plants, such as vegetables, are not ornamental (they may even be perceived to be an eyesore), this conception is incongruent with ancient gardening practices. And yet, Hermione Ricketts's complaint over useful and useless plants was not unfamiliar to ancient gardeners and garden owners. In fact, as will be shown below, we hear echoes of ancient Roman authors Martial and Horace in Ricketts's choice of adjectives. The *Cambridge Dictionary* defines ornament as "an object that is decorative rather than useful," yet this modern definition may not be entirely helpful when thinking about ornament in antiquity (*Cambridge Dictionary of American*

English 2008). Our modern term, ornament, is derived from the Latin *ornamentum,* used by various Roman authors to mean apparatus, equipment, accoutrement, attire, dress, furniture, trappings, mark of honor, décor—to be without ornament, *sine ornamentis*, is to be naked (Lewis & Short, "*ornamentum*"). For Romans, like encyclopedist Pliny the Elder, the idea of *ornamentum* is deeply rooted in artistry, labor, and production (Carey 2007: 32). This definition is markedly different from our own. For plants to act as *ornamentum* suggests that some level of craftsmanship is necessary, and that the plants may also have a covering, adorning, or screening function such as attire and dress on human bodies.

GERMINATING DESIGNS IN THE NEAR EAST: EARLY MESOPOTAMIAN AND EGYPTIAN PLEASURE GARDENS

The Tigris-Euphrates Valley in Mesopotamia is home to the earliest evidence of agricultural cultivation, followed shortly thereafter by the Nile and Indus Valleys. Some of the earliest cultivated plants in Mesopotamia include dates (*Phoenix dactylifera*), olives (*Olea europaea*), grapes (*Vitis vinifera*), figs (*Ficus carica*), and pomegranates (*Punica granatum*). In Egypt's Nile Valley the earliest agricultural cultivation focused on dates, followed by sycamore figs (*Ficus sycomorus*) and pomegranates (Janick 2005: 3). It is out of this arboricultural tradition that pleasure gardens arise. While these early gardens have left little trace archaeologically, surviving garden representations help us reconstruct their development. For example, an eighteenth-century fresco painting from the tomb of Nebamun in Thebes shows a rectangular pond, filled with fish, ducks and geese, and floating lotus blossoms (*Nymphaea lotus*) (see Figure 7.1). The pool is ringed with low-growing papyrus plants (*Cyperus papyrus*), and a ring of fruit trees stands over the pool. With the exception of the date palms, the trees are similarly shaped, making identification difficult, but the fruit they bear is clearly articulated: pomegranates and sycamore figs. In this orchard, date palms are by far the most numerous trees, and unlike the other trees, they are depicted at various stages of growth; some occupy the understory, while others are tall. A woman in the far corner carries fruit collected from the garden to baskets that are already over-full. Similarly, the "Investiture" panel from Mari Syria depicts a date palm and a scene of humans performing arboricultural labor, in this case two climbers scaling the tree trunk to gather the dates. In both examples date palms dominate the gardens; in Egypt they do so by their greater number, and in Mari by their great height and large scale, dominating the figures and other scenes. While date palms already had a long tradition of cultivation in Egypt by the time the tomb frescoes were painted, their representation in Syria is particularly significant as Mari is located beyond their natural growing habitat, and thus it would have been nearly impossible for the date palms to produce fruit. Yet the presence of scaling climbers reaching for the successfully cultivated fruit celebrates the botanical knowledge and rule at Mari that made such a feat possible (Luciani 2010: 114; see also Chapter 8 of this volume and Figure 8.3). In fact, the Mari scenes repeatedly represent the cultivation of exotic, and thus expensive and labor-intensive plants, in favor of native species (Luciani 2010: 115).

These early pleasure gardens feature many shared motifs: central pools or channels around which the garden is centered, fruit-laden trees, and representations of garden labor. In these images and likely in the real gardens, the plants and their bounty serve

FIGURE 7.1 Fresco from the tomb of Nebamun, shows a pool in a garden that might have belonged to Nebamun. The pool is full of birds, lotus flowers, and tilapia fish, while papyrus grows along the edge. Around the pool are date palms, dom-palms, sycamore figs, and mandrakes. Thebes, Egypt, 1300 BCE. Photo by World History Archive. Courtesy of Alamy Stock Photo.

as important a role in ornamentation as the performed human labor. Additionally, the representations illustrate a significant horticultural innovation: utilizing the shade of tree canopies to create a cooler microclimate in the understory, making it possible to cultivate non-native plants such as saffron-producing crocus (*Crocus sativus*) or the edible poppy (*Papaver somniferum*) (Luciani 2010: 99, 114; Semple 1929: 421). This is further supported by recent archaeological excavations in Egypt at Amarna and at Dra Abu el-Naga, Luxor, where tamarisk trees (*Tamarix nilotica*) have been found growing over tomb waffle gardens (a grid-shaped garden, farmed by packed mud to capture water) planted with a wide variety of edible plants, such as coriander (*Coriandrum sativum.*), melon (*Cucumis melo*), cucumber (*Cucumis sativus*), and caraway (*Carum carvi*) (Farrar 2016: 11, 33; Galán 2018). At Luxor, the pristine preservation conditions have even led to the discovery of purple blossoms, now more than four thousand years old (Galán 2018). These tomb gardens (located in front of tomb entrances) and their representations functioned as funerary vegetal ornamentation and as productive vegetable and herb gardens. Unlike Miami Shore Village's gardens, here these two functions are not mutually exclusive; instead, the gardens were "at once useful and beautiful" (Semple 1929: 421).

Growing and Cultivating Empire in Ornamental Gardens

Distinguished Assyriologist Leo Oppenheim suggested that Neo-Assyrian gardens changed significantly from utilitarian to ornamental gardens under the rule of Sargon II (722–705 BCE) and Sennacherib (705–681 BCE) (Amrhein 2015: 93; Oppenheim 1965: 331–3). This new change in the role of gardens and the kinds of plants that were cultivated is a direct result of imperial expansion. Perhaps one of the best-known gardens, built by Sennacherib, sought to miniaturize and recreate the environment and plants of two culturally and imperially significant places: Mount Amanus in modern-day Syria and Chaldea in southeastern Mesopotamia. Sennacherib boasts that within this garden he planted trees native to Chaldea, as well as cypress (*Cupressus sempervirens*), sissoo (*Dalbergia sissoo*, a hardwood native to the Indus Valley), and olive trees. He also planted a botanical garden and even replicated Mount Amanus itself, as well as its forests, by importing trees known to naturally grow on the mountain. A bas-relief from the palace of his grandson, Ashurbanipal, may illustrate this very garden (British Museum, London n. 124939) (see Figure 7.2). The representation of plants is highly abstracted, so that it is impossible to identify the trees definitively, but there is clear differentiation between taller, fruitless trees, with conical canopies, and shorter and more squat, multi-branched trees, with fruit-laden branches—perhaps olive trees. While it may be tempting to identify the taller, conical trees as cypress, earlier Mesopotamian, Near Eastern, and Egyptian depictions of fruit trees show a variety of trees having a conical shape, suggesting that representational tree morphology should not be heavily relied upon for tree identification. It may be the case that select tree species, like date palms, warranted more naturalistic representation of tree morphology while others were depicted in a more standardized

FIGURE 7.2 Sennacherib's Mount Amanus garden, Palace of Ashurbanipal, North Palace, room H, British Museum. Photo by Heritage Image Partnership Ltd. Courtesy of Alamy Stock Photo.

fashion. Alternatively, a great variety of trees may actually have been shaped, by pruning and training, into similar, conical forms. In any event, it is quite likely that the garden also featured cedars (*Cedrus libani*), junipers (*Juniperus drupacea*), stone pine (*Pinus pinea*), and an additional, unspecified pine species (*Pinus* sp.), as these are known to have grown abundantly in the area of Mount Amanus and to have been highly prized for their aromatic timber (Thomason 2005: 175). Importantly, beyond miniaturizing and replicating, an ornamental Neo-Assyrian garden served as a botanical map. Exotic flora and fauna from the entire empire, brought as tribute, were collected in these gardens. Indeed, one may think of them as "universal gardens," functioning as microcosms of the known world (Amrhein 2015: 92; Novák 2001: 445).

Climate, Ornament, and the Arts

The growing horticultural knowledge that accompanied imperial expansion and assimilation of non-native plants also resulted in the development of specialized garden structures that even further manipulated the local climate into mimicking vastly different environments. While Sennacherib claims to have replicated Mount Amanus, it is unclear how he could have recreated an alpine climate, but another garden type, the sunken garden—developed in the Near East by Sargon II, Sennacherib, and Assurbanipal, but also popular in Egypt—successfully controlled moisture, temperature, and wind. A sunken garden is one that is cut into the earth so that it sits below the natural ground level, and is generally protected from wind by walls and colonnades. During the especially arid summer season, the lowered and protected position of a sunken garden would have trapped cool, damp air, which naturally settles into low-lying areas. The lower position and tall walls would also have provided protection from wind, thereby slowing the evaporation of moisture from soil and plants. In many ways, the sunken garden is a more effective structural solution for mitigating arid conditions than planting in the cooler, moister, understory of trees. This new structure would have allowed gardeners to cultivate plants with distinctly different moisture, temperature, and climate requirements—plants that were not native and likely not accustomed to the local arid climate. The excavations in sunken gardens in Egypt, such as at Amarna (1346 CE), where temperatures range from 48 degrees Celsius (118° F) to below freezing, have yielded tree cavities and signs of flower beds, but botanical analysis has not identified the species of plants that may have benefited from the more controlled and moister sunken garden microclimate (Wilkinson 1998: 147). Based on the surrounding surviving decoration, it is likely that these kinds of gardens featured cornflowers (*Centaurea cyanus*), poppy, hollyhock (*Alcea*, unidentified species) or mallow (*Althaea offcinalis*), giant reed (*Arundo donax*), nut grass (*Cyperus rotundus*), and papyrus (Wilkinson 1998: 160–4). It is possible that some subtropical plants from Central Asia or the Indus Valley may have been cultivated in such spaces. Even more commonly cultivated plants, such as pomegranate trees, may have decorated sunken gardens, where their high water and moisture needs would have been better met and thus would have likely resulted in more bountiful harvests.

Some scholars have suggested that these ornamental sunken gardens were closely related in design and function to royal carpets and to carved patterns in stone floors, which acted as a kind of permanent, interior sunken garden (Polinger-Foster 2004: 211–16). The relationship between real plantings and representations in textiles and carving is suggestive of connections between various types of specialized laborers, such as gardeners, designers, weavers, and sculptors, to name but a few. Beginning with Sargon

II, floral patterns, featuring rosettes, lotus flowers, flower buds, and garlands became increasingly common on threshold carvings (Polinger-Foster 2004: 213). While these carvings are in stone, their primary role as floral floor decoration is highly evocative both of real carpets and real gardens. Royal courts were adorned with textiles such as hangings, carpets, and floor coverings, and the throne was commonly placed in front of or on top of floor coverings. A fragmentary inscription listing the inventory of Assyrian king Tukulti-Ninurta I's (1243–1207 BCE) textiles includes two items that hint at the presence of gardens: one is described as being decorated with images of fruit trees, possibly pomegranate, lotus blossoms, and rosettes, as well as a second item with kings, towers, animals, cities (Barrelet 1977: 58; Polinger-Foster 2004: 212–13). One can imagine the playfulness of designers as they translated garden designs into patterns to be stepped on. Unlike real gardens, these could be trampled. Moreover, in contrast to hard, bare stone floors, soft floor coverings—whether leather with appliqué, woven, or pile—replicated the experience of walking on plush, cultivated, turned earth.

While developed in the Near East and Egypt, the decoration of floors with garden motifs remained popular in antiquity across the Mediterranean, and indeed, remains popular today. Nearly a thousand years after Assyria's fall, the Egyptian king Ptolemy II Philadelphus (283–246 BCE), is said to have covered the floor of his banquet hall so thoroughly with real blossoms that the floor turned into a magnificent winter-time meadow (Ath., *Deip.* V. 197e). The motif remained equally popular during the Roman period, as evidenced by a large body of garden-themed mosaics from across the Mediterranean. Many of these mosaics, depicting sprays of flowers and cut branches, eternalize Ptolemy's freshly strewn blossoms (Malek 2018: 317–40). The evocation of real garden beds in interior spaces was no doubt further heightened by the use of floral perfumes and incense that would have lent garden representations fragrance. Modern "carpets" of wooly-thyme (*Thymus pseudolanudinosus*, meaning fake-wool, a more recently developed cultivar), a low-growing, fuzzy-haired groundcover, exemplify the persistence of the garden carpet idea. Beyond the sensory experience of fuzzy, wool-like twigs/threads under one's feet, even the name, wooly-thyme, evokes textile production.

BLOSSOMING IDEALS: PLANTING THE WORLD IN THE CHINESE GARDEN

The predominant concepts explored in the ancient Near Eastern gardens (miniaturization, replication, and garden as a collection or microcosm) also defined the Supreme Forest, the great imperial hunting park of the Qin and Han Dynasties in China, established in the third century BCE by Qin Shi Huang. Little is known of the park's dimensions, features, flora, and fauna during the Qin period, but a decree issued after the fall of the Qin, in the second year of the following Han Dynasty, is illustrative. The decree returned all Qin imperial gardens and parks, including the Supreme Forest, to the common people for agricultural cultivation. Yet the legislation was short-lived, and sometime before 138 BCE the Supreme Forest was reinstated and enclosed by the Han, and by 138 BCE was greatly expanded by Emperor Wu. This new, enlarged Supreme Forest was enclosed by a 300-kilometer-long wall (roughly 200 miles), and featured between thirty-six and seventy structures (depending on the source), as well as woods, meadows, streams, swamps, waterfalls, islands, marshes, hills, valleys, pools, ponds, water gardens, and an encyclopedic collection of flora and fauna from the far reaches of the empire, so that the entire world was held within (Allsen 2006: 42). Over three thousand plant species have been identified in the Supreme Forest—too many to list here (Guangping et al. 2012: 61–193). The forest featured ferns, palms, evergreen and deciduous trees and shrubs, trees for timber, fruit trees, ornamental trees, flowering shrubs, and many other types of plants (Guangping et al. 2012: 61–193). A quick glance at the variety of *Prunus* found in the forest serves as a helpful example of the encyclopedic nature of the plants: three varieties of cherries are documented (*Prunus pseudocerasus*, *Prunus japonica*, *Prunus cerasus*), two varieties of plums (*Prunus mume*, *Prunus salicina*), and two varieties of peaches (*Prunus persica*, *Prunus davidiana*) (Guangping et al. 2012: 110–63). Moreover, many of these were exotic, subtropical, and tropical plants from the south and from Central Asia, such as the recently introduced Iranian grapes, mandarin oranges (*Citrus reticulata*), camphor trees (*Cinnamomum camphora*), and Lychee tree (*Litchi chinensis*). Their presence in the forest suggests that Han gardeners must have employed sophisticated horticultural practices, such as some sort of green-houses or portable planters that could be moved to warmer areas, allowing plants to survive winter and perhaps even to produce fruit (Guangping et al. 2012: 53–6; Schafer 1968: 328). By trying to replicate nearly every environment and climate, and collecting three thousand species of plants, the park drew on "the magic belief that by artificially making a replica of something one wields power over the real object" (Guangping et al. 2012: 53–6; Hunt 2012: 12; Ledderose 1983: 166). Here, as in Sennacherib's gardens, the bounty of the empire, exemplified in non-native plants, new horticultural knowledge, new materials and raw resources, served as natural ornament.

After the decree returning the park to the people, and the subsequent reinstallation of the imperial grounds, we again find a parallel to Ricketts's frustration with "useless grass." The late second-century BCE poet, Ssu-ma Hsiang-ju, reports that Emperor Wu's advisors criticized the forest as "wasteful," and recommended that he instead use the preserve as cultivated, agricultural land and for firewood for his subjects (Snyder 2016: 53–4). The forest consumed resources in the form of horticultural labor, care of pavilions, and care of animals in the hunting preserve without producing consumable goods for the common people. Moreover, this example illustrates that our definition of ornamental plants and ornamental gardens need not be restricted to formal, flower-based creations, since a park or other type of planted space may also be perceived as ornamental.

While Sennacherib's and Emperor Wu's gardens utilized the resources, products, labor, knowledge, and craftsmanship afforded by conquest to cultivate botanical ornamentation, imperial expansion also fostered plant trade with distant, unconquered lands (Milburn 2016: 28). Writing in or before 217 CE, Cao Di, then soon to be emperor of the Wei Dynasty, wrote and commissioned rhapsodies commemorating the importation of the "*midiexiang*," an aromatic plant from Rome (Milburn 2016: 26–7). Scholars have theorized the identity of this plant, suggesting that it may be rosemary (*Rosmarinus officinalis*); however, there is much disagreement and this identification is far from secure (Milburn 2016: 28–9). While many tropical plants from Nanyue (modern Vietnam) had been brought to Chang'an, the Han capitol, in 111 BCE as trophies of conquest, the rhapsodies describe *midiexiang* as possessing agency and as willingly choosing to escape the barbaric West in favor of the more civilized Wei Dynasty (Milburn 2016: 35). Interestingly, in keeping with the Roman concept of *ornamentum* as attire or dress, the rhapsodies describe *midiexiang* as being worn on the body, with branches tucked into layers of fabric, thereby covering the body in aromatic perfume (Milburn 2016: 36–42).

While the Supreme Forest condensed the ruled world into a forest-sized microcosm, *penjing*, the Chinese art of miniaturized, tray-sized landscapes and dwarfed trees compressed the world even further. Inversely, the level of skill and knowledge required to further miniaturize the world was even greater than that required to cultivate the microcosm of the Supreme Forest—indicative of the political rule that made such knowledge possible (Bailey 2005: 30; Bowan 2017: 14). According to legend, *penjing* dates to the Qin and Han Dynasties (221 BCE–220 CE), although the earliest surviving representation of *penjing*, found in the Qianling Mausoleum during excavations in 1972, is significantly later, dating to 706 CE (Bailey 2005: 30; Hu 1987: 7). The Mausoleum wall painting depicts a court attendant holding a shallow bowl featuring two plants (perhaps fruit trees), each shorter than a forearm, amid lumps of stone (Bowan 2017: 14) (see Figure 7.3). Small enough and light enough to be held by a single figure, the mobility of these miniaturized spaces is central to the mural motif. While today some 160 varieties of trees are cultivated in *penjing* form, chosen for their transplantation and pruning

FIGURE 7.3 Earliest representation of *penjing* (bonsai), Qianling Mausoleum, 706 CE. Courtesy of Kimberly Wlczak.

resilience, longevity, tiny leaves, and woody structure, it is difficult to reconstruct the early history of this practice with any certainty. This is made further difficult by evidence of distinct regional differences in preferred tree species in later periods, suggesting divergent histories (Hu 1987: 20).

RIPENING ORNAMENT: GRECO-ROMAN NATURAL *ORNAMENTUM*

Modern plans of ancient Greek and Roman cities tend only to depict walls, streets, paved courtyards, and other structures, creating the impression that urban spaces were barren, devoid of green spaces. Yet this could not be further from the truth. While we know much less about Greek gardens and their plants than their Roman counterparts, a brief glance at just one Greek city, Athens, illustrates the important role of urban green space. For example, during the Classical and Hellenistic periods (508 BCE–31 BCE), the altar of the Twelve Gods was ornamented by a small grove of olive trees and laurel (*Laurus nobilis*) which grew within its walls (Carroll-Spillecke 1992: 86). Likewise, the temple of Hephaestus was surrounded by two rows of plantings in planting pots, theorized by Thompson to have been composed of laurel and pomegranate (Thompson 1937: 423–5) (see Figure 7.4). Similarity, the Agora, an open space used for assemblies and markets, is described as shaded by the canopies of plane trees (*Platanus orientalis*) and poplar (unknown species) which are thought to have grown along pathways, roads, and at drains (Carroll-Spillecke 1992: 86; Thompson 1982: 8). In contrast to the wooded civic and religious heart of the city, the residential areas within the city walls appear to bear little evidence of plantings directly in the ground, as courtyards were generally paved, though perhaps household courtyards featured portable, potted plants. Instead, Greek cities like Athens, were ringed by intensely cultivated suburban gardens composed of privately tended or rented fruit orchards, vegetable gardens, and vineyards (Carroll-Spillecke 1992: 86, 89). The gymnasia (places for physical and intellectual training), located in the suburbs, were likewise decorated with plants. For example, one of these gymnasia, the Academy, founded by Plato in the fourth century BCE and attended by philosophers, was shaded by elm, poplar, plane trees, and olive trees (Carroll-Spillecke 1992: 91; Giesecke 2007).

Thus far we have established that the Greek urban fabric did include green spaces, primarily in the form of trees, but were they perceived to be ornamental? Scholars have generally argued that in comparison to other ancient Mediterranean cultures, the Greeks did not grow ornamental gardens, and that instead they valued intensive cultivation in productive gardens. Yet this conception has recently been questioned (Foxhall 2007: 219–46). Textual sources suggest that trees in particular appear to have held a prominent functional (shade producing) and ornamental role. Thus, the Greek essayist Plutarch says of the general Cimon's planting the Agora and Academy with trees that he was "the first to beautify the city" (Plut., *Cim.* 13.8). Similarly, the Persian prince Cyrus' garden at Sardis is admired in part for its beautiful trees, but also for the precision of the gardeners in measuring and mapping the tree placement and their alignment with one another (Foxhall 2007: 221). While Cyrus' garden does not survive, a preference for orderly plantings is supported by archaeological excavations, such as the double rows of planting pits, each aligned with the columns, surrounding the Temple of Hephaestus in Athens. As such, the excavations and textual sources suggest that Greek ornamental gardens did not differ from productive gardens: it was the manner in which plants were planted

FIGURE 7.4 Planting pits each featuring a single planting pot, surrounding the colonnade of the Temple of Hephaestus, Athens, third century BCE. Courtesy of American School of Classical Studies, Athens.

and arranged in orderly rows, as well as their usefulness, that made them pleasing and ornamental (Foxhall 2007: 221).

In contrast to the more limited Greek evidence, the concept of plants as ornament explodes in gardens and across all media in Rome in the first century BCE. At the end of the Republic, amid extreme elite competition in public works, the general Pompey the Great constructed Rome's first public park. Completed in 62 BCE, the Porticus of Pompey featured a large grove of plane trees (Gleason 1994: 13). As the first emperor, Augustus, took power and renovated the city, transforming Rome from a city of brick into a city of marble (Suet., *Aug.* 29), he also transformed the city into a multi-media garden, planting new public gardens and veneering the cityscape in vegetal reliefs (Tally-Schumacher and Niemeier 2016). Thus, after the fall of the Republic and the commencement of the Imperial Age, Livia, Augustus' wife, gifted to the Porticus Liviae (a multi-use public structure) a grapevine that grew so large it spanned the facades of two buildings, thereby transforming it into a garden space. The abundant growth of the vine was the direct product and visualization of the fecundity, fertility, and success of the new Augustan peace (Clarke 2003: 26; von Stackelberg 2009: 91). In addition to housing the temples to Jupiter and Juno, a curia (an assembly place), a library, and a school, the Porticus of Octavia, Augustus' sister, also featured gardens in which paintings and sculpture were displayed (although we do not know what these gardens looked like) (Hemelrijk and Woolf 2013: 30). Augustus' own house was famously decorated by two laurels on either side of the main door. The laurels were so significant, in part because they served to further connect his private home with the adjoining Temple of Apollo, that the motif of a door flanked by two laurels was minted on contemporary coins, and quickly became synonymous with Augustus' home, and even with Augustus himself. The Mausoleum of Augustus, begun shortly after the Battle of Actium in 31 BCE, also featured plants as a prominent form of ornamentation. The geographer Strabo, writing in the first centuries BCE and CE, describes the tomb as an artificial mountain covered with evergreen trees, *aethalési déndrōn* (Strab., *Geograph.* V.3.8). The choice of evergreen plants instead of fruit trees or other culturally significant plants was likely intentional. Unlike deciduous trees, evergreens do not lose their foliage and thus remain green year-round, thereby defying the signs of the passage of time and seasonal change. As such, evergreens serve as an ideal visualization of the idea that Augustan rule, fertility, and fecundity know no physical or temporal boundaries. Strabo does not identify specific evergreen species, but based on contemporary representations, such as the garden painting from Livia's and Augustus' Villa at Prima Porta, it is possible that the mausoleum mound was planted with recently imported Norway spruce (*Picea excelsa*), and more traditional evergreens such as cypress, stone pine, and even non-conifer evergreens, like box (*Buxus sempervirens*) (Caneva and Bohuny 2003: 151).

Potted Plants: Adonis Gardens and Ollae Perforatae

While Greek homes appear not to have had gardens, ceramic planters have been found at Olynthos, attesting to smaller, more portable domestic cultivation. Planters have also been found buried in the planting pits surrounding the Temple of Hephaestus at Athens, where they likely held fruit trees that had been propagated via air layering (a method in which a pot with soil is placed on a fruit tree branch, promoting root growth, and allowing transplantation) (Carroll-Spillecke 1992: 86; Thompson 1937). Ceramic planters also played an important role in Greek religious celebrations. For example, Adonis gardens, miniaturized single-pot-sized gardens planted with fast growing seeds, were part of an important annual Adonis festival celebrated by women (Carroll-Spillecke 1992: 86; Reitzammer 2016). The tiny gardens, once sprouted, would be placed on the rooftop of houses, and left to wilt, marking the death of Aphrodite's consort's death. Intentionally short-lived and miniaturized in scale if not in the treatment of plants, the Adonis gardens nonetheless subtlety evoke another form of miniaturized, potted gardens: Chinese *penjing*.

It is in Roman gardens, however, that the use of ceramic planters blossoms (Macaulay-Lewis 2006 a, b). Found in various contexts, some of the most famous examples come from Villa Oplontis in the Bay of Naples. During the 1970s excavations numerous *ollae perforatae* (perforated ceramic vessels used as planters, buried directly in the ground) were found in one of the large peristyle gardens (garden 59) (Bergmann 2002: 87–120; 2016: 96–110). The planters were arranged in two parallel rows, following the contours of the three excavated walls of the garden. Both rows of planters were aligned with the columns of the colonnade. One row, placed further into the garden, featured larger planters, likely used for shrubs or small trees. Another row of planters was placed directly at the colonnade's edge, with each planter placed in front of a column and tilted towards it, as one would do to guide a climbing plant like ivy (*Hedera helix*) up the column (Jashemski 1979: 293–5; Macaulay-Lewis 2006b: 7). Using plants to ornament columns appears to have been a popular motif in gardens, paintings, and in garden discourse. A fresco painting from triclinium 14, also from Villa Oplontis, depicts a column that sprouts gilded acanthus leaves at the base, and is bound by a swirling, bejeweled, and gilded vine running up its shaft, mirroring the real vines in garden 59 (Gleason 2014: 1080). The Roman statesman Cicero, writing to his brother Quintus, describes how his gardener "clothed" the intercolumniations (the spaces between columns) and his villa foundations in ivy (Cic., *QFr.* 3.1.5). It is significant to note Cicero's word choice, "to clothe," as it is indicative of the complimentary relationship between architecture and plants—the structure is nude, incomplete, even barbaric, and uncultured without vegetal dress. This idea has deep roots; two seventh-century BCE Assyrian kings, Esarhaddon and Ashurbanipal, also describe plants as clothing for buildings, thereby expressing the cultured state of their rule (Amrhein 2015: 96).

Snipped, Cut, Pruned: Flowers, Fruit, and Leaves

Cut flowers, fruit, and leafy branches were also important ornaments used to "clothe" buildings. Tracing the history of cutting plants for garlands, Pliny the Elder tells us that the earliest chaplets (garlands or wreaths worn on heads) were made from olive, oak (*Quercus robur* or perhaps *ilex*), laurel, and myrtle branches (*Myrtus communis*) (*HN* XXI.1–3). Though archaeological and literary evidence reveals his assessment to be untrue, floral garlands being of very great antiquity, Pliny goes on to write that both Romans and Greeks began creating garlands with flowers, versus greenery, only later, after the 100th Olympiad (fourth century BCE). He states that this custom resulted from a competition between a painter, Pausias of Sicyon, and the female garland maker, Glycera—a competition in which the painter's pigments could not compete with the full palette Nature provides in flowers. As floral garlands, commonly composed of roses (*Rosa gallica*, or other varieties), lilies (*Lilium candidum*), and violets (possibly *Viola reichenbachiana*), would have only lasted a day before wilting and losing their fragrance, Pliny says that they serve as a reminder that "that which is the most beauteous and most attractive to the eye, is the very first to date and die" (*HN* XXI.1; Jashemski 1963: 115). In Athens, garlands were commonly woven by women and girls, and were sold at the Myrtle Market. As in Rome, the flowers were likely grown in specialized gardens beyond the city walls (Ar., *Thesm.* 445–60; Plin., *HN* XXI.1; Jashemski 1963: 116).

Garlands were commonly used to decorate a variety of structures. The forum scenes on a fresco from the Praedia Julia Felix at Pompeii depict green garlands hung in between the columns of that city's forum. The quotidian nature of the scenes suggests that perhaps the civic center may have been thus adorned regularly, not only during celebrations and

FIGURE 7.5 Garland composed of grape bunches and leaves (*Vitis vinifera*), laurel leaves (*Laurus nobilis*), quince fruits (*Cydonia oblonga*), pine cones and pine sprigs (possibly *Pinus pinae*), wheat heads (possibly *Triticum dicoccum*), pomegranate fruits (*Punica granatum*), and poppy heads (*Papaver somniferum*), Villa of P. Fannius Synistor at Boscoreale, 50–40 BCE. Photo by Artokoloro. Courtesy of Alamy Stock Photo.

festivals. Moreover, the lack of colorful but quickly wilting blossoms suggests that the garlands would have remained fresh and green for a significant time, making them ideal for prolonged ornamentation in a public space. We find similar representations of garlands hung in private domestic spaces, such as the Villa of P. Fannius Synistor at Boscoreale, which suggests that similar, real garlands also decorated the house (see Figure 7.5). Unlike the more plain, green civic garlands, these are composed of grape bunches and leaves, laurel leaves, quince fruits (*Cydonia oblonga*), pine cones and pine sprigs (possibly *Pinus pinea*), wheat heads (possibly *Triticum turgidum*), pomegranate fruits, and poppy heads, suggesting that garlands used in private contexts, or at least representations of these, might be more sumptuous than their civic cousins. The hanging of fresh, real garlands no doubt lent perfume to the more eternal sculptured and painted versions that so often decorated public and private spaces.

Similarly, representations of garlands and real garlands decorated Roman tombs, cinerary urns, and sarcophagi. In fact, at Pompeii garland hooks may be found on tomb exteriors, likely used during the *Rosalia* and the *Parentalia* when rose and violet garlands were hung in commemoration of the dead (Campbell 2015: 11; Jashemski 1963: 115). Analogous to Roman traditions, Greek tombs and funerals were similarly ornamented with garlands and wreaths. Greek funerary rites included washing the body and ornamenting it with floral wreaths and garlands. Exquisite gold leafed oak and myrtle funerary crowns have been found in Aigai (300–330 BCE), the first capital of Macedonia. Made of gold, these elite crowns eternalize the more ethereal (and likely more common) real flower wreaths buried with the dead and worn by mourners at the funeral banquet (Retief and Cilliers 2006: 54). Relatedly, a fifth-century white ground lekythos depicts a woman placing a similar wreath onto a funerary stele (Louvre MNB 3059).

Ornamentum: *Widowed, Unproductive, Dwarfed, and Abortive Trees*

Although there was a proliferation of ornamental plant use in Roman daily life, there were nonetheless critics of these new ornamental plants, critics who, like Hermione Ricketts, rejected "useless" plants in favor of "productive" gardens. The plane tree is perhaps one of the most prominently cited and criticized "ornamental" ancient plants. In describing the introduction of the plane tree to Italy, Pliny the Elder proclaims, "Who can fail to be astonished, perfectly reasonably, at the fact that a tree has been introduced from an alien world just for its shadow" (*HN* 12.6). Indeed, Pliny's surprise underscores a key Roman criticism of "unnatural" plants, namely, their un-productiveness, or as Ricketts would say, "uselessness." The problem for Pliny and others is twofold. First, the tree is useless in that it does not produce edible fruit, nor is it used for timber: the tree has no function other than providing full shade. Second, while many gardeners and their owners clearly prized the deep, cool shade of plane canopies, the fullness of the coverage disrupted traditional Roman agricultural practices. As was the practice in Near Eastern gardens, Roman gardeners utilized the understory to grow plants in dappled shade, and vines were commonly trained along tree trunks. But the full shade of plane trees, unlike elm, was too dark to support plant life underneath, earning the plane tree the epithet "*caelebs,*" which can be translated as unwed, single, widowed, or divorced (Landgren 2013: 91; Hor. *Carm.* 2.15.4). In describing the changing Roman landscape and the influx of ornamental plants, the poet Horace and the satirist Martial lament that plane trees have driven out elms, that perfumed beds of violets and myrtle are now grown in former olive orchards, and that laurels block the hot midday sun (*Carm.* 2.15.4; Mart., *Epigr.* 3.58.1–7). These new shaded, perfumed, and unproductive gardens are a clear departure from the agronomic prescriptions of Rome's founding fathers. But the criticism should not be misunderstood as a lack of interest in these new ornamental plants and gardens. In fact, the demand for new ornamental plants was so great that *topiarii*, a new kind of specialized gardener and designer, bred myrtle into new cultivars: some with broad, small and thick leaves—a feat unparalleled in the cultivation of other plants (Landgren 2004: 77). While some sources describe these new plants as widowed, single, unproductive, they were far from useless. Rather, the value of these new plants was found in their widowed and unproductive state, as "only a person of tremendous taste, vast wealth, and highly cultured insouciance would devote the most fertile productive locations to the purely aesthetic and pleasurable" (Purcell 1996: 136). Thus these new unproductive plants served as significant markers of wealth.

It was not just the planting of purely ornamental plants that was criticized, but the unnatural treatment of them as well. Pliny the Elder describes the stunting or dwarfing of plane trees, stating that, "we find in Italy some plane-trees, which are known as *chamæplatani*, in consequence of their stunted growth; for we have discovered the art of causing abortion in trees even, and hence, even in the vegetable world we shall have occasion to make mention of dwarfs, an unprepossessing subject in every case"

(*HN* 12.6). Pliny's use of *abortus*, an untimely birth, again accentuates the unnatural and unreproductive quality of these new ornamental practices. Even the term "stunted growth," *coactae brevitatis*, from *coactus*, to force, is suggestive of the idea that the tree's will or natural desire has been coerced against nature (Plin., *HN* 12.6; Landgren 2013: 91). Pliny's choice of words in describing the miniaturization as dwarfing (*pumilionum*) is equally insightful, indicating that Pliny perceives the horticultural manipulation to be a kind of abomination (Landgren 2013: 91). While Pliny tells us that dwarfing of planes was accomplished by pruning (*recidere*) and planting (*serere*), it is unclear how exactly the trees were manipulated into smaller stature. "Planting" may refer to growing plants in pots rather than directly in the ground, and "pruning" may refer to pruning of roots (Landgren 2013: 92). This type of tree pruning and miniaturization is certainly evocative of Chinese *penjing*, discussed earlier. Although there is no evidence directly linking Roman first century CE dwarfing to Chinese *penjing*, one cannot help but wonder at the horticultural coincidence, particularly as there is evidence for trade between China and the Mediterranean during this period, including trade in plants.

Dwarfing and pruning are not the only horticultural practices associated with new ornamental plants. The Greek poet Phillipus' first century BCE epigram 9.247 rather oddly features a speaking plane tree that is able to stand straight and tall even in the strongest winds thanks to having been watered with wine:

> I am a fine plane tree that the furious blasts of the south wind uprooted and laid low on the ground. But after a bath of wine I stand again erect, vivified both in summer and winter by rain sweeter than that of heaven. By death I lived, and I alone, after drinking the juice of Bacchus which makes others bend, am seen to stand straighter.
>
> (*Greek Anthology* 9.247)

The juxtaposition of a tree that, though inebriated, stands erect with an equally intoxicated but bent-over (wilted?) human is certainly comical to the point of defying belief, yet intriguingly, the practice of watering an ornamental tree with wine is echoed in numerous authors. Pliny the Elder tells us that at one point the plane tree achieved such a prestigious level, that it was watered with wine, for this was thought to be beneficial for the roots, and Martial mentions Julius Caesar's own plane tree, which grew magnificently due to regular libations of wine (Plin., *HN* 12.4; Mart., *Epigr.* 9.61.5–16). The poet Ovid mentions that planes take pleasure (*gaude*[*n*]*t*) in wine (Ov., *Rem. am.* 141). Even Roman author Macrobius, writing in the fifth century CE about a first-century BCE court case between the speaker and politician Hortensius and Cicero, describes Hortensius attempting to leave a court to rush home and water, or rather to "wine" his plane tree (Macrob., *Sat.* 3.13.3). On one level, these passages illustrate a new, performative arboriculture, whereby useless ornamentals were tended to by extravagant and ornamented practices. As such, the plane tree is doubly wasteful: nothing can be harvested from under its dense canopy, and by drinking wine the tree consumes the "useful" produce of other plants. On another level, however, these passages are evidence of ornamental plants stimulating horticultural innovations in soil science. While watering with wine has been dismissed as unnecessary and as a "display of virtuoso," it is nonetheless true that wine, grape must (freshly pressed grape juice with seeds, skins, and fruit stems), and the waste products of wine production are all rich in nitrogen, one of the main macronutrients necessary for healthy plant growth and a key nutrient in fertilizer (Bell and Henschke 2005; Moldes et al. 2007; Purcell 1996: 136). For instance, recent studies have shown that waste from wine production can be recycled successfully as fertilizer for cash crops, suggesting that we might question whether watering with wine was only a performative display of wealth, or whether it might also have been used intentionally as a fertilizer (Mulidzi 2007: 103–9).

A BUDDING FIELD: MESOAMERICAN NATURAL ORNAMENT

Looking westward to the New World, it is more challenging to definitively find evidence for specifically ornamental uses of plants that predate 500 CE. Without texts that identify and differentiate between ritual, agricultural, economic, medical, and ornamental uses of plants, we more acutely face the problem that plants likely had multiple functions. For example, in the Mediterranean basin, pomegranate trees, their flowers, and fruit, are simultaneously ornamental, bear religious connotations, possess an economic value, and were grown as food, all of which is substantiated by a wide range of literary and archaeological sources. Thus, we cannot simply look for pollen evidence of non-edible plants in New World gardens and label the pollen's sources ornamental, as the edible or medicinal plants may have also served as natural ornament. Moreover, part of the challenge in studying pre-500 CE Mesoamerican gardens stems from the fact that until recently archaeological excavations have centered on architectural spaces and gray spaces (such as paved plazas), with green open spaces receiving little to no attention (Stark 2014: 376). The lack of study does not mean that gardens, even ornamental pleasure gardens, did not exist. For example, analysis of the constructed and open spaces at Cerro de las Mesas (300–600 CE), located in Veracruz, Mexico, suggests that the five major palatial platforms have sufficient surrounding space for gardens, but these areas have not yet been explored (Stark 2014: 380–1). The elite status of these palaces suggests that we might expect to find different kinds of gardening practices there than in lower-status green areas.

Although there has been little archaeological investigation of ancient Mesoamerican green spaces, there is rich evidence for ornamental pleasure gardens, cultivation, sacred groves, and medical uses of plants primarily from the late Pre-Columbian period and after Columbus' arrival in 1492 (König 2005: 79–97). For example, Aztec palatial and temple pleasure gardens featured "ornamental, aromatic, and medical plants" maintained by a large body of laborers (Granziera 2001: 186). These elite pleasure gardens principally held plants that were non-functional and were purely ornamental, as elites received food as tribute from lower-class gardens (Granziera 2001: 188). Much like the Qin and Han Supreme Forest or Sennacherib's gardens, the Aztec gardens featured plants from the furthest regions of the empire. Furthermore, the royal gardens were not just a botanical microcosm; plants were brought by their native gardeners who ensured proper care, so that the gardens were a collection of peoples and their knowledge as much as collections of plants (Granziera 2001: 191). The sophisticated horticultural knowledge and network of gardeners could not have developed overnight, which is suggestive of a long gardening tradition. Still, pre-Aztec (i.e. pre-thirteenth century) gardens remain an open question (Evans 2000: 206–28; Stark 2014: 376).

But this does not mean that these earlier gardens are unknowable; we simply need to employ different methods to reconstruct pre-500 CE Mesoamerican horticultural practices. One such method is tracing garden knowledge linguistically. For instance, the analysis of the origin of forty-one plant names found in prehistoric Mesoamerican proto-languages spoken in Mexico and north Central America and north Central American, allows us to identify key periods in botanical knowledge development (Brown 2010). While the oldest names are dated to 5000 BCE (five plant names originating in proto-Otomanguean, a highland language), there is a clear linguistic botanical "explosion" between 2000–1000 BCE, with the number of plant names jumping suddenly to twenty-one (Brown 2010: 89). Between 1250–450 BCE another nineteen plants were deemed culturally significant enough to warrant new names (Brown 2010: 89). It has been argued convincingly that the creation and retention of plant names within a culture directly correlates to the importance that a given plant is awarded; insignificant plant names change over time while important plants retain the same name even in later, related cultures (Berlin et al. 1973). Moreover, an increased interest in plants, leading to a greater botanical vocabulary, was undoubtedly related to greater horticultural knowledge and thus more developed and sophisticated gardens. And so we might speculate whether the botanical explosion between 2000–1000 BCE also saw the development of edible and useful plants as natural ornament. If ancient Mediterranean gardens and orchards are described as at once ornamental and useful, there is no reason to doubt that maize (*Zea mays*), avocado (*Persea americana*), guava (*Psidium guajava*), pineapple (*Ananas comosus*), papaya (*Carica papaya*), beans (*Phaseolus vulgaris*), sunflowers (*Helianthus annuus*), cotton (*Gossypium hirsutum*), and other Mesoamerican plants were not equally appreciated.

In terms of other methods of examining ancient Mesoamerican gardens, there are serious preservation and methodological limitations. For example, while it is customary to take pollen samples from soil during archaeological excavations, thereby capturing and identifying local ancient plants, the results of such samples have created an incorrect portrayal of Maya forest gardening (a sustainable method of gardening in tropical forests where cultivation occurs at various elevation levels under tree canopies) (Ford 2008). Many of the native trees do not produce wind-born pollen, meaning that tree pollen does not make its way into soil. Instead, wind-born pollen of other plants, such as grasses, are found in the samples. Pollen analyses that did not take these factors into account suggested that the ancient Maya forest gardens led to massive deforestation (Ford 2008: 193).

In the face of such methodological limitations, examining contemporary gardening practices adopted for native plants may help us reconstruct those of the past. Accordingly, observing people living at Naranjal, located on top of an ancient Maya site in the Yucatan peninsula, is invaluable for reconstructing gardening practices over the last three millennia of habitation here (Fedick et al. 2008). The area is known for its lack of soil: the limestone bedrock is exposed at the surface and is pocketed all over with cavities ranging from a few centimeters to many meters deep. Upon seeing this strange pock-marked limestone landscape, the Spanish bishop Diego de Landa (1566) wrote, "Yucatan is the country with the least earth that I have seen, since all of it is one living rock and has wonderfully little earth ... And among the stones and over them they [Maya] sow and all their seeds spring up and all trees grow and some so large and beautiful that they are marvelous to see" (Tozzer 1941: 186). Similar "tree cavities" in limestone bedrock have been found at other ancient Maya sites, including Chunchucmil (Fedick et al. 2008: 302). De Landa's 400-year-old description remains accurate today: trees—even fruit trees with their deeper roots—are cultivated here (Fedick et al. 2008: 296). In the shallower cavities, multiple herbaceous plants are grown together. Based on the geological limitations of the area, it is likely that these modern techniques mirror those of the ancient Maya. Certainly, using the naturally occurring cavities as a kind of "container garden" is the only way enough cultivation could have occurred here to sustain a long-term settlement. While the initial purpose of the cavity gardens may have been functional (i.e. to produce sustenance), the juxtaposition of bare, living stone, with fertile, fruit-bearing trees, at times in bloom, at times heavy with fruit, must have also been perceived as a magnificent joining of nature and human craft, and thus as a kind of ornamentation. While the cavities have not yielded ancient Maya botanical materials due to poor preservation conditions, linguistic analysis of proto-languages in the same region has made it possible tentatively to reconstruct this landscape. Envisioning the larger, tree-sized cavities, we might find cacao (*Theobroma cacao*), hog plum (*Spondias* spp.), papaya (*Carica papaya*), pineapple (*Ananas comosus*), guava (*Psidium guajava*), and avocado (*Persea americana*).

Beyond linguistic evidence and contemporary gardening practices, ancient Maya ceramics perhaps most clearly illustrate the role plants played as natural ornament. Many Maya artifacts, dating to 250 CE–900 CE, are adorned with sculpted and painted plants, leaves, and flowers. A few plants—cacao, corn, calabashes (*Bignoniaceae*, unknown species), and water lilies (*Nymphaeaceae*, unknown variety)—feature most prominently on ceramics (Zidar and Elisens 2009: 119). Further, some vessels replicate plant shapes, such as gourds, squash, and other plants (see Figure 1.8 of this volume). For example, a group of burial urns and incense burners from Mexico, Guatemala, and Belize, thought to have been used during death and sacrifice rituals, feature prominent spikes on the body of the vessel, clearly mimicking the characteristic trunk spikes commonly found on trees in what was formerly known as the Bombacacae family (now classified as the *Bombacoideae* subfamily of Malvaceae), such as the Kapok tree (*Ceiba pentandra*). *Bombacoideae* flowers, including Guiana chestnut (*Pachira aquatica*) and Shaving brush tree (*Pseudobombax ellipticum*), are also found on both monochrome (primarily used as everyday vessels) and polychrome ceramics (used for ceremonial and ritualistic purposes) (Zidar and Elisens 2009: 125). Even Kapok tree seeds (*Ceiba pentandra*) and their silky strands are found decorating ceramics (Zidar and Elisens 2009: 128). The profusion of vegetal ornamental motifs on both everyday utilitarian wares and also on more specialized ceremonial vessels illustrates the great cultural value of *Bombacoideae* trees, but also shows that plants as natural ornament were an integral part of society across a wide spectrum of activities.

CONCLUDING REMARKS

The Cambridge Dictionary would lead us to believe that ornamental plants are primarily decorative not useful, yet in antiquity these categories were not exclusive and plants served multiple purposes. Moreover, the modern definition suggests a static perception of what is ornamental, when in fact ornamentation is deeply rooted in cultural norms, and thus means something different to each culture. Thus, when Sennacherib plants trees native of Chaldea on his Mount Amanus replica, they are ornamental not because of an intrinsic aesthetic quality, but because they commemorate his victory over Chaldea. Likewise, the *midiexiang*, brought from Rome to China, was perceived to have had an ornamental function in part because of its distant origin. Romans likely thought very differently about it. Additionally, while ancient texts help us reconstruct culturally specific conceptions of ornamentation, it is the more fragmentary evidence that prompts to us to question our definition and assumptions about ornamentation. For example, the fragmentary nature of the Mesoamerican material prompts us to reexamine the kinds of methods we employ to gather data, and how such methods, like tree-pollen analysis, may provide an incorrect assessment of ancient gardening practices. In some ways, without textual sources defining ornamentation, one cannot rule out any plant's possible ornamental function.

From the Supreme Forest to Augustus' landscaping of Rome, the chapter has primarily focused on elite examples of plants used as natural ornamentation. But in no way should this suggest that ornamentation is only the purview of elites. Rather this is merely a reflection of the material record. Less expensive materials, such as those utilized by non-elites, are also often more perishable. Thus, the funerary gold wreaths of the elites survive, while the more common, real flower wreaths have perished. Similarly, textual sources, written by elites for elites, preserve their voices at the exclusion on others. But perhaps more importantly, until recently, archaeologists have only been interested in exploring the lives and the material record of the elites: palaces, villas, and monumental structures. It is only more recently that scholars have begun to develop a methodology for studying more invisible populations. As a result, our conception of plants as natural ornament in antiquity will likely change as we begin to explore how non-elites may have defined ornamentation and what kinds of plants they may have cultivated.

Finally, while we began with Hermione Ricketts's useless lawn grass, ornamental plants are far from useless. Even Pliny's, Martial's, and Horace's unproductive, widowed, and abortive plane trees, myrtles, violets, and box, while not utilized in traditional agriculture, served a significant function. Possessing no earning potential as food, fuel, or raw material, these plants represented a level of wealth so great, that fertile, productive land could be "wasted" on non-essential cultivation. As such, these new ornamental plants helped construct their owner's social identity. Moreover, it is no coincidence that Rickett's criticism of the city ordinance echoes ancient Roman authors, as our conception of ornamental plants does in fact have roots in Roman botanical thought. While some Roman authors were critical of these new plants, Roman elites and their gardeners were clearly deeply invested in this trend, to the degree that it spurred the propagation of new, specialized ornamental varieties. It is these new ornamental varieties, the various myrtle types with differently shaped leaves, that form the basis of modern, Western ornamental plants.

CHAPTER EIGHT

The Representation of Plants

ALLISON THOMASON, JOANNA DAY, AND ANNETTE GIESECKE

THE ANCIENT NEAR EAST

The ancient Near East is home to the first experiments in plant cultivation and agriculture, beginning as early as twelve thousand years ago, so it is appropriate to begin a survey of the representation of plants in the art of this time and space. By 3500 BCE, large, densely populated urban city-states and plains fed by canals peppered the region of Mesopotamia between two rivers, the Tigris and Euphrates, in what is now southern Iraq (see Figure 8.1). As complex civilizations evolved, the Mesopotamians developed many different relationships with their vegetal environments, including farms and fields, subsistence gardens, and orchards. Notably, Mesopotamia is also the legendary home to two of the most famous botanical landscapes of the ancient world, the Garden of Eden and the Hanging Gardens of Babylon, one of the Classical Seven Wonders. These gardens were imagined by later authors as both "naturalistic and fantastic" (Luciani 2010: 100). Clearly, the rivers and their canals supported lush plant growth, whether cultivated or naturally occurring.

Plants sustained by these various environments, or landscapes, were variously represented in accordance with artistic developments and conventions coupled with cultural ideologies and pressures of patronage. For example, entire landscapes could be indicated artistically by repeating a limited number of plant specimens—often cypress and date palm—or by single exemplars standing alone, sometimes in relation to figures in narratives. Plants could be represented in abstract form, as a decorative element on pottery, luxury vessels, and jewelry, but they could also be portrayed more naturalistically, as context demanded. Plants often, or even predominantly, appeared in symbolic contexts, representing deities or divinely empowered regents.

Warka/Uruk Vase

The earliest cities in southern Mesopotamia were dominated by large temple complexes that controlled the agricultural surplus of the surrounding fields. By 3500 BCE, the elites of these cities, such as Uruk, Ur, Nippur, and Kish, recorded vegetal offerings to the gods and goddesses on clay tablets, and one of the first pictographic signs was that for "grain" (Sumerian *ŠE*). The period known for the invention of writing—the Uruk period—gave rise to "an explosion of visual imagery" on cylinder seals and other works of art, such as ceremonial vessels carved in stone (Miller et al. 2016: 56). The best known of these is the Warka/Uruk Vase (see Figure 8.2), a large (96 cm) cylindrical, footed alabaster vessel

FIGURE 8.1 Map of ancient Mesopotamian archaeological sites. Courtesy of the Oriental Institute of the University of Chicago.

excavated in the 1930s from a votive offering deposit in the temple precinct of Inana, goddess of fertility and war, at Uruk.[1]

The vase is decorated with a series of bands or registers with the narrowest at the bottom of the vase and the widest, presumably containing the most important scene, at the top. The bottom register contains a narrow, wavy line representing flowing water. The next band up portrays two alternating, schematically rendered, "long-stemmed" types of plants, one with a brush-like and the other with a trident-shaped crown. The remaining bands depict alternating rams and ewes progressing to the left; nude, shaven males carrying various baskets and vessels; and, at the top, a procession of male figures carrying various items and led by a larger male wearing a skirt and rolled cap. The males thus are to be understood as advancing around and "up" the vase to a culminating scene in which a female figure wearing an elaborate headdress stands waiting. The scenes on the vase have been variously understood as: a hierarchical ordering of the Mesopotamian cosmos and/or society (Suter 2014); an actual ritual in honor of Inana (Bahrani 2002); and a "sacred marriage" between a priest or other member of the elite and the goddess that would ensure fertility and abundance for humanity as well as domesticated plants and animals (Winter 2003). Identification of the plants on the lower register has also sparked

FIGURE 8.2 Warka Vase and reconstruction, Uruk. Photo by Jennifer Mei. Creative Commons 4.0. Courtesy of Wikicommons.

debate. Most scholars agree that the two alternating species represent two different types of grain, most likely wheat and barley, which were already agricultural staples by this time in Mesopotamia. Others argue that while the brush-like plant represented is indeed wheat or barley, the trident-shaped plants are date palm saplings (Miller et al. 2016: 3).[2] The latter interpretation has gained acceptance in recent years given the importance of date palm cultivation for its fruit as early as 4500 BCE (Miller et al. 2016: 8; Zohary et al. 2012). Understood in context with other images on the vase, these plants are widely believed to symbolize the idea of abundance provided by the goddess and ensured by the ruling elites (Winter 2006). Furthermore, these vegetal elements are perfectly at home in the "larger symbolic environment" of early image systems known from seals, script, and other works of art from the Uruk period in Mesopotamia (Miller et al. 2016: 55; Ross 2014: 301).[3]

Eventually the Mesopotamian city-states, including Uruk, were consolidated politically and militarily into kingdoms by individuals and dynasties, such as that of King Hammurabi of Babylon (1790–1750 BCE), known for his famous "Law Code." The elites of the Mesopotamian world dominated artistic and textual production, creating literature and works of art legitimizing their reigns and disseminating messages to humans and gods alike about their military power, justness, and capabilities of producing abundance. These took the form of wall paintings, sculpture in the round and in relief, architectural decoration, molded ceramics, as well luxury arts such as metal objects and jewelry.

"Investiture Painting" at the Amorite Palace of Mari

The palace at Mari up (north along) the Euphrates River from Sumer was occupied and modified over several generations by various Mesopotamian kings. From about 2000 BCE until its demise in 1760 BCE, it was the capital of the Amorites, whose network of alliances stretched far and wide in the ancient Near East. The palace's structure, decoration, and cuneiform archives demonstrate that Mari had diplomatic ties with the kingdoms of Alalakh and Qatna in Syria, Assyria and Eshnunna in northern Iraq, and Babylon in southern Iraq. Inside the building, a central courtyard served as the entrance to the official reception suite of the palace. This courtyard (court 106) was called the "Court of Palms" by the excavator, Jean-Claude Margueron, due to the presence of several fragments of artificial and potted date palm trees that had presumably decorated the courtyard.[4] One wall of the courtyard had an awning supported by wooden posts that protected a tempera fresco from the elements (von Ruden 2014: 69) (see Figure 8.3). The central scene of the so-called "Investiture Painting" depicts an ornately dressed king led in procession by deities and standing in front of a goddess, probably Ishtar (the Akkadian, Babylonian, and Assyrian equivalent of Inana), who hands him the "rod and ring," which are the traditional symbols of Mesopotamian kingship.[5] Scenes to either side contain images of hybrid, sphinx, or griffin-like creatures as well as deities, identified as such by their distinctive helmets with bulls' horns.

The elements of this enigmatic composition that are of greatest interest to plant historians are the four plants symmetrically flanking the central scene, one specimen of each distinct type on either side. Being realistically enough portrayed to enable secure

FIGURE 8.3 Drawing of "Investiture Painting," Palace of Mari, from archives of André Parrot, and Archaeological Mission of Mari/Ministry of Culture, France. Courtesy of the Mission Archéologique Française de Mari, Archives A. Parrot.

identification, the two tall trees farthest from the central scene depict mature date palms, replete with fronds and fruit, and gardeners climbing the characteristically cross-hatched trunks. The sytlized inner plants, however, are less obviously identifiable; they have tall stems with stalks terminating in fruits or open flowers arranged symmetrically at the top. They have been identified as papyrus or "papyriform" (Feldman 2006: 84) and also as immature date palms (Luciani 2010: 113). Since the excavation of the palace in the 1930s, scholars have noted their resemblance to plants known in wall paintings from elite buildings in the Aegean (Knossos, Thera), Egypt (Tell ed-Daba'), the Levant (Tell Kabri, Tall Burak), and Syria (Alalakh, Qatna) (Parrot 1937; von Ruden 2014). This observation has led to the consensus that the entire painting, with its vegetal motifs, spirals, and multi-colored frames, was part of a *koinē* (common language) of shared styles and motifs created by various craftsmen throughout the Levant and Aegean area in this time period (Feldman 2006; von Ruden 2014: 74). Regardless of precise identification, plants depicted at Mari were hybrid and polyvalent, incorporating at the same time an international aesthetic connecting Mari with points further west, and a more regional and traditional iconography of fertility and abundance drawn from southern Mesopotamia down the Euphrates River.

The "Sacred Tree": Symbolism and Symmetry

The symmetricity noted in the case of Mari becomes a common and distinctive feature of plant representation in Mesopotamia, and is thus a notable feature of the prominent "Sacred Tree" motif.[6] Images of a large-scale plant, appearing singly and with symmetrically arranged parts—and usually flanked by identical human, animal, or mythical figures—are known in Mesopotamian art from objects excavated at the Royal Cemetery at the site of Ur (*c.* 2600 BCE). The motif appears often in glyptic art of cylinder seals throughout Mesopotamian history, but reaches its peak of popularity in the Neo-Assyrian period during the first millennium BCE (Collins 2006; Giovina 2007; Porter 1993; Richardson 1999–2001). The plant is depicted in many ways, but has certain persistent traits: a single columnar stem rising vertically from the ground, water, or a mountain and having symmetric or alternating leaves or fronds, usually terminating in some type of flower or fruit (MacDonald 2002: 113). Occasionally, this plant is rendered as a short bush or shrub, but the arrangement of the numerous leaves and flowers is usually symmetric. The type of tree represented is the subject of much debate, as is the meaning and significance of the motif. On glyptic art of the third and second millennia BCE, where it is most common at that time, the motif lacks sufficient detail to identify the type of plant. However, by the first millennium with the Middle and Neo-Assyrian periods (*c.* 1200–600 BCE), the tree gains detail and is represented in portable and monumental art as a "columnar plant" with "a palmate canopy" (MacDonald 2002: 114). An early example of the sacred tree in Assyria is incised into two ivory objects, a comb and a round pyxis (lidded cylindrical box), from Tomb 45 at the royal city of Ashur (*c.* 1350 BCE). In the case of the pyxis, a date palm is associated with a seated bird, thus sharing iconography with the earlier "Investiture Painting" at Mari discussed earlier. Since the Assyrians have an Amorite origin, it is likely that this motif of bird-in-stylized-date palm was a traditional motif in northern Mesopotamian art. There are also iconographic links to glyptic and ceramic decoration from the culture of Mittanni, to the immediate west of Assyria. The highly stylized tree in Neo-Assyrian art is often associated with the chief deity, Ashur, whose winged disk hovers above the plant and flanking figures. Whether the tree can be identified with an actual species such as a date palm (the most common choice), a

FIGURE 8.4 Neo-Assyrian, *c.* 883–859 BCE. Mesopotamia, Nimrud (ancient Kalhu). Gypsum alabaster relief. New York, The Metropolitan Museum of Art. © Gift of J. Pierpont Morgan, 1917. Courtesy of The Met Museum.

pomegranate tree (Matthews et al. 2013), fir tree (Collins 2006: 103), or a "stylized lotus shoot" (MacDonald 2002: 117) remains to be determined, but more important from the Mesopotamian standpoint is the significance and meaning of the tree in relation to cosmology and royal ideology.

Owing to its biblical importance, there have been nearly as many interpretations of the sacred tree as we have depictions of it![7] The most common interpretation offered by scholars holds that the tree is a date palm, representing fertility and abundance, especially connected with Inana/Ishtar as on the Warka Vase discussed earlier (Porter 1993: 138). In the Neo-Assyrian context, the king or mythical genii hold various objects up to its fruits in a ritualized activity, indicating that they are pollinating it. The most frequently cited example with the king present is a relief from the throne room of Ashurnasirpal II's Northwest Palace at Nimrud in which the sacred tree terminates in palmettes.[8] This activity of pollination thus connects the fertility of the universe given by the god Ashur and Inana/Ishtar with the king's ability to engender and maintain it, thus being intimately entwined in Assyria with royal ideology and legitimation (Winter 2003; Collins 2006) (see Figure 8.4). When the sacred tree is combined with genie figures and placed on either side of doorways in the palaces, it can also have an apotropaic function, warding off evil and protecting the inhabitants within the rooms of the palace. Other interpretations of the meaning of the sacred tree in Assyria include the suggestion that the sacred tree is connected to the ideology of gender (Collins 2006), an ancestor cult (Richardson 1999–2001), or the sacerdotal role of the king in Assyrian cosmological thought (Ataç 2010: 126–9). As a plant form in Mesopotamia, the highly stylized sacred tree clearly held deep meaning, symbolism, and cosmological significance, connecting the gods to humans in myriad ways. Thus, the motif is a prime example of the polysemic nature of Mesopotamian artistic representations of plants.

Plants in Assyrian Palace Gardens

Less stylized plants, whose species are clearly identifiable through the details depicted by Assyrian artists, suggest a high degree of observation and appreciation of the natural environment. Royal inscriptions claim that as a part of their ideology of expansion and abundance, the Assyrian kings created botanical gardens in their capital cities at Nimrud and Nineveh from plants "captured" while on military campaign abroad (Foster 1999; Thomason 2005; Winter 2003; Wiseman 1976):

> By divine will, vines, all kinds of fruit trees, olive trees, and aromatic trees flourished greatly in (those) gardens. Cypress trees, musukkannu trees, (and) all kinds of trees [gr]ew tall and sent [out] shoots. ... I created a marsh to moderate the flow of water for (those) gardens and had a canebrake planted (in it).
>
> (Grayson and Novotny 2012: no. 138)

We are fortunate to have numerous depictions in relief of these artificially constructed installations. Although the highly propagandistic images and inscriptions of these gardens cannot be taken at face value, there is no reason to believe that royal gardens were completely fabricated ideas.[9] Rather, the royal representations in text and image are instructive regarding the symbolic and ideological importance of the gardens as well as helpful in imagining what plants might have been gathered in them.

The most famous scene depicting plants in a garden setting dates to the last years of the Assyrian period, and was found in the North Palace of Ashurbanipal at Nineveh (*c.* 650–612 BCE). In this scene, which was smaller in scale than reliefs discussed earlier, the reclining king and seated queen dine beneath a grape arbor, both attended by servants. The scene has been called the "Garden Party" owing its garden setting and has been cited as a direct or indirect influence on banqueting scenes outside of Mesopotamia, as far away as Etruria (Rathje 1994).[10] This garden contains neat rows of alternating conifers and date palms, both species that do not grow easily or naturally in the environment of Assyria (northern Iraq/Kurdistan region). The conifers were depicted in earlier Neo-Assyrian narratives within their native environment, the landscapes in the area of Syria to the west which the Assyrians had incorporated into their empire as early as the ninth century BCE (Thomason 2001).[11] In the Assyrian idiom, the conifers are shown in section, having straight vertical trunks and branches that alternate somewhat asymmetrically on either side of the trunk. The branches and needles are numerously detailed and sculpted, indicating thick luxuriousness and their ability to produce shade. Each trunk terminates with a brush-like branch showing the further vertical growth potential of the tree.

As for the date palms, they are clearly identifiable in Assyrian reliefs, with their characteristic hatched trunks, widespread shade-giving fronds, and ripe fruits. While Assyrians were certainly aware of this plant, as indicated in the sacred tree imagery discussed earlier, the natural climate and environment of Assyria was not conducive to wild date palm growth. Rather, date palm seeds and saplings had to be imported at least initially, and along with marsh reeds, were associated in texts and campaign reliefs with their native southern Mesopotamia/Babylonia. Thus, in the politically and ideologically charged setting of the royal garden scene, the conifers and date palms indexically represent their captured native lands, which were successfully incorporated into the Assyrian heartland by the kings at the behest of the gods (Thomason 2001, 2005). This appears to be the case, too, in other reliefs from the palace, among them one that depicts a hillside park containing an aqueduct, walkways, sculptures, and a variety of trees from throughout the empire: well-ordered conifers, date palms, and deciduous trees with radial branches.[12] The image likely represents the botanical gardens that Ashurbanipal and his grandfather Sennacherib installed at Nineveh near to their palaces and is described elaborately in royal inscriptions. Depictions as appear on this relief affirm that the luxurious vegetation portrayed in both natural landscapes and artificially constructed installations served as a symbol of the beneficence and abundance brought by the royal father and grandson to the Assyrian heartland, and the pictorial representations match the titulary and other textual evidence from the king's royal inscriptions.[13] Furthermore, there may be more liminal significance to the royal garden enclosures, as their descriptions match those of the sacred garden enjoyed by the Mesopotamian gods Ashur and Marduk and their consorts as they relax and engage in love-making and other pleasurable activities.[14]

That vegetal abundance ensured by the Assyrian kings was also closely aligned with fertility in the animal and human world is clearly indicated by iconographic linkage. Another Nineveh relief shows a lion and lioness resting peacefully in a luxuriant garden space (the Assyrians kept live lions in their gardens as targets for royal hunts) (see Figure 8.5). In this image, specific and individual plants stand out as starkly different from the stock conifer trees that appear in many other scenes. Here, with intricate detail, artists have carved three daisy or aster flowers, which wrestle for space with climbing grapevines; two tall, blooming lilies, which frame the lions; and a broad-leafed, flowering, low plant, most likely a mandrake/mandragora.[15] These relatively large flowers bloom between alternating straight date palm trunks and conifer trees. As in other landscape scenes, grapevines figure prominently, here weighed down by clusters of fruit and flowers whose pestles reach out to touch the leonine couple. The dual symbolism of pollination and male-female coupling to represent fertility in abundance is hard to miss here. When considered in conjunction with royal inscriptions that claim the kings are "virile lions" (Ashurnasirpal II), or that they made animals "multiply abundantly" in their gardens (Sennacherib), the connections made in the royal palatial programs to fecundity and the kings' abilities to maintain the bounty provided by the gods are obvious. The famed "Garden Party" scene, with which we began this discussion of Assyrian gardens, contains similar connotations of coupling fertility, as a bowed grapevine replete with clusters hangs over the heads of Ashurbanipal and his queen.

FIGURE 8.5 Lion and lioness in a garden. Palace of Assurbanipal, Nineveh. London, British Museum. Courtesy of Trustees of the British Museum.

Such symbols of fecundity, in this case associated with female human fertility, can be found on jewelry and other portable objects from the palaces at Nimrud. In the early 1990s excavators uncovered four richly furnished tombs, with inscriptions naming royal wives of the kings Ashurnasirpal II, Tiglath-Pileser III, and Sargon II, under the floors of Ashurnasirpal II's Northwest Palace at Nimrud. This area became a royal burial center for females for nearly two centuries. Among the finds were gold crowns and other objects festooned with hanging, grape cluster-shaped stone pendants, and inlaid plaques with fruited date palms connected to woven gold tassels ending in pomegranate-shaped beads. The tombs also included several large headdresses or crowns made with hammered gold decorations consisting of flowers with long narrow petals that alternate with pestles loaded with granulated gold "pollen," as well as one with more orthogonally-shaped rosettes. Thus, the imagery in the palatial reliefs is reinforced in the objects that accompanied royal consorts after their deaths. Such objects certainly adorned the queens' bodies in death, but also might have been worn on ceremonial occasions in life, allowing the royal wives to assert their central roles in the display of the fecundity and beneficence of the royal house (Gansell 2016; Svärd 2015).

Babylonian Hanging Gardens and Persian Paradeisos

Any discussion of plants in the cultural history of Mesopotamia would seem incomplete without a consideration of the so-called Hanging Gardens in the city of Babylon, home to the successors of the Neo-Assyrian Empire, the Neo-Babylonian Dynasty (612–539 BCE). Though ancient authors such as Berossus (third century BCE), Pliny (23–79 CE), and Strabo (c. 64 BCE–*c*. 21 CE) describe these gardens structurally as containing elaborate artifices to support and lift the vegetation as if it appeared to grow from above, thus "hanging" in the air, there is no mention of the actual plantings (Joseph., *Ap.* 1.141, *Ant.* 10.226; *HN* 19.19.49; Strab. 16.1.5). In fact, no contemporary Neo-Babylonian pictorial or literary representations of such botanical gardens exist, nor is there currently archaeological evidence of their existence (Dalley 2004, 2013).[16] Nonetheless, etymological evidence connects the Hanging Gardens to another well-known Near Eastern royal garden type, the Persian *paradeisos*. The Classical authors who mention the Hanging Gardens of Babylon continuously referred to them in Greek as *paradeisos*, a word from which our English word "paradise," with its liminal and sacred connotations, is derived. The Greeks developed their word from a more "mundane" Persian word for a royal garden and game park attached to a palace compound in the Persian capitols (Susa, Persepolis, and Pasargadae in modern Iran) (Lackenbacher 1990; Reade 2000: 200; Stronach 1990; Tuplin 1996: 81). Though traces of these have been found, their reconstruction remains hypothetical. In addition, there are very few landscape elements in Persian reliefs, and what images we do have of plants appear very schematic and geometrically arranged. The staircase and platform of the great Apadana reception complex at Persepolis contains the most large-scale plant images from the Achaemenid era.[17] In the reliefs supporting the Apadana platform, tributaries and courtiers bring baskets of pomegranates and other fruits to the Persian king, who greets them holding a single lotus flower (see Figure 8.6). Spaced regularly between the tributaries and lined up vertically along the registers are conifer trees with symmetrically arranged branches forming a dense teardrop-shaped canopy. On the staircases nearby, the same teardrop-shaped conifers spring neatly in pairs from terraced steps while below them animals fight next to rows of straight plants with plumed tops and short lateral leaves, perhaps representing lilies or lotuses. Despite their stock and schematic appearance, the plant images arranged in perfect rows leave the impression that the action in the reliefs took place in an environment full of vegetation, whether in the actual *paradeisoi* or not.

In this context, it is important to point out that the metaphorical and mythical idea of a garden paradise makes its way into the Old Testament, where in close parallels to the Assyrian inscriptions, the Garden of Eden is described as a lush space filled with "all kinds of beautiful trees" and "good fruit" (Gen. 2–3).[18] Unfortunately, the iconographic repertoire of plants from the Iron Age Levant, where the Hebrew Bible was composed, is limited in comparison to other areas of the Near East, with almost no extant large-scale narratives having vegetal elements. Most images represent single plants or repeated symbols, such as lotuses, generic rosettes, and pomegranates. These symbols can be found especially on portable works of art such as ceramic incense stands from the south or ivory furniture plaques found throughout the northern and southern regions. On the ivory figurines, the plants and fruits are most often associated with female figures and represent sexuality, love, or beauty.[19]

Egyptian Plants in Tombs and Temples

From the abundant pictorial, archaeological, and textual evidence available, it is clear that plants and gardens were of great practical and symbolic importance in ancient Egypt. In royal iconography, the papyrus plant and lotus flower, the two main vegetal symbols of ancient Egypt, represent the northern and southern regions of the Nile respectively, and they are often shown linked together at their stems to indicate the unity of the cosmos and the state. These prominent icons are known from the earliest of Egyptian royal objects, including the Narmer Palette (3200–3000 BCE), where the defeat of the northern

FIGURE 8.6 Persian King Darius I, Persepolis Apadana. Iran, Tehran, National Museum of Iran. Photo by Tuul and Bruno Morandi. Courtesy of Alamy Stock Photo.

regions by the southern king, Narmer, is represented by a Horus falcon hooking the nose of an anthropomorphized papyrus plant. These same two plants find architectural manifestation in the bundled papyrus and lotus capitals on Egyptian columns, particularly in temples of the New Kingdom (*c.* 1570–*c.* 1069 BCE), but present as early as the Third Dynasty (*c.* 2667–2648 BCE) in Zoser's pyramid complex at Saqqara. In addition, certain deities were associated with individual plant species in texts and images. For example, the goddesses Isis and Hathor had close links with the sycamore fig tree (*Ficus sycomorus*), and the date palm with Re and Min (Daines 2008; Wilkinson 1998). Osiris, the god of the underworld, was most often associated with willow (*Salix mucronata*) or tamarisk trees (*Tamarix nilotica*, *T. aphylla*, *T.* spp.) and it is generally believed that his skin is painted dark green in two-dimensional images and on statues to emphasize this arboreal connection. Individual gods' links with specific trees could be represented in various ways: their temple gardens could contain them, they could be depicted holding the tree or sitting in it, and they could represent the deities indexically or symbolically. These trees were important not only for the their ability to bring life and bear fruit, but also to provide pleasant and cooling shade in the sometimes harsh desert climate of Egypt, hence their divine associations.

As for gardens, these are well documented both in and for secular and temple contexts. Most temple complexes had artificially constructed lakes surrounded by arbors and planting beds, according to Egyptian religious texts and archaeological excavations, as for example at Amarna. Pictorial evidence for such a temple garden—albeit a unique one—can be found in a complex of rooms adjacent to Tuthmosis III's Festival Hall in the temple of Amun-Re at Karnak, Thebes. There, walls are covered with images carved in relief representing a "botanical garden" full of diverse plants and animals (Beaux 1990; Wilkinson 1998: 137). The plants are often shown as individual specimens growing from the ground line, but also appear as cuttings of fruit and flowers without stems suspended in space in a typical Egyptian style (see Figure 0.6 and the Introduction in this volume for further discussion). The garden rooms were constructed in the twenty-fifth year of Tuthmosis III's reign, upon his return from military campaign in Syria. The vegetation of that landscape must have thrilled the Egyptian king and his court, as many of the plants depicted represent species that were "exotic" and not native to Egypt, some even fantastical in appearance (Cashman 2014: 143). The same cosmological meanings of the world "out there" captured and represented in microcosm (discussed above in a Mesopotamian context) must have held for Tuthmosis III's garden, too, showing the king's ability to control and maintain the order of nature for the gods.

The numerous paintings of gardens from tombs of the elite are vibrantly rendered and "naturalistically" populated with fish, fowl, gods, and humans, both the deceased and their servants (see also Figure 7.1 and further discussion in Chapter 7 of this volume). Representations typically show a formal complex with a rectangular or T-shaped pool containing lotus (*Nymphaea lotus* or *maculata*) and rimmed with papyrus (*Cyperus papyrus*). Around the water feature are planted rows of shade and fruit trees, including tamarisks (*Tamarix aphylla*) and willows (*Salix mucronata*), sycamores (*Ficus sycomorus*), date palms, acacias (*Acacia* spp.), and perseas (*Mimusops laurifolia*); as well as flowers such as lilies (*Lilium candidum*), poppies (*Papaver rhoeas*), and cornflowers (*Centaurea depressa*) which would have offered pleasant scents and a mosaic of colors. Plants are largely shown in profile but can also—as with the water lily—be shown as viewed from above, in bird's eye "perspective." Egyptian pictorial conventions called for a certain degree of abstraction—trees, for example, were not only represented in section, but often took the shape of a rounded cone, regardless of species—yet individual species are generally nonetheless rendered realistically enough to be recognizable, except in the case of highly abstracted, hybrid floral ornament. The most complete and detailed examples of garden images, in this case in two dimensions, come from courtiers' tombs of the Eighteenth Dynasty (Nebamun for example), but earlier examples can be found in tombs from the Old Kingdom (Wilkinson 1998).

Gardens depicted in tomb contexts might have represented actual ones owned and maintained by elite Egyptians, thus marking the status and prestige of the deceased and his family in the earthly world and afterlife. Furthermore, the formal gardens depicted in tomb scenes had important symbolic value in that water features represented the sacred lake in the afterlife across which every human's soul, or *ba*, must travel to reach eternity (Wilkinson 1998). Made available in effigy, the gardens' plantings, in turn, could continue to be enjoyed by the deceased in the afterlife not only for their beauty but also for their life-giving bounty, as there were utilitarian uses for all garden plants. Texts from all periods indicate that various species provided fuel, perfume, food, and medicines, and there is no shortage of working gardens represented in tombs: grapes are harvested and wine is made, waterlilies are picked, and plants are watered, for example (Daines 2008: 17; Wilkinson 1998).

Not all plants appearing in tomb paintings were depicted in garden settings, however. Indeed, an important group of paintings show scenes of activities occurring "in nature." In tomb scenes from the Old Kingdom through to the New Kingdom, the deceased often chose to portray himself, his family members, or servants fishing and fowling in the Nile. The best example of such a scene occurs, once again, in the tomb of Nebamun, whose artists must have been from royal workshops. In one scene, the courtier stands with feet widespread on a boat attended and aided by family members and servants. The boat makes its way through a thick stand of papyrus plants with blue stems and red-tipped flowers; water lilies bow beneath it. While the stand of papyrus plants seems fairly schematic here, the artists have taken care to show the individual plants in the water under the boat, and these details, along with the fish, birds, and insects, bring vivid liveliness to the image. Taken together, the elements of the scene convey sentiments of rebirth and eternal life, symbolic associations deliberately intended for the context of the tomb.

BRONZE AGE AEGEAN PLANT REPRESENTATION

The Bronze Age civilizations of the Aegean (*c.* 3300–1200 BCE)—Minoans on Crete and Mycenaeans on mainland Greece—have left a rich but contrasting iconographic record. The natural world is prevalent in Minoan imagery, with seemingly realistic plant motifs occurring across a range of media such as frescoes, ceramics, and jewellery. Conversely, Mycenaeans favored different themes like hunting and processions, and when plants did feature, they were highly stylized. The Cycladic islands are a third important source for Bronze Age Aegean floral iconography. Of particular significance is the town of Akrotiri on the island of Thera (modern Santorini); preserved beneath debris from a volcanic eruption, it is an unparalleled source for in-situ frescoes as well as a multitude of other artifacts, many of which show botanical inspiration. Given that the vast majority of Aegean Bronze Age plant imagery is Minoan or Cycladic, the majority of examples cited here will come from Crete and Akrotiri; Mycenaean material will be referred to when relevant, but the comparative paucity of floral imagery indicates a fundamental difference between the cultures with regard to depicting plants.

In terms of chronology, the Late Bronze Age (*c.* 1600–1200 BCE) with its wide array of decorated material culture witnessed the zenith of floral imagery, although in the preceding Middle Bronze Age (*c.* 1900–1600 BCE) flowers were depicted on certain classes of artifact, such as ceramics.[20] These are the periods that the examples below are drawn from. It is possible to seek origins of this floral imagery in the Early Bronze Age (*c.* 3300–1900 BCE), although botanical motifs, like all representational art, were not so common in the third millennium BCE. It is worth noting, too, that the antecedent Neolithic cultures of the Aegean did not represent plants, or at least nothing survives of such imagery. This does not mean that the plant world was not of importance at this time, however, and palaeobotanical studies inform us about the changing dynamic between these early agricultural communities and their natural environment.

Stone Crocuses

While many of the Aegean Bronze Age images of plants are well known, this study is structured around a relatively understated floral artifact—a Minoan "blossom bowl"—a stone vessel with carved petals on the exterior (see Figure 8.7). This bowl, like most of the other *c.* 150 examples, is carved in serpentinite, a soft stone available on Crete. It dates to the Neopalatial era (*c.* 1700–1450 BCE), a time when the "Minoan palaces" flourished on the island.[21] These complex structures served as centers for administrative, religious, and craft activities, and many of the florally decorated artifacts come from them or from other sites associated with an elite class. Stone carving was one of a number of crafts that produced items for elite consumption during the Neopalatial and earlier, but in notable contrast to contemporary Egypt, there is little evidence for botanically inspired architectural motifs in Crete or the wider Aegean. Neither do the Minoan stone vessels with exquisitely carved relief scenes feature flowers, apart from one famous exception: the Zakros rhyton (Shaw 1978: figs. 7–8), a conical, chlorite vessel thought to be for pouring libations via its pierced base. On it is depicted a shrine in a rocky landscape, with goats and clumps of flowers. Although the flowers are only one minor element within a larger composition, the three petals and blade-like leaves mean they can be identified as crocuses, one of the most common flowers in Minoan art. Crocus clumps in a rocky

FIGURE 8.7 Serpentine blossom bowl. New York, The Metropolitan Museum. © The Cesnola Collection, by exchange, 1911. Courtesy of The Met Museum.

landscape occur in frescoes, too, as well as on ceramics, and may even be identifiable on some gold rings despite their small scale. In fact, crocuses were depicted in a number of different styles in Minoan and Cycladic art throughout the Middle and Late Bronze Age, ceramics in particular showing a wide variety of representations, from a simple three-petalled flower with prominent stigmas to the "crocus and festoons motif" to the rarer crocus heads that float across the vessel surface, no longer anchored to a ground line (Day 2011a: table 1).

There has been much debate about the meaning(s) of the crocus in Minoan and Cycladic society, given its predominance as a motif. Interpretations tend to link its importance to the fact that saffron can be produced from the dried stigmas of *Crocus sativus*, a triploid mutant of *Crocus cartwrightianus*, native in Greece. Saffron can be used as a dye for food, clothes, and skin, but it also has potent pharmacological properties. The plant may have been venerated in some way by the Minoans and their Cycladic neighbors, and it perhaps brought wealth to communities that produced the spice (Day 2011a). The fresco cycle in Xeste 3, Akrotiri, showing women of different ages engaged in collecting the crocus flowers and offering them to a deity may support these claims (e.g. Doumas 1992: figs. 116–30; Rehak 2002). Interestingly, while Mycenaean inscriptions list quantities of saffron, demonstrating an administrative or commercial interest in the crop (Day 2011b), the crocus does not occur in Mycenaean art, reinforcing its apparent link to Minoan culture specifically.

Exotica and Gardens

Returning to the blossom bowl, it is important to note Egyptian influence on Minoan stone-carving traditions from the Early Bronze Age, when Egyptian vessels were first imported and local Cretan imitations were produced also (Bevan 2007). In fact, stone vessels are only one of a number of artifact types indicating cultural contact between the two civilizations throughout the Bronze Age, and this adoption and adaptation of Egyptian material culture and iconography extends to include plant imagery. A number of questions lie at the heart of this transfer of botanical motifs. Were actual Egyptian plants grown in the Aegean, or was the imagery brought by traveling artists? Were they local adaptations of foreign motifs, and indeed, what meanings could have been associated with such exotic plants in their new home?

The so-called "Nilotic frieze" from room 5 of the West House in Akrotiri shows a lush riverine landscape, with papyrus plants, palm trees, a griffin, and a large feline (Doumas 1992: figs. 30–4). There has been much debate over this scene: does it represent Egypt, or Libya, or an imaginary world with mythical beasts, or a generic exotic land, thereby standing for the travel undertaken and knowledge accrued by the house's owner (Morgan 1988: 88–92; Younger 2011: 166)? One interpretation sees it as an Aegean adaptation of the Egyptian trope of fowling in a riverine setting, human protagonists replaced by real and mythical felines (Morgan 1988: 146–50, 160), while others suggest that it represents a real Cycladic landscape (Warren 1979), a claim most recently supported by palaeobotanical research (Vlachopoulos and Zorzos 2014). Papyrus features elsewhere too: in the fresco in the House of the Ladies at Akrotiri (Doumas 1992: figs. 2–5) and the Knossian Birds and Monkeys scene (see Figure 8.8), for example, as well as on seals and pottery (Betts 1978; Hood 1994: fig. 18) and terracotta burial chests, in imagery of rituals and perhaps an afterlife (Watrous 1991). It is one of the few recognizable plant motifs that also occurs in Mycenaean imagery, such as the more stylized spiral and papyrus

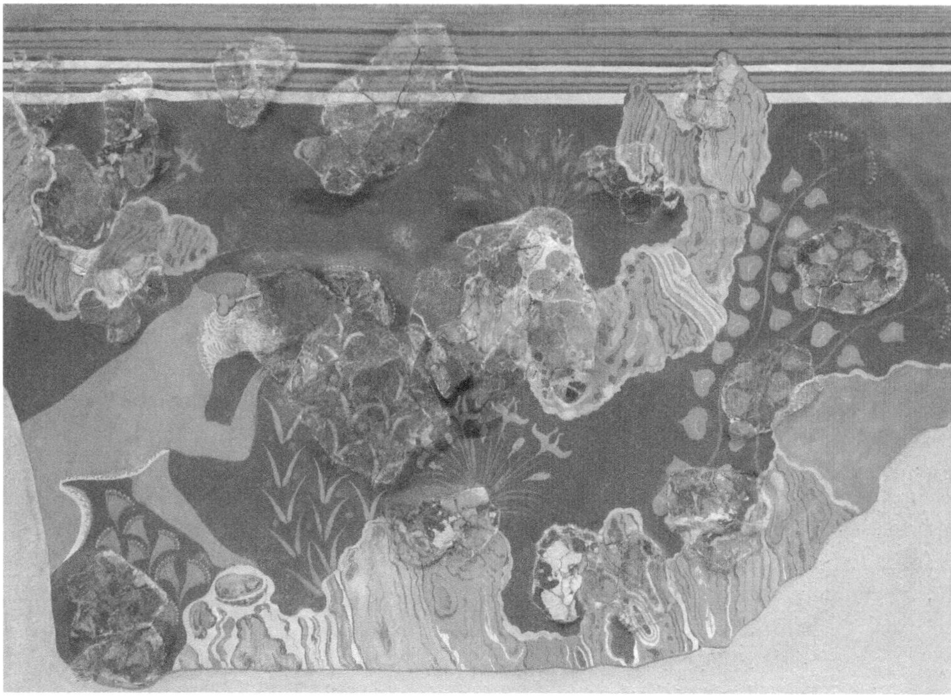

FIGURE 8.8 Fragment of the Birds and Monkeys fresco from House of the Frescoes, Knossos. Heraklion Archaeological Museum. Courtesy of Joanna Day.

friezes found at Mycenae, Tiryns, Orchomenos, Thebes, and Argos in mainland Greece (Immerwahr 1990: 142). In its most stylized Minoan version, papyrus was combined with a lily to create the so-called "waz-lily" (Evans 1921–35: II:776). Yet *Cyperus papyrus* is not native to the Aegean, although it can and does grow there; it may have been introduced from Egypt already by the Late Bronze Age, cultivated as an ornamental specimen plant rather than on a large scale for its various useful products (Morgan 1988: 21–4, 149–50, 160; Warren 1976).

If Aegeans were growing papyrus ornamentally, does this then indicate the presence of Bronze Age pleasure gardens (as opposed to kitchen gardens that provided comestibles)? Some scholars have suggested that various other frescoes apart from the Nilotic frieze (e.g. Amnisos, and the Birds and Monkeys from House of the Frescoes, Knossos [Shaw 1993]; see Figure 8.8) may represent gardens because of the unrealistic mixture of plant species and the apparent use of flower pots, as well as the inclusion of "pet" monkeys. If these cultures were similar to their eastern Mediterranean neighbors, we might expect to find gardens planted with foreign species as demonstrations of rulers' power as well as for religious purposes (Day 2010). There are no incontrovertible archaeological remains of gardens associated with any Minoan or Mycenaean palace, however, and the potential garden imagery most likely depicts imagined or symbolic landscapes rather than reality (Day 2006).

It should also be noted that iconographic transfer was not all one way, and Aegean-style fresco imagery and technique has been uncovered at Tel Kabri in Israel, Alalakh in Syria, and Avaris in the Nile Delta of Egypt. Depictions of reeds/grasses at Alalakh, chains of crocuses and irises at Tel Kabri, and myrtle, reed, and ivy from Avaris show that botanical motifs were amongst the Aegean material consumed by eastern Mediterranean elites in the Middle and Late Bronze Age. As in the case of the blossom bowls, scholarly debate has centered on whether these were made by itinerant craftsmen or locals with Aegean training, and whether Aegean people necessarily accompanied such imagery (Niemeier and Niemeier 1998).

Identification and Naturalism

Attempts to identify species of plants depicted in Aegean Bronze Age art have a long history because, at first glance at least, many artifacts do appear to depict real-world species. The crocus is one such plant, and the lily is also easily recognizable. This ease of identification has spawned a general belief that Minoan plant imagery is predominantly naturalistic in contrast with highly stylized Mycenaean representations, yet the first comprehensive botanical study recognized four main categories to organize the material: certain; probable; uncertain; unidentifiable (Möbius 1933).[22] More recent scholarship has seen a spectrum of naturalism that moves from near naturalistic to essentialist (Warren 2000) and derivation from earlier geometric and abstract shapes (Walberg 1987).

Blossom bowls neatly exemplify this identity crisis, having been identified by scholars as representing either crocuses (Bevan 2007: 131), lilies (Baumann 1993: 189), or poppies (Nugent 2012: 593), although the vessels all share the same morphology. A better-known example of conflicting identification is related to the Spring Fresco from Delta 2 in Akrotiri (Doumas 1992: figs. 66–71). Here lilies are painted in a rocky landscape, but scholars are torn over whether the red *Lilium chalcedonicum* or the white *Lilium candidum* is shown: the painted flowers combine the recurved petals of *chalcedonicum* with the upright flower heads of *candidum* (Rackham 1978). A similar debate occurred over the plants from room 1 of the House of Ladies, Akrotiri, with scholars unable to agree whether they are sea lilies (*Pancratium maritimum*) or papyrus (*Cyperus papyrus*) (Doumas 1992: figs. 1–5; Porter 2000; Warren 1976). The term "hybridization" is often used in relation to such confusing plants, suggesting that characteristics of several species were deliberately combined by painters into a novel form. Even if some have preferred to see these artistic hybrids as mistakes rather than deliberate (Rackham 1978: 756), or simply relating to the color of the background wall, i.e. painting a white flower on a white wall would not work (Warren 2000: 373), it appears that a faithful copy of a real plant was not necessarily the goal of floral imagery. Moreover, looking closely at some of the landscape scenes reveals that the various flowers shown in bloom simultaneously do not correspond with any actual growing season and instead may reflect a more timeless, idealized landscape (Chapin 2004: 57–8)—although "seasonality," like "species," may be another modern concept unthinkingly foisted upon a Bronze Age society.[23] It is worth considering, therefore, why imaginary flowers and floral landscapes might have been a desirable goal for artists and patrons.

Fragrance and Multisensory Experience

Returning to the blossom bowl, how might its floral affordances have been desired and exploited in Late Minoan Crete? Flowers are inherently multisensory—colorful, often scented, sometimes edible, soft, and moist to touch—as well as relatively ephemeral, blooming for a short time, wilting, dying and perhaps being reborn year upon year. Floral artifacts of all materials bring to mind at least some of these characteristics, and thinking about their use-contexts illuminates how and why "floweriness" may have been desirable. For example, Kamares Ware pottery of the Minoan Protopalatial era (*c.* 1900–1700 BCE) is decorated with complex motifs, many botanical in inspiration. The black cups, jugs, and bowls feature swirling, repeated patterns in white, orange, and red, making this a very striking-looking tableware (see Figure 8.9). It is well-nigh impossible to recognize species amidst this profusion of florally inspired imagery, but this probably didn't matter. These vessels, found mainly in palatial contexts, would have been used at elite feasts and would have contributed to the sensory experiences—eating, drinking, tasting, smelling, seeing, touching, talking, singing, listening, intoxication—that mediated social interactions and structures on the island at this time (Day 2013; Hamilakis 2013: 161–90). Floral decoration not only enhanced some of these physical experiences, but may have operated on a metaphorical level also, bringing imaginary scents to an already sensuous event. Blossom bowls could be interpreted as similarly synaesthetic objects, the fragrance of their contents (thought to be spices or unguents) enhanced by the metaphorical smell exuded by the calyx-shaped vessels (Day 2013).

A similar approach can be used to comprehend Minoan palatial floral frescoes, such as the Birds and Monkeys in the House of the Frescoes at Knossos, the Floral Fresco from the Unexplored Mansion at Knossos, the Akrotiri Spring Fresco, the Amnisos lilies, and Ayia Triadha room 14. Their appearance on the walls of small rooms with restricted access suggests that the paintings were viewed by a privileged few. Usually interpreted

FIGURE 8.9 Kamares Ware pottery, details of decorations by Sir Arthur John Evans, *The Palace of Minos at Knossos*, 1921 edition. © DEA Picture Library. Courtesy of Getty Images.

as having a ritual significance linked to nature and fecundity, they may have served as a backdrop for religious activities that occurred in these rooms (e.g. Blakolmer 2014; Marinatos 1984) or outdoors, underlining the link between the elite and the divine world (Chapin 2004). They might also serve as a constant "fragrant" offering to a deity, their blossoms permanently open and even imparting an imagined scent to whatever activities happened in these rooms. Indeed, fragrance is a key element in ritual, often functioning as a signal of the presence of deities or used as a medium for communication between human and non-human realms, e.g. incense (Howes 1987). Explicitly floral links to scent in a ritual context can be extrapolated from the seal stone from Myrsinochori in Messenia that shows a female smelling a flower (CMS I.279), and the bunches of flowers held by women processing in the frescoes from Thebes, Pylos (Immerwahr 1990: 115–18, fig. 32e, pl. 57) and Akrotiri (Vlachopoulos 2014: pls. LXI, LXIV).

Plants and Ritual

As can be gleaned from above, ritual was deeply embedded in Aegean Bronze Age life, and, at least for the Minoans, the natural world played an important role in practices and beliefs.

Mountain tops and caves were the main venues for ritual activities, as attested by offerings pushed into crevices or pools and the ceramic and faunal remains of feasts and/or sacrifices. One example of such a shrine embedded in the natural world has already been mentioned above: the Zakros rhyton, which perhaps shows a peak sanctuary. Gold rings have been a rich field for studies of Minoan religion and its links to nature, their intricate engraving often depicting humans interacting directly with trees and plants, for example a ring from the Isopata tomb, near Knossos, showing four females (perhaps dancing) amidst clumps of flowers. Other rings or sealings (e.g. CMS I.126; I.219; II.3.114; XII.264) show figures pulling branches of trees downwards, an action that may have led to an epiphany of a divinity. Trees also occur on boats (e.g. CMS II.3.252; VI.280; VSIA.55), a motif suggested to represent the transport of sacred trees to built shrines, which may be what trees pictured in boxes were intended to depict (e.g. CMS I.119; I.126; II.3.326; see also the ivory pyxis from Mochlos [Soles 2016: pls. LXXXI–LXXXII]). In most cases, tree species are difficult to identify in glyptic art, although some are thought to be olives, palms, or figs (Moody forthcoming). The challenge has been interpreting this iconography. Who or what was being worshipped by the Minoans? A Great Goddess whose domain was the fecundity of the natural world was first proposed early in the twentieth century and is still a popular interpretation (Goodison and Morris 1998: 113). More recent work has shied away from identifying a specific deity, preferring to see the Minoans entangled in a relationship with features of the natural world, giving agency to sacred rocks, trees, and places (e.g. Day 2012; Herva 2006). This would go some way towards explaining the apparent involvement of aspects of the natural world, including plants, in diverse rituals, and hence their prominence in art, too.

PLANT REPRESENTATION IN THE GRECO-ROMAN WORLD: FROM THE GREEK DARK AGE TO LATE ANTIQUITY

Geometric Greece and the Eclipse of Botanical Representation

The end of the Greek Bronze Age and its artistic celebration—particularly on Crete and the Cyclades—of plants and landscape, was likely precipitated by a convergence of environmental and political factors that may have included drought, earthquakes, social instability, and civil war. Massive upheaval attended by abandonment of settlements and displacement of populations extended from mainland Greece to the Cyclades, Crete, and the Near East. By about 1050 BCE, many Mycenaean palace centers had been destroyed and abandoned. The disappearance of palace culture necessarily entailed loss of support for luxury arts such as monumental architecture, sculpture, and painting. Travel and trade became limited, and the art of writing was lost. The period from the eleventh to the eighth century BCE, the Dark or Iron Age, is marked by a material culture consisting necessarily of utilitarian objects. As, over time, conditions became more settled and populations grew more prosperous, the aesthetics of such objects could again become a concern. In the case of ceramic vessels, which constitute the bulk of Greece's archaeological remains, a basic repertoire of triangles, circles, and wavy lines evolved into geometric ornament applied to increasingly well-proportioned vessels in such a way as to emphasize the structural components (neck, shoulder, belly, and foot) of the vessels themselves. The onset of the Geometric period, around 900 BCE, was marked by a pronounced shift from circular to rectilinear ornament in vase painting and a tendency to cover the entire surface with bands of repetitive ornaments including meanders, zigzags, triangles, chevrons, checkerboards, and dots. Mimesis of natural forms was not a driving impulse, and if any of these patterns were meant to represent plants, they were abstracted beyond recognition (Boardman 1998a: 7–141; Giesecke 2007: 59–60). In fact, a century would pass before identifiable figures drawn from the natural world began to appear with some regularity: horses, deer or antelope, waterfowl, and ultimately humans. A few isolated plants also appeared, especially notable because of their rarity. For example, two deer rear up heraldically on either side of a tree on a hydria from Chalcis (Coldstream 1977: 193, fig. 61b).

Interestingly, this motif, the so-called "Sacred Tree," is not only distinctive but also distinctly non-Greek, hearkening to the Near East and thus foreshadowing the renewal of active cultural influence or confluences and exchange. As contact with the Near East and Egypt increased, a new painting style developed under the influence of imported objects, likely ivories, textiles, and other portable items (Boardman 1980: 77–109; Payne 1971: 7–66; Rassmussen and Spivey 2005: 60–73). In the new Orientalizing style of the seventh century BCE rectilinear patterns gave way to more fluid Eastern-inspired ornaments such as intertwined cables, running spirals, rosettes, lotuses, and palmettes, filling vacant space and forming the boundaries of registers enclosing the animals—now among them panthers, lions, dogs, goats, roosters, and boars—ambling rhythmically around the pots they embellish. Such decoration, and its tapestry-like effect, was perfected in Corinth. Heroes from the realm of myth would also make an appearance on pots of this period under Eastern influence, and it was their deeds, as well as the deeds of ordinary men and women, that ultimately prevailed over the "animal style" as decoration favored by Greek pot painters and their patrons, with the finest wares coming from Athens. Importantly, even then, however, Eastern floral decoration persisted, the lotus being a major element.

Botanical Motifs and Personal Adornment in Archaic Greece: Phrasikleia and the Lotus

> I am the grave marker of Phrasikleia. I will always be called "*korē*" (maiden), since this is the fate allotted me by the gods in place of marriage.
>
> (IG I^3 1261)

FIGURE 8.10 Statue of a maiden, Parian Marble, found in Merenda, Attica. Grave Marker of Phrasikleia. Sculptor, Aristion from Paros, 550–540 BCE. Athens, Greece, National Archaeological Museum. Photo by Kisler Creations. Courtesy of Alamy Stock Photo.

The statue known as Phrasikleia (*c.* 550/30 BCE) was discovered on May 18, 1972, at the town of Merenda in a region outside of Athens where olives, figs, and grapes are still grown (see Figure 8.10). The agricultural land from which she emerged had already for some years been undergoing periodic excavation, as deep ploughing had unearthed rich funerary finds establishing the area as an ancient cemetery containing burials from the Geometric through the Classical periods. Phrasikleia was not found alone; the pit enclosing her also contained the statue of a young male, a *kouros*, affectionately dubbed her brother. Placed so as to face each other, both were largely intact, together constituting an extraordinary find (Kakavoyiannis 2007). Possibly carved slightly later, the *kouros* lacked an identifying inscription. The *korē*, meanwhile, was identifiable, as her inscribed base, re-used as a column capital for the Byzantine church of the Panaghia about 200 meters (650 ft) south of the burial pit, had been found and published at an earlier date (see quote above).

Phrasikleia, if we are right to call her this, is clad in an ornately patterned gown still bearing significant traces of the vibrant original colors that animated her marble form. The garment, a *peplos*, is purple-red, suggesting that it was dyed with the precious pigment derived from the murex shell, and it is decorated with floral rosettes, swastika-squares, and bands of meander. Her elaborate gown is suitable for a memorial to a young woman of elite status, as is the jewelry with which she is adorned: a crown or headdress, necklace, and earrings, all based on vegetal forms. A string of tiny pomegranates paired with *aryballoi* (perfume containers) adorn her neck, but stylized lotus is her jewelry's dominant motif. Closed lotus buds alternate with buds beginning to flower on her headdress, and closed buds, mirroring the single bud she holds in her left hand, dangle from hoops that pierce her ears. That Phrasikleia is so heavily adorned with flowers is in itself fully appropriate, since, as Greek sources attest, flowers and young, unmarried women are closely associated in Greek thought, both in life and in death (Stieber 2004: 145–78). The picking of flowers as a metaphor for maidens' transition to marriage and loss of virginity may be of particular relevance here—though Phrasikleia, like Demeter's daughter Persephone, would be wed to Hades, god of the Underworld, her wedding crown doubling as a burial crown.

What is most striking about Phrasikleia's jewelry, then, is not so much its vegetal character; indeed, leaf, flower, bud, and seedpod forms have a long history in Greek jewelry especially in later periods (Alexander 1941; Higgins 1980; Platz-Horster and Vickers 2002). Rather, it is the predominance of the lotus, likely *Nymphaea caerulea* or *Nymphaea lotus*, a plant that is associated not with Greece, where it did not grow natively, but with Egypt. Dominance of the lotus motif is underscored by the garment's rosettes, which plausibly have been identified as stylized or abstracted lotus seedpods or lotus flowers shown as viewed from above (Riegl 1992: 56). Lotus ornament was already of great antiquity when Phrasikleia was created, and its origins, like the flower upon which it is based, are to be sought in Egypt, whence it was introduced to the ancient Near East, Greece, Etruria, and the Greek colonies of southern Italy (Rawson 1984: 199–222). In Egypt, representations of lotus flowers were ubiquitous, appearing in artistic media ranging from tomb paintings, columns, furniture, and musical instruments to smaller luxury items such as lamps, ivories, and jewelry (Hepper 2009: 11–16). This is not surprising, given the flower's importance as a symbol of upper (southern) Egypt, of the sun, and—as its petals close at night and reopen in the morning—by extension, of the cycle of death and resurgent life. The Egyptian lotus's use in a Greek funerary context would imply that the flower's Egyptian symbolism as well as its form had been assimilated into Greek culture.

While Phrasikleia is a unique creation, unparalleled in her particulars by any extant find, she nonetheless serves as a fitting introduction to the representation of plants in Greece from the Dark or Iron Age through the Classical period more generally. In the sphere of representation, the eclipse of landscape by the human form, together with the reduction of plants to ornament and frame for human action (and human interaction with the divine), would characterize this period. At the same time, the cultural importance of plants was in no way diminished. Rather, as will be shown below, this shift was accompanied by a recalibration of the relation between humans and their natural environment, the essential force and mystery of the plant world still acknowledged and powerfully divinized.

Classical Greece and the Crystallization of Botanical Ornament

Since Athenian wares came to dominate the Greek pottery market in terms of quality and quantity produced, it is meaningful to examine an Athenian vase and its botanical ornament in some detail. A vessel that demonstrates an interesting range of types of vegetal motifs is a mixing bowl for wine and water (*krater*) whose paintwork is attributed to the so-called Painter of the Wooly Satyrs and is currently housed in the Metropolitan Museum, New York (see Figures 8.11a and b). The *krater*, which had been broken into many pieces, is thought to have been found in Italy and is dated roughly to the middle of the fifth century BCE (Richter 1936: 126–9). The exquisite figured scenes and ornaments that cover most of its surface have been painted in the Red Figure Style. In this style,

FIGURE 8.11a and b Terracotta volute-*krater* (bowl for mixing wine and water), *c.* 450 BCE. Attributed to the Painter of the Woolly Satyrs (name piece). Around the body: Amazonomachy (battle between Greeks and Amazons). On the neck (front view), battle of Centaurs and Lapiths; (back view), youths and women. *c.* 450 BCE. Greek, Attic. © Rogers Fund, 1907. Courtesy of The Metropolitan Museum of Art.

decorative schemes were outlined in paint or through incision, and the area outside the outline was filled in with black, leaving the figures red, the original color of the clay. Details were painted rather than incised, as they had been in the earlier Black Figure Style (the reverse of Red Figure, with decoration painted in black silhouette on the clay's red background surface) thus allowing more flexibility in rendering the human form, movements, and expressions. The larger picture field on the vessel's belly is decorated with a single scene likely inspired by contemporary large-scale painting: the legendary battle between the Greeks and the Amazons, a tribe of female warriors by some accounts from what is now central northern Turkey (Hdt. 4.110.1; Strab. 12.8.6). The fact that the figures have been placed at different levels on the uneven ground and, through use of foreshortening and three-quarter views, appear to move freely in space lends the composition a depth that establishes a tension with the vessel's essential solidity. The scene is clearly dominated by the human form, but plants do appear, assuming a subtle but important supporting role as they do on the majority of Greek pots (Hurwit 1991: 53). A diminutive tree and shrub as well as a few flowering plants serve to establish an outdoor setting. Here botanical accuracy, well within the painter's grasp, was not elected. A recognizable palm, for example, might have suggested the agency of Apollo and set scene on Delos, the island of Apollo's birth (Miller 1979). An olive could have intimated the patronage of Athena or possibly indicated the setting as Athens, where the olive first appeared (Giesecke 2014: 76–80). In this scene, the landscape is not a thematically or metaphorically loaded one, however.

The foliate wreaths that appear as the Greek warriors' shield devices and adorn several of their helmets are a different story. In this case botanical identification can at least be attempted, as there are a finite number of plants with lanceolate leaves (much longer than wide, with the widest part lower than the middle and a pointed apex) that were commonly used for wreaths. The most important of these are laurel (*Laurus nobilis*), olive (*Olea europaea*), and myrtle (*Myrtus communis*). In many instances it is possible to make a positive identification; for example, a *krater* in Agrigento (4688) attributed to the Kleophon Painter depicts a scene of sacrifice to Apollo who, seated in his temple, looks on. Those making sacrifice wear wreaths of myrtle and Apollo, appropriately, wears a wreath of laurel, his sacred plant, and holds a laurel staff too (Kunze-Götte 2006: 40–1 and fig. 18a–c). In this scene, the myrtle leaves are shorter, less elongated, and more pointed; the laurel leaves, by contrast, are longer, more widely spaced, and alternate.

Both wreaths of myrtle and laurel were polyvalent, appropriate as marks of distinction and signifiers of a variety of ritual contexts (see also Chapter 6 of this volume for further discussion). However, myrtle, fragrant and sacred to both Aphrodite and Hera, had special associations with fertility and marriage, thus seeming less appropriate for use in the context of the New York vase's battle. Laurel, meanwhile, would signify the presence of Apollo, who as god of reason, enlightenment, and high culture would presumably support the efforts of the Greeks in their effort against the "barbarian" and, by definition, uncouth foe. In her publication of the Painter of the Wooly Satyrs' vase, Gisela Richter reasonably identified the wreaths as laurel. A case might also be made for the wreaths to be made of olive, suggesting the tutelage of Athena, appropriate enough if the Greeks, as is most likely, are Athenians. Interestingly, laurel (or olive) also appears on the ornately patterned garment of what must be the Amazon queen Antiope (or Hippolyte). The foliate pattern on her *chiton* (tunic) belies her presumed barbarism and elevates her to heroic status, making her a formidable adversary for Theseus, the Athenian king. Whether appearing as wreaths or as decorative motifs on garments, the plants depicted are cultural symbols that mediate or define interaction between humans as well as human relations with the divine.

Utilitarian objects crafted from plant sources, in this case the wood of trees, appear on this vase alongside plants used as culture symbols; among them are spears, arrows, staffs, columns, furniture, and chariot components. In these instances, too, the plant world has been appropriated for human use. Meanwhile, heavily stylized plant forms constitute the chief form of ornament on the *krater*: on its lip, neck, handles, and shoulders. Ivy shoots (*Hedera helix*), easily recognizable as such, run along each curving handle. Ivy is a fitting choice on a symbolic as well as design level since, together with the grapevine, it is closely linked to Dionysus, god of wine (Giesecke 2014: 66–72). The vessel is, after all, a mixing bowl for wine and water. The tongue pattern on lip and shoulder and the scrolling palmette frieze are more challenging to decode botanically, but all may well be derived directly from the lotus, which brings us back to Phrasikleia and her exotic lotus accoutrements. As remarked earlier, lotus ornament is Egyptian in origin, and the alternation of open flower with bud had been firmly established as a convention for depicting the lotus, whether in ornamental borders or gathered in bouquets, by the period of the New Kingdom (sixteenth century BCE). Egyptian artists also produced a variation of the lotus and bud border in which other plants or "pseudo-plants"—that have come to be identified as palmettes—were substituted for the buds (Rawson 1984: 204). This variation enjoyed great popularity not only in Egypt but, through a series of political and commercial encounters, also in the Near East, particularly Assyria, where it was further developed, with the palmette finding a special, additional application as a key structural element of the Assyrian Sacred Tree. Presumably via small luxury items from Phoenicia and the Syrian coast, lotus and palmette patterns found their way in the Archaic period into the repertoire of Greek pot painters both in eastern Greece—Rhodes and Melos, for example—and on the mainland, where they became a ubiquitous and continuously evolving decorative feature (Boardman 1980: 77–84; Hamlin 1919; Rawson 1984: 149–222; Rykwert 1994).

Palmette and Pausian Scroll: Transfigurations and the Rise of the Roman Empire

Throughout the Greek world the versatile palmette ornament found application in a variety of contexts and media, for example as an architectural ornament used for friezes, temple columns (hence, ultimately, the Ionic capital's scroll), and akroteria for roofs as well as on finials for grave stones. The lotus-palmette was exported west to Italy, and it was re-exported east, traveling along the Silk Road(s) to be adopted and further modified in Achaemenid Persia and, later, the Hellenistic kingdoms as well as India, Central Asia, and China. As on the Painter of the Wooly Satyrs' *krater*, a particular transformation of the lotus-palmette ornament saw the near-complete repression of the lotus (with only the slightest trace remaining) and its domination by palmettes seated on scrolling, interlocking base "stems." Obfuscation of its original symbolism likely attended the lotus's diminution, but it would be wrong to assume that the motif had no importance. At a minimum, the *krater's* ornate, carefully painted (lotus and) palmette band established the vessel as a luxury item and articulated its well-proportioned body, reaffirming the solidity conceptually challenged by the spatial depth of the Amazon scenes.

Its original, Egypto-Assyrian vegetal forms and symbolism forgotten, lotus-palmette ornament could be and was re-naturalized—botanically reinvented—in Greek hands. Around the middle of the fifth century BCE, it underwent a process of "acanthusization," acanthus leaves being added as palmette bases and at junctures of stems or "tendrils," into which the increasingly stylized, elongated stems of these pseudo-plants evolved (Jacobsthal 1925). It is in this context that the Corinthian capital, distinguished by its layers of unfurling acanthus leaves, emerged, contrary to first-century BCE Roman architect Vitruvius' very specific and very much later account (*De arch*. 4.1.9–10) of its origins: inspiration of the Corinthian sculptor Kallimachos by a basket of funeral offerings that had become overgrown with and enveloped by acanthus leaves (Ebeling 1924; Scahill 2009).

Acanthusized floral scrolls used as ornament burgeoned and multiplied, intermingling with grape vine and sprouting new varieties flowers. These floral scrolls achieved particularly exuberant expression on the necks and sides of so-called Ornate Style

FIGURE 8.12 Marble relief portraying a farmer on his way to market. First century CE. Munich, Staatliche Antikensammlung. Courtesy of De Agostini Picture Library, Bridgeman Images.

pottery of Apulia from the middle of the fourth century BCE. Quite plausibly an organic development from the lotus-palmette, the Apulian floral ornament, often "sprouting" distinctive female heads identified in one instance as Aura (a fecund breeze), became associated with the name of Pausias as its "inventor"; the master-painter Pausias of Sicyon was said to have fallen in love with Glycera, a young woman attributed with being the first to have plaited chaplets of flowers that he imitated in paint (Plin. *HN* 35.135). Pausian scrolls, possibly of Sicyonian inspiration, would appear in the Greco-Roman world as mosaic borders, on decorative tableware and furnishings, on frescoed walls, and perhaps most importantly, as carved architectural ornament from the fourth century BCE through the Hellenistic and Roman periods.

Botanical identifications are challenging in the case of early Pausian scrollwork. Acanthus and grape vine appear regularly on Apulian vases, for example, but flower species can only very tentatively be identified as rock rose (Cistaceae), dianthus, campanula, and lily (Trendall 1989: 265). On the much later Ara Pacis (dedicated in 9 BCE), the emperor Augustus' monument to prosperous peace in the newly forged Roman Empire, botanical variety and verism, an effect enhanced by the presence of small animals and insects, increased, albeit coupled with a high degree of fantasy (Caneva 2010; Castriota 1995). These qualities were founded in a deepening interest in landscape and the natural world both in literature and the visual arts that characterized the Hellenistic period (323–31 BCE). In this period, Greek culture spread through the lands that were conquered by Alexander the Great and, after his death, were divided ultimately between four ruling dynasties: the Seleucids, governing much of the former Persian Empire; the Attalids, governing Pergamon; the Antigonids, governing Macedonia; and the Ptolemies, governing Egypt. It was a time when cities grew in size and number, the challenges of urban life occasioning both nostalgia for a rustic past and yearning for escape to the unspoiled countryside. It was also a time when ethnographic, geographic, and diverse forms of scientific knowledge were assiduously gathered and variously displayed. In the arts, which flourished under royal patronage, there was a new interest in realism, exoticism, and individualism, coupled with notions of vegetal super-abundance as a reflection of good government (Guest 2016: 151). A late first century BCE marble relief depicting a peasant driving a cow to market illustrates such tendencies (see Figure 8.12). Heavily burdened with the produce of his—or some other's—allotment, he passes by the crumbling walls of an ancient shrine built around a gnarled, almost leafless plane tree. Presumably of still greater antiquity than the encircling shrine, its limbs now press hard against and extend beyond the sanctuary's gateway. Notably, the scene is not dominated by its human actor as it would have been in the Greek Classical period; rather, setting and actor are to natural scale.

Influence of Greek Hellenistic art on that of Rome was very great. What treasures the Romans did not acquire via trade or carry off in the course of their Eastern campaigns, they both imitated faithfully and creatively re-cast, relying on the expertise of foreign, subject craftsmen as well as their own. As for the Roman Ara Pacis, its vegetal imagery can be seen as a stylistic and symbolic development of Hellenistic art from Attalid Pergamon that employed botanical accuracy (and fancy) in an imagery of State-ensured abundance. Recognizable plants on its exterior panels include laurel, rose, poppy, wheat, and oak, which feature among what may be as many as ninety "real" or imaginary specimens (Caneva 2010; Castriota 1995). Importantly, the easily identifiable plants are all connected with and symbolic of deities whose support of the Augustan regime is broadcast in Augustan monuments and literature alike: laurel and acanthus were sacred to Apollo; rose to Venus (Greek Venus); poppy and wheat to Ceres (Greek Demeter); oak to Jupiter (Greek Zeus); and grapevine to Bacchus (Greek Dionysus) (see also Figure 0.7 as well as the Introduction and Chapter 6 of this volume). Later imperial monuments likewise boasted vegetal scrollwork, but there was a general trend towards stylization and an inclination to populate the scrollwork with putti, griffins, masks, and miniaturized large animal species represented "out of scale." As for the grapevine scroll specifically, this was a favored motif on sarcophagi, since Dionysus, the god of wine, was also associated with burgeoning plant life and (re-)generation. Indeed, it was because of these associations that looping grapevines became the dominant form of vegetal scroll in late antiquity both in Italy and in the Roman provinces, though now linked with Christ in place of Dionysus (Toynbee and Ward-Perkins 1950).

Plants at Center Stage: Roman Garden Frescoes

The most striking representation of plants in the Greco-Roman world, and the most strikingly realistic, occurs in garden paintings that, judging from remains at Pompeii, Herculaneum, and other sites on the Bay of Naples preserved by the 79 CE eruption of Mt. Vesuvius, were a regular feature in the decorative schemes of Roman houses. The majority of garden paintings known to date appear on walls forming a backdrop to landscaped areas, as in the case of the House of the Marine Venus and the House of Orpheus, both at Pompeii. Some, however, adorn the walls of interior living spaces, as in of the House of Livia at Prima Porta (just outside Rome), where garden paintings adorn all four sides of a semi-subterranean dining room (for further discussion, see the Introductory chapter of this volume), and the House of the Fruit Orchard at Pompeii, where garden paintings cover the walls of two bedrooms. Among the best preserved and thus important "interior" garden paintings are those discovered in the House of the Golden Bracelet, an elaborate Pompeiian town house (see Figure 8.13). Unearthed in 1974, the frescoes covered three sides of a *diaeta* (room for entertaining) on the ground level of this three-story house. The room, entirely open on one side and located adjacent to the summer *triclinium* (dining room), looked out over both a formal garden with symmetrical planting beds and an elaborate water feature—a pergola-covered pool rimmed with twenty-eight jets of water—and, beyond that, the sea. At the same time, the garden paintings illusionistically dissolved the diaeta's walls, so that guests would have had the impression of a complete garden immersion while still being under cover. On these walls appear a profusion of plants: massed shrubs and trees that form a dense, verdant backdrop to a variety of multi-colored flowers. Among the shrubs and trees that feature here are viburnum (*Viburnum tinus*), oleander (*Nerium oleander*), palm (*Phoenix dactylifera*), plane (*Platanus orientalis*), laurel (*Laurus nobilis*), pine (*Pinus pinea*), strawberry tree (*Arbutus unedo*), and rose (*Rosa gallica*). Flowers shown growing in the foreground include violet (*Viola odorata*), vinca (*Vinca major* or *minor*), poppy (*Papaver somniferum*), Madonna lily (*Lilium candidum*), spotted rock rose (*Tuberaria guttata*), corn marigold (*Glebionis segetum*), and chamomile (variously identified as *Matricaria recutita*, *Anthemis arvensis*, *Anthemis cotula*, and *Anthemis nobile*). Among these blooms grow Solomon's seal (*Polygonatum multiflorum)* as well as the climbers giant bindweed (*Calystegia sylvatica*, or perhaps *sepium*) and variegated ivy (*Hedera helix* "Chrysophylla") (Ciarallo 1991: 29–31; 2006). That what is depicted here is no wild, unruly landscape but rather a cultivated space is clearly signaled by the plaque-supporting herms (squared stone pillars topped with a carved bust or head), scalloped

FIGURE 8.13 Garden fresco, House of the Golden Bracelet, Pompeii. First century CE.

fountains, flower stakes, and theater-mask *oscilla*, hanging ornaments that would swing in the breeze. It is indicated, too, though more subtly, by very deliberate planting rhythms, for example a single "specimen" plane tree located behind each of the fountains and the even spacing to either side of foreground oleander. Beyond a doubt, the paintings represent a garden that mimics and complements the live plantings that they faced. They do not, however, present a true mirror image of the villa's garden. Rather, they depict an idealized landscape in which all plants bloom or produce flowers simultaneously regardless of seasonal propriety; for example, viburnum, shown in bloom, produces its flowers in early spring, while the strawberry tree, here hung with berries, produces its fruit in summer to early fall.

A combination of verism and idealism—more or less "realistic" depiction of plant species coupled with an impossible alignment or coincidence of their life cycles—is typical of Roman garden painting. This "paradisical" condition could be achieved through the agency of the gods to whom the gardens' plants were linked; as already noted, different plants were viewed as sacred to, even embodiments of, certain deities, with a special role being played by Dionysus, guarantor of burgeoning plant life (see also the Introductory chapter of this volume for further discussion). It is doubtless through Dionysus' magic that this garden's statuary is curiously animate: herm heads gaze soulfully at the viewer, reclining nymphs shown on the *pinakes* (plaques) rouse themselves from slumber, and suspended theater masks stare out in surprise or horror, mouths agape. All of the statuary, incidentally, have ties with this god, being members of his usual mythical entourage: woodland-dwelling fauns, part human and part goat, and maenads, the god's "possessed" female celebrants. Himself a shape-shifter able to blur distinctions between real and unreal, Dionysus was not only god of the vintage and of vegetal life but also of the theater, which fact is important to the genesis of garden painting as a "genre." Garden paintings should logically be viewed as a development of the so-called Second or Architectural Style of Roman wall painting, which came into vogue around 100 BCE. Practitioners of the Second Style employed *trompe l'oeil* devices pioneered by Greek painters of the fifth and fourth centuries BCE illusionistically to pierce solid walls that offered "vistas" into various landscapes, among them grottoes, temple precincts, and peristyle courts, all framed by diverse architectural members. Vitruvius indicates that at least some of these Greek painterly innovations were driven by the production of scenery for the stage (*De arch*. 7. Pr. 11). Roman fresco painting, for its part, can be seen as scenery for the theater of daily life. Whether appearing on walls of grand seaside villas on the Tyrrhenian coast, of modest town houses in Pompeii, or of villas and houses throughout the Roman empire, such frescoes projected an undeniable degree of luxury and, in the case of pure garden painting proper, of a felicitous immersion in bountiful, numinous landscapes.

Plant Motifs into Late Antiquity

Among the most captivating visual creations of the Late Antique period (the period of transition from Classical antiquity to the Middle Ages, *c*. 234–mid-seventh century CE, more properly the subject matter of Volume 2 of this series) are mosaics. In earlier Hellenistic and Roman mosaics, botanical motifs had figured prominently in borders to central picture fields, taking the form of Pausian Scrolls and garlands. The stag hunt mosaic from Pella in Macedonia (late fourth century BCE; Dunbabin 1999: 14–15, figs. 12, 13) with its floral scroll and the "foyer" threshold mosaic from the House of the Faun at Pompeii (late second century BCE), which depicts a garland of flowers, fruit, and leaves punctuated by two tragic masks (late second century BCE; Naples Archaeological Museum Inv. No. 9994), are much-discussed examples. But, with the passage of time, the role of botanic motifs expanded. The Roman provinces—especially Britain, North Africa, the Iberian peninsula, Greece, Syria, Asia Minor, Palestine, and Transjordan—have yielded a wealth of Late Antique mosaics, a great many of which feature mythological or secular themes; among the latter, representations of the hunt, combats in the arena, and the vintage are recurrent, and plants are often incorporated. Now, at this later period, mosaicists were also creating floors that featured vegetal designs such as sweeps of massed small flowers

FIGURE 8.14 Figures of the Apostles with the crown of martyrdom and palm trees, detail of the early sixth-century CE mosaics of the dome, Arian Baptistery, Ravenna, Italy. © DEA/A. DAGLI ORTI. Courtesy of Getty Images.

(florets), scattered branches, and scrolling shoots of acanthus or grape vine twining in "organized" whorls over their whole expanse. These mosaics have been described as floral counterparts of the geometric carpet pattern that then was enjoying particular popularity, and the effect they produced as being of an immersive garden experience (Dauphin 1997: 14; Malek 2018: 317–40).

Mosaics laid in religious buildings constructed by Christian and Jewish communities are an interesting case of the continued migration and adaptation, in the Mediterranean region, of botanical motifs and their symbolism to accompany cultural shifts. As mentioned earlier in this chapter, grapevines are a case in point. Grapes and wine, which had been a staple of the Mediterranean artistic repertoire for millennia, came to symbolize the blood of Christ, but the vine and its branches could also represent Christ and his followers or, in Judaism, the people of Israel, as presumably on the mosaic from the synagogue at Ma'on-Nirim, Israel (Dunbabin 1999: 193, fig. 205). The date palm was another plant that saw continued representation in all manner of media, including mosaic; while remaining a symbol of peace, victory, and humanity's flourishing, it could be linked with the prosperity of Christ's followers as well as with Christ's victory over death in Christian contexts (e.g. Ps. 92: 12; Rev. 7:9) (see Figure 8.14). A final example in closing, the pomegranate, still a symbol of fertility and of renewed life, appears to either side of the bust of Christ on a mosaic pavement (mid- to late fourth century CE) from a villa discovered at the village of Hinton St Mary in Dorset, southern England, where it signifies the Resurrection (Dunbabin 1999: 95). Plants and flowers, both "realistic" and stylized, would continue to bloom eternally, and in proliferation, in the mosaics of Byzantium and the Islamic world. Indeed, botanica, so central to our spiritual and physical wellbeing, would capture artists' imaginations in all media through the ages.

NOTES

Chapter 2

1 Adapted from Shaughnessy's translation (2010: 69). For the single repeated word represented by my two words "soft, fresh," and obviously intending the girl as well as the peach, Shaughnessy has "yummy." Perhaps he's right after all.

2 Other such plants are described in Theophrastos' *History of Plants*. The peach does not appear in that text, but a description is excerpted and attributed to Theophrastos by Athenaios, *Deipnosophists* 82e (Dalby 2003: 252).

3 "The despairing prayer ... of the last women to be chosen at the dance." Adapted from Shaughnessy's translation (2010: 69). The "slant basket" in double entendre implies the vagina (see also Harper 1998: 413).

4 Sima Xiangru, "Sir Vacuous" in Wen xuan (Knechtges 1987: 58–9).

5 Zhang Heng, "Southern Capital Rhapsody" lines 123–30 in Wen xuan (Knechtges 1982: 320–3).

6 Qu Yuan, "Li sao" lines 25–29 in Chu ci (Hawkes 1985: 69). Chinese or Sichuan pepper, *Zanthoxylum* spp., is quite different from black pepper and relatives and from chilli pepper and relatives. Both will be discussed below.

7 Qu Yuan, "Li sao" line 110 in Chu ci (Hawkes 1985: 71). Zuo Si, "Shu Capital Rhapsody" line 128 in Wen xuan (Knechtges 1982: 350–1).

8 Sima Xiangru, "The Imperial Park" lines 201–12 in Wen xuan (Knechtges 1987: 90–3).

9 The lemon, *Citrus* x *limonium*, necessarily originated later because it is a hybrid of bitter orange and citron. It cannot be traced with certainty before sixth-century India. Lemons found by some, including me, in ancient Roman sources must certainly be interpreted as citrons.

10 Sima Xiangru, "Sir Vacuous" line 71 in Wen xuan (Knechtges 1987: 58–9) in the marshy Yunmeng hunting park near Yiyang in Hunan; later, growing near Nanyang, according to Zhang Heng. Others have thought that the hybridization took place in New Guinea (Denham 2004; Grivet 2004; Lebot 1999). In New Guinea and Melanesia sugar has been something approaching a staple; elsewhere a spice, and even now a luxury food.

11 Carakasaṃhitā, *Sūtrasthāna* 5.76cd–77 (Zumbroich 2007: 118). Both plants have been identified in early Chinese literature, the areca palm in Sima Xiangru's "The Imperial Park" (second century BCE) and the betel vine in Zuo Si's "Wu Capital Rhapsody," describing Nanjing in the third century CE. "The Imperial Park" line 222 in Wen xuan (Knechtges 1987: 90–3); "Wu Capital Rhapsody" line 176 in Wen xuan (Knechtges 1982: 384–5). Cloves and nutmeg, listed among aromatics in this quotation, both originate in the Moluccas in what is now eastern Indonesia. Both have been uncertainly recognized in texts of the Roman Empire, and there are persistent claims about the use of cloves in early China, but the *Carakasaṃhitā* (unfortunately impossible to date precisely) provides the earliest secure evidence that these two spices, so important later, had become commodities in the long-distance spice trade.

12 Sūpeyyapaññaṁ madhuraṁ, "sweet herbs for curry": the Cowell translation (1895: 4); Fausbøll's Pali text (1877: 99). The herbs, *sūpeyyapaññaṁ*, are literally "leaves for soup." The status of this soup as staple food is confirmed by the link between the base word, Sanskrit and Pali *supa* "soup," and its familiar compound, Sanskrit and Pali *sūpakāra* "soup-maker i.e. cook."

My thanks, as so often, to the Internet Archive. If it had not uploaded Fausbøll's edition of the Jātakaṭṭhakathā in January 2017, from an old printed copy at the Archaeological Survey of India, I would not have been able to write this note in February 2017.

13 Pliny often misreads his originals. Each of these three statements could be accepted for the banana if allowance is made for misunderstanding (the fruit grows from the crown; it is sweet but has no juice; a bunch would be enough for four). But the fruit of the Indian fig grows from the bark, the mango has very sweet juice and a single jackfruit is enough for four, so the passage could be even more confused than at first appears.

14 Winstedt (1909: 321, 351). The name *nargellia* (argellia in some manuscripts) is close to Sanskrit *nārikela* and Prakrit *nāriela* "coconut."

15 In the first century CE long pepper was exported from the port of Barygaza (modern Bharuch: Casson 1989: 199, 210) at the mouth of the same river Narmadda on which the prehistoric site of Navratoli once flourished.

16 Adapted from Lionel Casson's translation (1989: 90–3, 241–2). The practice described is surely a form of silent barter.

17 Archaeological reports of "pistachio" from prehistoric Europe and the Levant are inaccurate: "*Pistacia* sp." is what is intended, the fruits of *P. atlantica*, *P. terebinthus*, or *P. lentiscus*.

18 Linguistic evidence must be treated with caution. Historical reconstructions are sometimes abandoned or altered on the basis of new evidence. New meanings can be attached to words, and, in particular, new, preferable or more easily available foods are often given names that were previously applied to others (see note 20 for an example).

19 Brown (2010) gives precise absolute datings for the proto-languages to which he traces a series of Mexican plant names. In general the reconstructions appear reliable, the linguistic history valid, the datings less so. In the text, which is indebted to his work, the dates are presented with a level of uncertainty.

20 The tomato, *Solanum lycopersicum*, known to us by an Aztec name and much cultivated in later Mexico, originated in the central Andes and was first domesticated there in a form close to modern cherry tomatoes. It spread north through Mesoamerica, not without human help, and was domesticated a second time in Mexico around 3000 BCE. A name for it can be reconstructed to proto-Zapotecan, around 1250 BCE.

Chapter 3

1 Unless stated otherwise, all translations are mine.

2 The scientific plant names in this chapter are those given in the International Plant Names Index.

3 It is customary in references to Galenic texts to give the page in the relevant volume (here 14) and page (here 24) of Kühn's edition.

4 The story is also found in the writing of Aelian (third century CE): *De natura animalium* 1.38 and 7.39.

5 Pliny provides similar information on adulteration. This is because Pliny and Dioscorides had access to the same source(s). However, Dioscorides is usually more reliable.

Chapter 4

1 Most of the scores of documented bog bodies were naked; what limited clothing remains was mostly not plant-derived.
2 Although the word itself is Danish for "binding with needle" as a Viking textile with continuous circular looping, note *nålebinding* also existed from at least 300–500 CE in Coptic Egypt. Note the "Coppergate Sock" from Viking Yorvik in the medieval tenth century CE.
3 John Peter Wild has dozens of publications across decades of Roman textile research; note in addition to his Cambridge (1970) volume his 2002 (1–42) and 2008.
4 Hamilton Collection, GR 1772.3–7.172 (D550), Department of Greek and Roman, British Museum.
5 British Museum, from upper Thebes, # EA 6046, 38.7 cm in length, with a wooden handle.
6 British Museum, WA 124772 Nineveh South-West Palace Room VII, panel 11; Phoenician warships and freighters (Layard 1853: plates 13, 71. Wall reliefs, Nineveh, South-West Palace, Throne Room and Hall 6, slabs 14 and 53, reign of Sennacherib, 704–681 BCE).
7 Hyginus, *Fab.* 97 lists 1,154 ships (or 245); Apollodorus, *Epit.* 3.13–15 lists 1,013 ships.
8 British Museum (WA 121201), inlaid shell, red limestone, lapis lazuli, bitumen, 50.4 cm in length.
9 Assuming the Greek talent was 26 kg or 57 lbs.
10 Here Socrates speaks to Critobolus about Lysander and Cyrus (the Younger) and of "the many sweet scents which hung about them as they walked through the garden park." See also Boucharlat 2011: 557–74; Rezaeian 2012; Stronach 1978: 108; 1989: 475–502; Tuplin 1996: 80–131.
11 For example, Papyrus of Nakht, Eighteenth Dynasty (British Museum # EA 10471); Papyrus of Ani Nineteenth Dynasty (British Museum #EA 10470); Papyrus of Hunefer, Nineteenth Dynasty (British Museum # EA 9901); Greenfield Papyrus (British Museum # EA 10554), Twenty-second Dynasty; along with medical documents including the Edwin Smith Papyrus, Dynasty 16–17, *c.* 1600 BCE (New York Academy of Medicine); the Papyrus Ebers, *c.* 1550 BCE (University of Leipzig); the London Medical Papyrus, possibly Dynasty 16–17 (British Museum # EA 10059); and many others elsewhere like the mathematical Rhind Papyrus, Dynasty 16–17 (British Museum #EA 10057–8).

Chapter 6

The authors are profoundly grateful to Midori Hartman for bibliographic assistance when it was critically needed.

Chapter 8

1 Warka is the German name for the site of Uruk. The vase is referred to in scholarly literature by either name.
2 A less-accepted interpretation is that the "trident-shaped" plant represents sesame (Bahrani 2002).
3 Some similar cereal plants appear on ceramics and seal impressions from other Near Eastern sites in the Uruk period (*c.* 3500–2900 BCE), including the emerging cities in Syria on the Upper Euphrates River, and the sites in Susiana, Iran.
4 A tablet from the palace also refers to a "Court of Palms" as part of the palace (Luciani 2010: 103).
5 From the tablets in the archive, we know that several kings occupied the palace beginning with the Amorite king Shamshi-Adad I (1850 BCE) and ending with its final sacking and

abandonment by the Babylonian Hammurabi (1750 BCE). The most recent interpretation of the painting suggests that the king depicted here is Yahdun-Lim, one of the earlier Amorite occupants of the palace (Luciani 2010: 102).

6 Originally, with clear biblical undertones related to the portrayal of the sacred tree of life, which sheds its forbidden apple in the Garden of Eden (Gen. 2:21–4). For a recent discussion of (and additional bibliography about) symmetric compositions in Neo-Assyrian art of the first millennium BC, see Watanabe 2014.

7 For a complete accounting (with bibliography) of the appearance, placement, and meaning of the motif in Assyria, see Giovino (2007).

8 An example of the date palm and stylized tree from portable objects of the Neo-Assyrian period can be found on two cloisonné pendants made of gold and lapis lazuli from the Queen's tombs found in Ashurnasirpal II's Palace at Nimrud (Bahrani 2016: 254–5; Collins 2006: 100; Oates and Oates 2001: cover illus.).

9 In fact, they may have played an important religious role, serving as the setting for royal rituals that mimicked the gardens of gods known from literature, in which the gods Ashur and Marduk, and their consorts, frolicked and renewed the abundance of the universe.

10 The Assyrians might have come into contact with the iconographic motif of a seated, banqueting figure while campaigning abroad in Syro-Hittite areas. There are numerous stone reliefs from palatial, temple, and funerary contexts that show a single figure seated before a table laden with food. Here the vegetal world is represented as a single example, such as a lotus flower, held in the hand of the human or god figure, and little landscape setting is invoked pictorially.

11 In fact, royal inscriptions as well as administrative letters from the royal archives of Assyria discuss the court's oversight of transport of live conifer saplings from Syria (Parpola 1987: no. 226), intended for transplantation in the capitol city at Khorsabad.

12 Sennacherib's waterworks in the area northeast of Nineveh can be traced archaeologically at Jerwan and through CORONA satellite imagery (Dalley 2013; Reade 2000; Ur 2005). Another garden representation, known only from a drawing in the British Museum (Layard 1853: 232; Original Drawing IV, 77) as the relief is now lost, shows another garden image with trees built on terraces, aqueducts, and waterways (Dalley 2013).

13 The Assyrian court also used the symbolism of horticultural abundance in negative ways in scenes of military campaigns. On reliefs from many of the kings' reigns, vignettes showing Assyrian soldiers chopping down fruit trees neatly aligned in orchards reinforce common topoi in the royal inscriptions, which claim that the kings could literally and figuratively cut off the lifelines of enemy kings and cities as they claimed in numerous royal inscriptions and archival texts to have chopped down royal orchards of their enemies.

14 These gardens are attached not to palaces, but temples, and in particular a structure connected in Mesopotamian cosmology to the important New Year's Festival, the *bit akitu* (Pongratz-Leisten 1999).

15 Lilies and mandragoras are associated in the Hebrew Bible and Akkadian texts with fertility, sexuality, and love; both plants may be considered aphrodisiacs in the ancient world (2014b).

16 However, see Reade for arguments to the contrary (2000). There has also been a great deal of debate about the hydraulic mechanisms (Archimedes Screw) used to "lift" the water to the high terraces, but that subject is beyond the scope of this volume.

17 Tuplin 1996: 91. There are also a few orthogonal date palm trees flanking the king hunting on Persian royal seals.

18 Tuplin 1996: 81. The iconographic repertoire of plants from the Iron Age Levant, where the Hebrew Bible was first composed, is relatively limited in comparison to Mesopotamia,

with almost no extant landscape friezes. Most images represent single or stock plants, such as lotuses or rosettes on portable works of art such as ceramic incense stands or ivory furniture plaques.

19 For the latest summary of these connections, see Gansell (2014b).

20 Aegean Bronze Age chronology is complicated and absolute dates remain disputed. Those used in this article come from Dickinson (1994). Periods on Crete are referred to as "Minoan," i.e. Late Minoan (LM), those on the mainland as "Helladic," and those in the Cyclades as "Cycladic". A complementary chronological division based on the development of Minoan palaces is also in use for Crete: Prepalatial = EM I–MM IA; Protopalatial = MM IB–MM IIB; Neopalatial = MM IIIA–LM IB; Final Palatial = LM II (at Knossos only); Postpalatial = LM IIIA–LM IIIC.

21 This figure of 150 comes from Warren's (1969) survey of stone vessels, and should probably be increased to nearer 200 to allow for material excavated since that publication.

Two small limestone half-capitals from the Stepped Portico of the Central Court at Knossos represent flowers (Evans 1921–35: II: 814–15, supp. pl. XXX), and the stone slabs of the ceiling of the side room of the Mycenaean tholos tomb at Orchomenos were carved with rosettes and stylized papyrus (Hood 1994: fig. 58b), but there is little other architectonic use of plant imagery. A fresco from Akrotiri, however, may show a papyrus capital from a column of a shrine (Vlachopoulos 2007: 134, pl. 15.1, 15.3) (e.g. Crocus Panel at Knossos [Chapin and Shaw 2006]; room 14, Ayia Triadha [Jones 2007]); (e.g. the kymbai from Akrotiri [Immerwahr 1990: pl. VI]; the Anemospilia bull jar [Sakellarakis and Sapouna-Sakellarakis 1997: fig. 559]); (e.g. Sellopoulo [Popham, Catling, and Catling 1974: pl. 37a–b]; Poros [Dimopoulou and Rethemiotakis 2000]).

22 Hence debates about botanical species identification have not permeated studies of Mycenaean iconography to the same extent as they concern Minoanists.

23 Thanks to Jenny Moody for bringing this point about seasonality to my attention.

BIBLIOGRAPHY

Abdul-Nour, Hani. 2001. "Les incriptions forestières d'Hadrien. Mise au point et nouvelles découvertes." *Archaeology & History of Lebanon*, 14: 64–95.

Achaya, K. T. 1994. *Indian Food: A Historical Companion*. Delhi: Oxford University Press.

Achebe, C. 2009. *Things Fall Apart*. Toronto: Anchor Canada.

Ahmed, Mukhtar. 2014. *Ancient Pakistan: An Archaeological History*. CreateSpace.

Alcorn, J. B., et al. 2006. "Thipaak and the Origins of Maize in Northern Mesoamerica." In J. E. Staller et al. (eds.), *Histories of Maize: Multidisciplinary Approaches to the Prehistory, Linguistics, Biogeography, Domestication, and Evolution of Maize*, 600–608. Amsterdam: Academic Press.

Alexander, Christine. 1941. *Greek and Etruscan Jewelry*. New York: The Metropolitan Museum.

Alexiou, S., and W. Brice. 1972. "A Silver Pin from Mavro Spelio with an Inscription in Linear A: Her. Mus. 540." *Kadmos*, 11: 113–24.

Allevato, E., et al. 2009. "Woodland Exploitation and Roman Shipbuilding: Preliminary Data from the Shipwreck Napoli C" ("*L'utilizzo delle risorse forestali nell'industria navale romana. Primi dati dal relitto Napoli C*"). *Journal of Mediterranean Geography*, 112: 33–42.

Allred, Lance. 2013. "Diet, Ancient Near East." In Roger S. Bagnall et al. (eds.), *The Encyclopedia of Ancient History*, first edition, 2085–6. Chichester, UK: Blackwell Publishing.

Allsen, Thomas T. 2006. *The Royal Hunt in Eurasian History*. Philadelphia, PA: University of Pennsylvania Press.

Alpern, S. 1992. "The European Introduction of Crops in West Africa in Precolonial Times." *History in Africa*, 19: 13–43.

Ambrose of Milan. *Expositio evangelii secundum Lucam*. Edited by Marcus Adriaen. 1957. Corpus christianorum, series latina 14. Turnhout: Brepols.

Ambrósio Moreira, Priscila, et al. 2015. "The Domestication of Annatto (*Bixa orellana*) from *Bixa urucurana* in Amazonia." *Economic Botany*, 69(2): 1–9.

Amigues, Suzanne. 1996a. "Un cinnamone fantomatique." *Topoi Orient-Occident*, 6: 657–64.

Amigues, Suzanne. 1996b. "L'expédition d'Anaxicrate en Arabie occidentale." *Topoi Orient-Occident*, 6: 671–7.

Amigues, Suzanne. 2002. "La '*Prairie* d'Asphodèle' de l'*Odyssée* et de l'*Hymne homérique à Hermès*." *Revue de philologie, de littérature et d'histoire anciennes*, 76(1): 7–14.

Amigues, Suzanne. 2004. "Le silphium, état de la question." *Journal des Savants*, 191–226.

Amigues, Suzanne (ed. and trans.). 2006. *Théophraste. Recherches sur les plantes. Tome V. Livre IX*. Amigues. Paris: Les Belles Lettres.

Amrhein, Anastasia. 2015. "Neo-Assyrian Gardens: A Spectrum of Artificiality, Sacrality, and Accessibility." *Studies in the History of Gardens and Designed Landscapes*, 35(2): 91–114.

Anacreon. *Greek Lyric, Vol. 2: Anacreon, Anacreontea, Choral Lyric from Olympus to Alcman*. Edited and translated by David A. Campbell. 1988. Loeb Classical Library 143. Cambridge: Harvard University Press.

Anderson, J. K. [1996] 1998. "The Geometric Catalog of Ships." In Jane B. Carter and Sarah P. Morris (eds.), *The Ages of Homer: A Tribute to Emily Townsend Vermeule*, 181–191. Austin: University of Texas Press.

André, Jacques. 2010. *Les noms des plantes dans la Rome antique*. 2nd edn. Paris: Les Belles Lettres.

Andres, T. C. 2004. "Diversity in Tropical Pumpkin (*Cucurbita moschata*): Cultivar Origin and History." In A. Lebeda and H. S. Paris (eds.), *Progress in Cucurbit Genetics and Breeding Research*. Proceedings of Cucurbitaceae 2004, the 8th EUCARPIA Meeting on Cucurbit Genetics and Breeding July 12–17, 2004 Olomouc, Czech Republic: Palacký University in Olomouc Czech Republic.

Andrews, A. C. 1961. "Acclimatization of Citrus Fruits in the Mediterranean Region." *Agricultural History*, 35: 35–46.

Antisthenes. *Antisthenes of Athens: Texts, Translations, and Commentary*. Edited by Susan H. Prince. 2015. Ann Arbor, MI: University of Michigan Press.

Apollodorus. *The Library, Vol. 1: Books 1–3.9*. Edited and translated by James G. Frazer. 1921 Loeb Classical Library 121. Cambridge: Harvard University Press.

Apollonius Rhodius. *Argonautica*. Edited and translated by William H. Race. 2009. Loeb Classical Library 1. Cambridge: Harvard University Press.

Appadurai, A. 1986. "Introduction: Commodities and the Politics of Value." In A. Appadurai (ed.), *The Social Life of Things: Commodities in Cultural Perspective*, 3–64. Cambridge: Cambridge University Press.

Apuleius. *Metamorphoses (The Golden Ass), Vol. 1: Books 1–6*. Edited and translated by Arthur J. Hanson. 1966. Loeb Classical Library 44. Cambridge: Harvard University Press.

Aristophanes. *Clouds. Wasps. Peace*. Edited and translated by Jeffrey Henderson, 1988. Loeb Classical Library 488. Cambridge: Harvard University Press.

Aristophanes. *Aristophanes: Komödien, Bd. 1–4*. Edited and translated by Peter Rau. 2017. Darmstadt: Wissenschaftliche Buchgesellschaft.

Aristophanes. *Birds. Lysistrata. Women at the Thesmophoria. Aristophanes vol. III*. Edited and translated by Jeffrey Henderson. 2000. Cambridge, MA, and London: Harvard University Press.

Aristophanes. *Clouds. Wasps. Peace*. Edited and translated by Jeffrey Henderson. 1988. Loeb Classical Library 488. Cambridge: Harvard University Press.

Aristotle. *History of Animals. Books VII–X*. Edited and Translated by D. M. Balme. 1991. Cambridge, MA, and London: Harvard University Press.

Aristotle. *Minor Works: On Colours. On Things Heard. Physiognomics. On Plants. On Marvellous Things Heard. Mechanical Problems. On Indivisible Lines. The Situations and Names of Winds. On Melissus, Xenophanes, Gorgias*. Translated by W. S. Hett. 1936. Loeb Classical Library 307. Cambridge: Harvard University Press.

Arrian. *Anabasis of Alexander, Vol. II. Books 5–7. Indica*. Edited and translated by P. A. Brunt. 1983. Cambridge, MA, and London: Harvard University Press.

Arriaza, B., et al. 2016. "Microscopic Analysis of Botanical Residues from Cerro Esmeralda Burial in Northern Chile: State and Death Ritual Implications." *Interciencia*, 41(12): 844–50.

Artemidorus. *Artemidori Daldiani Onirocriticon Libri V*. Edited by Roger A. Pack. 1963. Leipzig: Teubner.

Artemidorus. *Oneirocritica: Text Translation, and Commentary*. Edited by Daniel E. Harris-McCoy. 2012. Oxford: Oxford University Press.

Atac, M. 2010. *The Mythology of Kingship in Neo-Assyrian Art*. Cambridge, UK: Cambridge University Press.

Athenaeus. *Deipnosophists: Vol. II*. Edited and translated by Charles Burton Gulick. 2002. Cambridge: Harvard University Press.

Athenaeus of Naucratis. *Athenaei Naucratitae Dipnosophistarum libri* 15, 3 vols. Edited by Georg Kaibel. 1985–92 (1887–90). Reprint, Stuttgart: Teubner.

Athenaeus of Naucratis. *The Learned Banqueters, Vol. 5: Books 10.420e–11*. Edited and translated by S. Douglas Olson. 2009. Loeb Classical Library 274. Cambridge: Harvard University Press.

Athenaeus of Naucratis. *The Learned Banqueters, Vol. 8: Books 13.594b–14*. Edited and translated by S. Douglas Olson. 2012. Loeb Classical Library 519. Cambridge: Harvard University Press.

Augustus. *Res gestae divi Augusti: Hauts faits du divin Auguste*. Edited by John Scheid, 2007. Collection Des Universités de France, Série Latine. Bd. 386. Paris: Les Belles Lettres.

Augustus. *Compendium of Roman History. Res Gestae Divi Augusti*. Edited and translated by Frederick W. Shipley. 1924. Loeb Classical Library 152. Cambridge: Harvard University Press.

Avanzini, Alessandra (ed.). 1997. *Profumi d'Arabia: Atti del convegno*. Rome: L'Erma di Bretschneider.

Avni, Raz, et al. 2017. "Wild Emmer Genome Architecture and Diversity Elucidate Wheat Evolution and Domestication." *Science*, 357(6346): 93–7.

Bahrani, Z. 2002. "The Performative Image: Narrative, Representation, and the Uruk Vase." In Erika Ehrenberg (ed.), *Studies in Honor of Donald P. Hansen*, 15–22. Winona Lake, IN: Eisenbrauns.

Bahrani, Z. 2016. *Art of Mesopotamia*. New York, NY: Thames and Hudson.

Bailey, Douglas W. 2005. *Prehistoric Figurines: Representation and Corporeality in the Neolithic*. London: Routledge.

Bakels, C., and S. Jacomet. 2003. "Access to Luxury Foods in Central Europe during the Roman Period: The Archaeobotanical Evidence." *World Archaeology*, 34(3): 542–57.

Balter, M. 2009. "Clothes Make the (Hu)Man." *Science AAAS*, 325: 1329.

Bar-Yosef, Ofer. 2011. "Climatic Fluctuations and Early Farming in West and East Asia." *Current Anthropology*, 52: S175–S193.

Barber, Elizabeth W. 1992. *Prehistoric Textiles: The Development of Cloth in the Neolithic and Bronze Ages*. Princeton: Princeton University Press.

Barber, Elizabeth W. 1994. *Women's Work: The First 20,000 Years: Women, Cloth and Society in Early Times*. New York: W.W. Norton & Company.

Barrelet, Marie-Thérèse. 1977. "Un Inventaire de Kar-Tukulti-Ninurta: Textiles Décores Assyriens et autres." *Revue d'Assyriologie et d'archéologie orientale*, 71(1): 51–92.

Bartoloni, Piero. [1988] 2000. "Ships and Navigation." In Sabatino Moscati (ed.), *The Phoenicians*, 84–91. New York: Rizzoli/London: I. B. Tauris.

Bauks, Michaela, and Martin F. Meyer (eds.). 2013. *Zur Kulturgeschichte der Botanik*. Trier: Wissenschaftlicher Verlag.

Baumann, Hellmut. 1993. *Die griechische Pflanzenwelt in Mythos, Kunst und Literatur*. 3rd edn. München: Hirmer.

Baumann, Hellmut. 1993. *The Greek Plant World in Myth, Art, and Literature*. Translated by William T. Stearn and Eldwyth R. Stearn. Oregon: Timber Press.

Baus, Karl. 1940. *Der Kranz in Antike und Christentum*. Theophaneia: Beiträge zur Religions- und Kirchengeschichte des Altertums, vol. 2. Bonn: Peter Hanstein Verlagsbuchhandlung.

Beadle, G. W. 1972. "The Mystery of Maize." *Field Museum of Natural History Bulletin*, 43 (10): 1–11.

Beaux, N. 1990. *Le cabinet de curiosités de Tutmosis III: Plants et animaux du "jardin botanique" du Karnak. Orientalia Lovaniensia* 36. Leuven: Peeters.

Beckby, Hermann. 1975. *Die griechischen Bukoliker: Theokrit, Moschos, Bion*. Beiträge zur Klassischen Philologie 49. Meisenheim am Glan: Hain.

Beek, Gus van. 1958. "Frankincense and Myrrh in Ancient South Arabia." *Journal of the American Oriental Society*, 78(3): 141–52.

Belfer-Cohen, Anna, and A. N. Goring-Morris. 2011. "Becoming Farmers: The Inside Story." *Current Anthropology*, 52: S209–S220.

Bell, Sally-Jean, and Paul A. Henschke. 2005. "Implications of Nitrogen Nutrition for Grapes, Fermentation, and Wine." *Australian Journal of Grape and Wine Research*, 11(3): 242–95.

Benz, Bruce. 2001. "Archaeological Evidence of Teosinte Domestication from Guilá Naquitz, Mexico." *Proceedings of the National Academy of Sciences*, 98: 2104–6.

Benz, Bruce F. 2006. "Maize in the Americas." In John E. Staller et al. (eds.), *Histories of Maize: Multidisciplinary Approaches to the Prehistory, Linguistics, Biogeography, Domestication, and Evolution of Maize*, 9–20. Amsterdam and Boston: Elsevier Academic Press.

Bergfjord, C., et al. 2012. "Nettle as a distinct Bronze Age textile plant." *Scientific Reports*, 2: 664.

Bergmann, Bettina. 2002. "Art and Nature in the Villa at Oplontis." *Journal of Roman Archaeology*, 47: 87–120.

Bergmann, Bettina. 2016. "The Gardens and Garden Paintings of Villa A." In Elaine K. Gazda and John R. Clarke (eds.), *Leisure and Luxury in the Age of Nero: The Villas of Oplontis Near Pompeii*, 96–110. Ann Arbor, MI: Kelsey Museum.

Berlin, Brent, et al. 1973. "Cultural Significance and Lexical Retention in Tzeltal-Tzotzil Ethnobotany." In Monro S. Edmonson (ed.), *Meaning in Mayan Languages: Ethnolinguistic Studies*. The Hague: Morton.

Berry, C. J. 1994. *The Idea of Luxury: A Conceptual and Historical Investigation*. Cambridge: Cambridge University Press.

Betts, A., et al. 1997. "Early Cotton on North Arabia." *Journal of Archaeological Science*, 21: 489–99.

Betts, J. 1978. "More Aegean Papyrus: Some Glyptic Evidence." *Athens Annals of Archaeology*, 9(1): 89–95.

Bevan, A. 2007. *Stone Vessels and Values in the Bronze Age Mediterranean*. Cambridge: Cambridge University Press.

Binford, Lewis R., and Chuan Kun Ho. 1985. "Taphonomy at a Distance: Zhoukoudian, 'The Cave Home of Beijing Man'?" *Current Anthropology*, 26: 413–42.

Bion. *Theocritus. Moschus. Bion*. Edited and translated by Neil Hopkinson. 2015. Loeb Classical Library 28. Cambridge: Harvard University Press.

Black, J. et al. 2006. *The Literature of Ancient Sumer*. Oxford: Oxford University Press.

Blake, M. 2006. "Dating the Initial Spread of *Zea mays*." In J. E. Staller et al. (eds.), *Histories of Maize: Multidisciplinary Approaches to the Prehistory, Linguistics, Biogeography, Domestication, and Evolution of Maize*, 55–68. Amsterdam: Academic Press.

Blakolmer, F. 2014. "Meaningful Landscapes: Minoan 'Landscape Rooms' and Peak Sanctuaries." In G. Touchais et al. (eds.), *PHYSIS. L'environnement naturel et la relation homme-milieu dans le monde Égéen protohistorique*. Aegaeum 37, 121–128. Liège: Peeters.

Blech, Michael. 1982. *Studien zum Kranz bei den Griechen*. Berlin: Walter de Gruyter.

Blench, Roger. 2004. "Fruits and Arboriculture in the Indo-Pacific Region." *Indo-Pacific Prehistory Association Bulletin*, 24: 31–50.
Boardman, John. [1964] 1980. *The Greeks Overseas*. London: Thames and Hudson.
Boardman, John. 1998a. *Early Greek Vase Painting*. London. Thames and Hudson.
Boardman, John. 1998b. "Reflections on the Origins of Indian Stone Architecture." *Bulletin of the Asia Institute*, NS 12: 13–22.
Boardman, S. 1999. "The Agricultural Foundation of the Aksumite Empire: An Interim Report." In M. Van Der Veen (ed.), *The Exploitation of Ancient Plant Resources in Ancient Africa*, 137–47. New York: Klewer Academic/Plenum.
Boardman, S. 2000. "Archaeobotany." In D. W. Phillipson (ed.), *Archaeology at Aksum, Ethiopia, 1993–97*, 2 vols, 127-128, 363-364, 412-414. London: British Institute in East Africa and the Society of Antiquaries of London.
Bohm, Robert Karl. 1973. "The Isis Episode in Apuleius." *The Classical Journal*, 68(3): 228–31.
Bonati, Isabella. 2016. *Il lessico dei vasi e dei contenitori greci nei papiri*: specimina *per un repertorio lessicale degli angionimi greci*. Berlin: De Gruyter.
Borza, Eugene N. 1987. "Timber and Politics in the Ancient World: Macedon and the Greeks." *Proceedings of the American Philosophical Society*, 131(1): 32–52.
Bostoen, K. 2014. "Wild Trees in the Subsistence Economy of Early Bantu Speech Communities: A Historical-Linguistic Approach." In C. Stevens et al. (eds.), *Archaeology of African Plant Use*, 129–40. Walnut Creek: Left Coast Press.
Bottéro, Jean, et al. (eds.). 2001. *Everyday Life in Ancient Mesopotamia*. Translated by Antonia Nevill. Baltimore: Johns Hopkins University Press.
Bottéro, Jean. 2004. *The Oldest Cuisine in the World*. Translated by Teresa Lavender Fagan. Chicago: University of Chicago Press.
Bötticher, Karl. 1856. *Der Baumkultus der Hellenen: nach den gottesdienstlichen gebräuchen und den überlieferten Bildwerken dargestellt*. Berlin: Weidmann.
Boucharlat, R. 2011, "Gardens and Parks at Pasargadae: Two 'Paradises'?" In R. Rollinger et al. (eds.), *Herodot und das Persische Weltreich*, 557–74. Wiesbaden, Harrassowitz.
Boudon, Véronique. 2002. "La thériaque selon Galien: poison salutaire ou remède empoisonné." In F. Collard and E. Samama (eds.), *Le corps à l'épreuve. Poisons, remèdes et chirurgie: aspects des pratiques médicales dans l'Antiquité et au Moyen Âge*, 45–56. Langres, France: Dominique Guéniot.
Boudon-Millet, Véronique. 2003. "Aux marges de la médecine rationnelle: médecins et charlatans à Rome au temps de Galien (IIe s. de notre ère)." *Revue des études grecques*, 116: 109–31.
Bowan, Terry. 2017. *A Bonsai Book: A Reference for Bonsai and Plant Aesthetics*. Lulu Electronic Publication.
Braun, Heike. 2010. *Geschichte des Gottesvolkes und christliche Identität: eine kanonisch-intertextuelle Auslegung der Stephanusepisode Apg 6,1–8,3*. Wissenschaftliche Untersuchungen zum Neuen Testament 2. Reihe 279. Tübingen: Mohr Siebeck.
Breed, Brian W. 2000. "Imitations of Originality: Theocritus and Lucretius at the Start of the Eclogues." *Vergilius*, 46: 3–20.
Bressani, R., et al. 1958. "Corn Nutrient Losses, Chemical Changes in Corn during Preparation of Tortillas." *Journal of Agricultural Food Chemistry*, 6(10): 770–4.
Breton, Jean-François. 1970. *IGLS* VIII, 3, *Les inscriptions forestières d'Hadrien dans le Mont-Liban* (BAH 104). Paris.
Bretzl, H. 1903. *Botanische Forschungen des Alexanderzuges*. Leipzig: Teubner.

Brickell, Christopher, and Judith D. Zuuk (eds.). 1997. *The American Horticultural Society A–Z Encyclopedia of Garden Plants*. New York: DK Publishing.

Bronner, Oscar. 1962. "The Isthmian Victory Crown." *American Journal of Archaeology*, 66(3): 259–63.

Broudy, Eric. 1993. *The Book of Looms: A History of the Handloom from Ancient Times to the Present*. Hanover: University Press of New England.

Brown, Cecil H. 2010. "The Development of Agriculture in Prehistoric Mesoamerica: The Linguistic Evidence." In John Statler and Michael Carrasco (eds.), *Pre-Columbian Foodways: Interdisciplinary Approaches to Food, Culture, and Markets in Ancient Mesoamerica*, 71–107. London: Springer.

Bru, Hadrien. 2011. *Le pouvoir impérial dans les provinces syriennes. Représentations et célébrations d'Auguste à Constantin (31 av. J.-C.- 337 ap. J.-C.)*. Leiden: Brill.

Brun, Jean-Pierre. 2000. "The Production of Perfumes in Antiquity: The Cases of Delos and Paestum." *American Journal of Archaeology*, 104(2): 277–308.

Bruno, Maria C. 2006. "A Morphological Approach to Documenting the Domestication of *Chenopodium* in the Andes." In Melinda A. Zeder (ed.), *Documenting Domestication: New Genetic and Archaeological Pradigms*, 32–45. Berkeley: University of California Press.

Bruno, Maria C. 2008. *Waranq Waranqa: Ethnobotanical Perspectives on Agricultural Intensification in the Lake Titicaca Basin (Taraco Peninsula, Bolivia)*. PhD thesis. Washington University in St. Louis, Missouri, USA.

Bruno, Maria C., and W. T. Whitehead. 2003. "*Chenopodium* Cultivation and Formative period Agriculture at Chiripa, Bolivia." *Latin American Antiquity*, 14: 339–5.

Buchheit, Vinzenz. 1986. "Resurrectio Carnis bel Prudentius." *Vigiliae Christianae*, 40(3): 261–85.

Buck, David D. 1975. "Three Han Dynasty Tombs at Ma-Wang-Tui." *World Archaeology*, 7: 30–45.

Caesarius of Arles. *Sancti Caesarii Arelatensis Opera: Sermones*. Corpus christianorum, series latina 104. Edited by G. Morin. 2nd edn. 1953. Turnhout: Brepols.

Cairns, Francis J. 1999. "Virgil Eclogue 1.1–2: A Literary Programme?" *Harvard Studies in Classical Philology*, 99: 289–93.

Callimachus. *Callimachus: Hymns and Epigrams. Lycophron: Alexandra. Aratus: Phaenomena*. Translated by A. W. Mair and G. R. Mair. 1921. Loeb Classical Library 129. Cambridge: Harvard University Press.

Cambridge Dictionary of American English. 2008. Cambridge: Cambridge University Press.

Campbell, Virginia. 2015. *The Tombs of Pompeii: Organization, Space, and Society*. New York: Routledge.

Caneva, Giulia. 2010. *Il Codice Botanico di Augusto: Roma—Ara Pacis*. Rome: Gangemi Editore.

Caneva, Guilia, and Lorenza Bohuny. 2003. "Botanic Analysis of Livia's Villa Painted Flora Prima Porta, Rome." *Journal of Cultural Heritage*, 4: 149–55.

(Cant.) *See* Song of Solomon.

Carey, Sarah. 2007. *Pliny's Catalogue of Culture: Art and Empire in the Natural World*. New York: Oxford University Press.

Carla, Filippo. 2013. "Barley." In Roger S. Bagnall et al. (eds.), *The Encyclopedia of Ancient History*, 1st edn., 1049–50. Chichester, UK: Blackwell Publishing Ltd.

Carlson, Deborah. 2002. "Caligula's Floating Palaces." *Archaeology*, 55(2): 26.

Carroll-Spillecke, Maureen. 1992. "The Gardens of Greece from Homeric to Roman Times." *Journal of Garden History*, 12(2): 84–101.

Carroll, Maureen. 2003. *Earthly Paradises: Ancient Gardens in History and Archaeology*. London: The British Museum Press.

Cashman, J. 2014. "Foreign Self and Familiar 'Other': the Impact of 'Global' Connectivity on New Kingdom Egypt." In O. S. LaBianca and S. A. Scham, *Connectivity in Antiquity: Globalization as a Long-Term Historical Process*, 130–138. London and New York: Routledge.

Cassianus Bassus Scholasticus. *Geoponica sive Cassiani Bassi scholastici de re rustica eclogue*. Edited by Henricus Beckh. 1895. Leipzig: B. G. Teubneri.

Cassiodorus. *Expositio psalmorum*. Edited by Marcus Adriaen. 1958. Corpus christianorum, series latina 97. Turnhout: Brepols.

Casson, Lionel (ed.). 1989. *The Periplus of the Erythraean Sea*. Princeton: Princeton University Press.

Casson, Lionel. [1989] 2012. *Periplus Maris Erythraei*. Princeton: Princeton University Press.

Casson, Lionel. 1950. "The Isis and Her Voyage." *Transactions of the American Philological Society*, 81: 43–56.

Casson, Lionel. 1984. *Ancient Trade and Society*. Detroit: Wayne State University Press.

Castillo, Cristina, and Dorian Q. Fuller. 2010. "Still Too Fragmentary and Dependent Upon Chance? Advances in the Study of Early Southeast Asian Archaeobotany." In B. Bellina et al. (eds.), *50 Years of Archaeology in Southeast Asia: Essays in Honour of Ian Glover*, 91–111. Bamgkok: River Books.

Castillo, Cristina, and Dorian Q. Fuller. 2015. "Bananas: The Spread of a Tropical Forest Fruit as an Agricultural Staple." In Julia Lee-Thorp and M. Anne Katzenberg (eds.), *The Oxford Handbook of the Archaeology of Diet*. Oxford: Oxford University Press.

Castillo, Cristina, et al. 2016. "Rice, Beans and Trade Crops on the Early Maritime Silk Route in Southeast Asia." *Antiquity*, 90: 1255–69.

Castriota, David. 1995. *The Ara Pacis Augustae and the Imagery of Abundance in Later Greek and Early Roman Imperial Art*. Princeton: Princeton University Press.

Cato. *On Agriculture*. In *Marcus Porcius Cato, On Agriculture. Marcus Terentius Varro, On Agriculture*. Revised edn. Edited and translated by W. D. Hooper and revised by H. Boyd Ash. 1935. Cambridge, MA and London: Harvard University Press.

Cato. *Marcus Porcius Cato: Über den Ackerbau: Herausgegeben*. Edited by Dieter Flach. 2005. Stuttgart: Franz Steiner Verlag.

Catullus. *Catullus. Tibullus. Pervigilium Veneris*. Translated by F. W. Cornish, J. P. Postgate, J. W. Mackail and revised by G. P. Goold. 1913. Loeb Classical Library 6. Cambridge: Harvard University Press.

Cavalieri, D., et al. 2003. "Evidence for *S. cerevisiae* Fermentation in Ancient Wine." *Journal of Molecular Evolution*, 57: S226–32.

CBC News. 2015. "Kanata Couple Angered by City Order to Change Veggie Garden." July 7. Available online: https://www.cbc.ca/news/canada/ottawa/kanata-couple-angered-by-city-order-to-change-veggie-garden-1.3141354#:~:text=An%20Ottawa%20couple%20who%20built,the%20wooden%20structures%20around%20it. (accessed March 28, 2021).

Chandler-Ezell, Karol, et al. 2006. "Root and Tuber Phytoliths and Starch Grains Document Manioc (Manihot Esculenta), Arrowroot (Maranta Arundinacea), and Llerén (Calathea sp.) at the Real Alto Site, Ecuador." *Economic Botany*, 60: 103–20.

Chaney, Ralph W. 1935. "The Occurrence of Endocarps of Celtis Barbouri at Choukoutien." *Acta Geologica Sinica*, 14: 99–118.

Chang, K. C. (ed.). 1977. *Food in Chinese Culture*. New Haven: Yale University Press.

Chapin, A. 2004. "Power, Privilege, and Landscape in Minoan Art." In A. Chapin (ed.), *XARIS. Essays in Honour of Sara A. Immerwahr*. Hesp. Supp. 33, 47–64. Princeton: The American School of Classical Studies at Athens.

Chapin, A., and M. Shaw. 2006. "The Frescoes from the House of the Frescoes at Knossos: A Reconsideration of Their Architectural Context and a New Reconstruction of the Crocus Panel." *Annual of the British School at Athens*, 101: 57–88.

Chavez, S. J. 2006. "Native Aymara and Quechua Botanical Terminologies of *Zea mays* in the Lake Titicaca and Cuzco Regions." In John E. Staller et al. (eds.), *Histories of Maize: Multidisciplinary Approaches to the Prehistory, Linguistics, Biogeography, Domestication, and Evolution of Maize*, 623–627. Amsterdam: Academic Press.

Cheddadi, Rachid, et al. 2015. "A History of Human Impact on Moroccan Mountain Landscapes." *African Archaeology Review* 32: 233–48.

(Ps.) Chrysostom, John. *Procopii Gazaei catena in Ecclesiasten necnon Pseudochrysostomi commentarius in eundem Ecclesiasten*. Edited by S. Leanza. 1978. Corpus christianorum, series graeca 4. Turnhout: Brepols.

Ciarallo, Annamaria, and Marta Mariotti Lippi. 1994. "The Garden of 'Casa dei Casti Amanti' (Pompeii, Italy)." *Garden History*, 21(1): 110–16.

Ciarallo, Annamaria. 1991. "Fauna e Flora." In Gianni Conti et al., *Il Gardino Dipinto nella Casa del Bracciale d' Oro a Pompei e il Suo Restauro*. Florence: Università Internazionale dell' Arte.

Ciarallo, Annamaria. 2002. *Il Gardino Pompeiano: Le Plante, L'Orto, I Segreti della Cucina*. Naples: Electa.

Ciarallo, Annamaria. 2006. *Elementi Vegetali nell' Iconografia Pompeiana*. Rome: L'Erma di Bretschneider.

Cicero. *Letters to Quintus and Brutus*. Edited and translated by D. R. Shackelton Bailey. 2002. Cambridge: Harvard University Press.

Cicero. *On the Republic. On the Laws*. Edited and translated by Clinton W. Keyes. 1928. Loeb Classical Library 213. Cambridge: Harvard University Press.

Cicero. *The Verrine Orations. Vol. I. Against Caecilius. Against Verres. Part 1. Part 2, Books 1–2*. Edited and translated by L. H. G. Greenwood. 1928. Cambridge, MA and London: Harvard University Press.

Clarke, John R. 2003. *Art in the Lives of Ordinary Romans: Visual Representation and Non-Elite Viewers in Italy 100 BC–AD 315*. Berkeley, CA: University of California Press.

Clement of Alexandria. *Stromata: Buch I–VI*. Edited by Otto Staehlin. 1906. Clemens Alexandrinus 2. GCS 15. Leipzig: Hinrichs.

Clement of Alexandria. *Stromateis, Books 1–3*. Translated by John Ferguson. 2005. Fathers of the Church Patristic Series 85. Washington, DC: The Catholic University of America Press.

Clement, Charles R. et al. 2010. "Origin and Domestication of Native Amazonian Crops." *Diversity*, 2: 72–106.

Coe, Sophie D., and Michael D. Coe. 1996. *The True History of Chocolate*. London: Thames and Hudson.

Coldstream, J. N. 1977. *Geometric Greece*. London: Ernest Benn Ltd.

Colledge, S. 2001. *Plant Exploitation on Epipalaeolithic and Early Neolithic Sites in the Levant*. Oxford: BAR International Series 986.

Collins, P. 2006. "Trees and Gender in Assyrian Art." *Iraq*, 68: 99–107.

Columella. *On Agriculture (De re rustica)*, vol. I. Edited and translated by H. B. Ash. 1941. Cambridge, MA: Harvard University Press.

Columella. *On Agriculture (De re rustica)*, vol. II. Edited and translated by E. S. Forster and Edward H. Heffner. 1954. Cambridge, MA: Harvard University Press.

Columella. *On Agriculture (De re rustica)*, vol. III. Edited and translated by E. S. Forster and Edward H. Heffner. 1955. Cambridge, MA: Harvard University Press.

Conan, Michael (ed.). 2007. *Middle East Garden Traditions: Unity and Diversity*. Washington, DC: Dumbarton Oaks.

Cordell, L. 1997. *Archaeology of the Southwest*. 2nd edn. London: Routledge.

Cordell, L. and M. E. McBrinn. 2012. *Archaeology of the Southwest*. 3rd edn. London: Routledge.

Cornutus, Lucius Annaeus. *De natura deorum*. Edited by Jean-Baptiste-Gaspard d'Ansse de Villoison and revised and commentary by Friedrich Osann. 1844. Gottingae: Library Dieterichiana.

Cowell, E. B. (ed.). 1895. *The Jataka*, vol. 1. Cambridge: Cambridge University Press.

Crawford, G. W., et al. 2006. "Pre-Contact Maize from Ontario, Canada." In John E. Staller et al. (eds.), *Histories of Maize: Multidisciplinary Approaches to the Prehistory, Linguistics, Biogeography, Domestication, and Evolution of Maize*, 549–556. Amsterdam: Academic Press.

Crawford, Gary W. 2011. "Advances in Understanding Early Agriculture in Japan." *Current Anthropology*, 52: S331–S345.

Crowther, A., et al. 2016. "Ancient Crops Provide First Archaeological Signature of the Westward Austronesian Expansion." *Proceedings of the National Academy of Sciences*, 113 (24): 6635–40.

Curtius, Ernst Robert. 1953. *European Literature and the Latin Middle Ages*. Translated by Willard R. Trask. Bollingen Series 36. Princeton: Princeton University Press.

Cyril of Jerusalem. *Cyrilli Hierosolymarum archiepiscopi opera*. Edited by Wilhelm Carl Reischl and Josef Rupp. 1967. 2nd edn. Hildesheim: G. Olms.

D'Andrea, A. C., et al. 2001. "Archaeobotanical Evidence for Pearl Millet (*Pennisetum glaucum*) in Sub-Saharan West Africa." *Antiquity*, 75(288): 341–8.

D'Andrea, A. C. 2008. "T'ef (*Eragrostis tef*) in Ancient Agricultural Systems of Highland Ethiopia." *Economic Botany*, 62(4): 547–66.

D'Andrea, A. C., P. R. Schmidt, and M. C. Curtis. 2007. "Palaeoethnobotanical Analysis and Agricultural Economy in Early 1st Millennium B.C.E. Sites around Asmara." In P. R. Schmidt et al. (eds.), *The Archaeology of Eritrea*, 207–16. Trenton, New Jersey: Red Sea Press.

Daines, A. 2008. "Egyptian Gardens." *Studia Antiqua*, 6(1): 15–25.

Dalby, Andrew. 2000a. *Dangerous Tastes: The Story of Spices*. London: British Museum Press.

Dalby, Andrew. 2000b. *Empire of Pleasures*. London: Routledge.

Dalby, Andrew. 2003. *Food in the Ancient World from A to Z*. London and New York: Routledge.

Dalley, Stephanie. 1994. "Nineveh, Babylon and the Hanging Gardens: Cuneiform and Classical Sources Reconciled." *Iraq*, 56: 45–58.

Dalley, Stephanie. 1993. "Ancient Mesopotamian Gardens and the Identification of the Hanging Gardens of Babylon Resolved." *Garden History*, 21(1): 1–13.

Dalley, Stephanie. 2013. *The Mystery of the Hanging Garden of Babylon: An Elusive World Wonder Traced*. Oxford, UK: Oxford University Press.

Dauphin, Claudine. 1997. "Carpets of Stone: The Graeco-Roman Legacy in the Levant." *Classics Ireland*, 4: 1–32.

Davies, D.R. 1995. "Peas." In J. Smartt and N. W. Simmonds (eds.), *Evolution of Crop Plants*, 2nd edn., 294–6. London: Longman, UK.

Day, Joanna. 2006. "Flower Lovers? Reconsidering the Gardens of Minoan Crete." In J-P. Morel et al. (eds.), *The Archaeology of Crop Fields and Gardens. Proceedings of the 1st*

Conference on Crop Fields and Gardens Archaeology, Barcelona (Spain), 1–3 June 2006, 189–95. Bari: Edipuglia.

Day, Joanna. 2010. "Plants, Prayers and Power: The Story of the First Mediterranean Gardens." In D. O'Brien (ed.), *Gardening: Philosophy for Everyone. Cultivating Wisdom*, 65–78. Oxford: Wiley-Blackwell.

Day, Joanna. 2011a. "Crocuses in Context: A Diachronic Survey of the Crocus Motif in the Aegean Bronze Age." *Hesperia*, 80(3): 337–79.

Day, Joanna. 2011b. "Counting Threads. Saffron in Aegean Bronze Age Writing and Society." *Oxford Journal of Archaeology*, 30(4): 369–91.

Day, Joanna. 2012. "Caught in a Web of a Living World. Tree-Human Interaction in Minoan Crete." *PAN: Philosophy, Activism, Nature*, 9: 11–21.

Day, Joanna. 2013. "Imagined Aromas and Artificial Flowers in Minoan Society." In J. Day (ed.), *Making Senses of the Past: Toward a Sensory Archaeology*, 286–309. Carbondale: Southern Illinois University Press.

De Moulins, Dominique. 1997. *Agriculture Canges at Euphrates and Steppe Sites in the Mid-8th to 6th Millennium B.C.* Oxford: BAR International Series 683.

De Romanis, F. 1996. *Cassia, cinnamomo, ossidiana. Uomini e mercanti tra oceano indiano e Mediterraneo*. Rome: L'Erma di Bretschneider.

Decker-Walters, D. S., et al. 1993. "Isozymic Characterisation of Wild Populations of *Cucurbita pepo*." *Journal of Ethnobiology* 13: 55–72.

Demuth, Stefan. 2012. *Das Mythische in der Natur: die Entstehung der Tier-und Pflanzenarten in der antiken Mythologie*. Regensburg: Schnell & Steiner.

Denham, Tim. 2004. "Early Agriculture in the Highlands of New Guinea: An Assessment of Phase 1 at Kuk Swamp." *Records of the Australian Museum*, suppl. 29: 47–57.

Deuteronomy. *The New Oxford Annotated Bible: New Revised Standard*. Edited by Michael D. Coogan. 2018. 5th edn. Oxford: Oxford University Press.

Dickinson, O. 1994. *The Aegean Bronze Age*. Cambridge: Cambridge University Press.

Die Fragmente der griechischen Historiker. "Apollodoros von Athen (244)." In Felix Jacoby (ed.), *Die Fragmente der Griechischen Historiker Parts I–III*. 2007. Available online: http://dx.doi.org/10.1163/1873-5363_boj_a244 (accessed January 31, 2019).

Diehl, M. W. 2009. "Early Agricultural Food Provisioning and Foraging." *Archaeology Southwest*, 23(1): 12–3.

Dillehay, Tom, et al. 2015. "New Archaeological Evidence for an Early Human Presence at Monte Verde, Chile." *PLoS ONE*, 10(11): e0141923.

Dimopoulou, N., and G. Rethemiotakis. 2000. "The 'sacred conversation' Ring from Poros." In I. Pini (ed.), *Minoische-Mykenische Glyptik Stil, Ikonographie, Funktion*. CMS Beiheft 6, 39–56. Berlin: Gebr. Mann Verlag.

Dinsmoor, William Bell. 1975. *The Architecture of Ancient Greece: An Account of its Historic Development*. New York: Norton.

Diodorus Siculus. *Library of History, Vol. 3: Books 4.59–8*. Edited and translated by C. H. Oldfather. 1939. Loeb Classical Library 340. Cambridge: Harvard University Press.

Dionysius of Halicarnassus. *Roman Antiquities, Vol. 3: Books 5–6.48*. Edited and translated by Earnest Cary. 1940. Loeb Classical Library 357. Cambridge: Harvard University Press.

Dionysius of Halicarnassus. *Roman Antiquities, Vol. 7: Books 11–20*. Edited and translated by Earnest Cary. 1950. Loeb Classical Library 388. Cambridge: Harvard University Press.

Dioscorides. *De materia medica*. Edited and translated by Lily Y. Beck. 2017. 3rd edn. Altertumswissenschaftliche Texte und Studien 38. Hildesheim: Georg Olms.

Dioscorides. *Pedanii Dioscuridis Anazarbei De materia medica libri quinque. Volumen I quo continentur libri I et II*. Edited by M. Wellmann. 1907. Berlin: Weidmann.

Dioscorides. *Pedanii Dioscuridi Anazarbei De materia medica libri quinque. Volumen II quo continentur libri III et IV*. Edited by M. Wellmann. 1906. Berlin: Weidmann.

Dioscorides. *Pedanii Dioscuridis Anazarbei De materia medica libri quinque. Volumen III quo continentur liber V, Crateuae et Sixtii Nigri fragmenta, Diocuridis libri de simplicibus*. Edited by M. Wellmann. 1914. Berlin: Weidmann.

Dixon, D. M. 1969. "The Transplantation of Punt Incense Trees in Egypt." *Journal of the Egypt Exploration Society*, 55: 55–65.

Doebley, J. 1990. "Molecular Evidence and the Evolution of Maize." *Economic Botany*, 44(3): 6–27.

Doumas, C. 1992. *The Wall-Paintings of Thera*. Athens: The Thera Foundation and Petros M. Nomikos.

Dresken-Weiland, Jutta. 2018. "Christian Sarcophagi from Rome." In Robin M. Jensen and Mark D. Ellison (eds.), *The Routledge Handbook of Early Christian Art*, 39–55. New York and London: Routledge.

Dunbabin, Katherine M. 1999. *Mosaics of the Greek and Roman World*. Cambridge and New York: Cambridge University Press.

Dyck, Andrew R. 2004. *A Commentary on Cicero*, De Legibus. Ann Arbor: University of Michigan Press.

Ebeling, H. L. 1924. "The Origin of the Corinthian Capital." *The Art Bulletin*, 6(3): 75–81.

Ecclesiastes. *The New Oxford Annotated Bible: New Revised Standard*. Edited by Michael D. Coogan. 2018. 5th edn. Oxford: Oxford University Press.

Edgell, H. Stewart, et al. 2004. "The Myth of the 'Lost City of the Arabian Sands.'" *Proceedings of the Seminar for Arabian Studies* 34, Papers from the thirty-seventh meeting of the Seminar for Arabian Studies held in London, 17–19 July 2003, 105–20.

Edwards, D. N. 1996. "Sorghum, Beer and Kushite Society." *Norwegian Archaeological Review*, 29(2): 65–77.

Ehret, C. 2014. "Linguistic Evidence and the Origins of Food Production in Africa: Where are We Now?" In C. Stevens et al. (eds.), *Archaeology of African Plant Use*, 233–42. Walnut Creek, CA: Left Coast Press.

Eitrem, Samson. 1915. *Opferritus und voropfer der Griechen und Römer*. Kristiania: J. Dybwad.

Ellison, Rosemary. 1984. "Methods of Food Preparation in Mesopotamia (ca. 3000–600 BC)." *Journal of the Economic and Social History of the Orient*, 27: 89–98.

Engemann, Jürgen. 1983. "Die imperialen Grundlagen der frühchristlichen Kunst." In Herbert Beck and Peter Bol (eds.), *Spätantike und frühes Christentum*, 260–266. Frankfurt: Das Liebieghaus.

Erdmann, Kurt, et al. 1954. "Art. Baum." In Theodor Klauser (ed.), *Reallexikon für Antike und Christentum*, vol. 2, 1–34. Stuttgart: Hiersemann.

Esperandieu, E. 1913. *Recueil général des bas-reliefs, statues et bustes de la Gaule romaine. Tome V*. Paris: Imprimerie Nationale.

Estreicher, Stefan K. 2019. "Wine." In Kai Brodersen et al. (eds.), *The Encyclopedia of Ancient History*, online reference book. John Wiley & Sons Ltd. Available online: https://onlinelibrary-wiley-com.udel.idm.oclc.org/action/doSearch?field1=Contrib&text1=estreicher&field2=AllField&text2=&field3=AllField&text3=&publication%5B%5D=10.1002%2F9781444338386&Ppub= (accessed March 23, 2021).

Eubanks, M. 2001. "An Interdisciplinary Perspective on the Origin of Maize." *Latin American Antiquity*, 12(1): 91–8.

Eubulus. *The Fragments*. Edited with a commentary by Richard L. Hunter. 1983. Cambridge: Cambridge University Press.

Euripides. *Bacchae. Iphigenia at Aulis. Rhesus.* Edited and translated by David Kovacs. 2003. Loeb Classical Library 495. Cambridge: Harvard University Press.

Euripides. *Cyclops. Alcestis. Medea.* Edited and translated by David Kovacs. 1994. Loeb Classical Library 12. Cambridge: Harvard University Press.

Euripides. *Euripidis fabulae*, vol. 1. Edited by James Diggle. 1984. Oxford Classical Texts. Oxford: Oxford University Press.

Euripides. *Trojan Women. Iphigenia among the Taurians. Ion.* Edited and translated by David Kovacs. 1999. Loeb Classical Library 10. Cambridge: Harvard University Press.

Evans, Arthur. 1921–35. *The Palace of Minos at Knossos.* 4 vols. London: MacMillan and Co.

Evans, Arthur. 1921. *The Palace of Minos: A Comparative Account of the Successive Stages of the Early Cretan Civilization as Illustrated by the Discoveries at Knossos*, vol. 1, *The Neolithic and Early and Middle Minoan Ages.* London: Macmillan and Co.

Evans, Susan Toby. 2000. "Aztec Royal Pleasure Parks: Conspicuous Consumption and Elite Status Rivalry." *Studies in the History of Gardens and Design Landscapes*, 20: 206–28.

Falk, Harry. 1989. "Soma I and II". *Bulletin of the School of Oriental and African Studies*, 52: 77–90.

Farrar, Linda. 1998. *Ancient Roman Gardens.* Stroud: Sutton Publishing.

Farrar, Linda. 2016. *Gardens and Gardeners of the Ancient World.* Oxford: Oxbow Books.

Fausboll, V. (ed.). 1877. *The Jataka Together With Its Commentary, vol. 1.* London: Trübner.

Fedick, Scott L., et al. 2008. "Adaptation of Maya Homegardens by 'Container Gardening' in Limestone Bedrock Cavities." *Journal of Ethnobiology*, 28(2): 290–304.

Fehrle, Eugen. 1910. *Die kultische Keuschheit im Altertum.* Giessen: Alfred Töpelmann Verlag.

Feldman, M. 2006. *Diplomacy by Design: Luxury Arts and an "International Style" in the Ancient Near East, 1400–1200 BCE.* Chicago, IL: University of Chicago Press.

Feldman, M. 2014. *Communities of Style: Portable Luxury Arts, Identity and Collective Memory in the Iron Age Levant.* Chicago, IL: University of Chicago.

Feliks, Jehudah. 2007. "Balsam." In F. Skolnik et al. (eds.), *Encyclopaedia Judaica*, vol. 3. 2nd edn. Farmington Hills, MI: Thomson Gale.

Festus, Sextus Pompeius. *Sexti Pompei Festi De verborum significatu quae supersunt cum Pauli epitome.* Edited by W. M. Lindsay. 1913. Leipzig: B. G. Teubneri.

(FGrH) *See Die Fragmente der griechischen Historiker.*

Filipczak, Dorota. 1993. "Is there no Balm in Gilead?—Biblical Intertext in *the Handmaid's Tale.*" *Literature and Theology*, 7(2): 171–85.

Finkel, Irving. 2014. *The Ark Before Noah: Decoding the Story of the Flood.* New York: Anchor/Doubleday.

Finley, Robert. 2015. "Marlene Creates' Boreal Poetry Garden." In Annette Giesecke and Naomi Jacobs (eds.), *The Good Gardener? Nature, Humanity, and the Garden*, 162–71. London: Artifice Books on Architecture.

Flemming, Rebecca. 2005. "Empires of Knowledge: Medicine and Health in the Hellenistic World." In Andrew Erskine (ed.), *A Companion to the Hellenistic World*, 449–463. Malden, MA: Blackwell.

Ford, Anabel. 2008. "Dominant Plants of the Maya Forest and Gardens of El Pilar: Implications for Paleoenvironmental Reconstructions." *Journal of Ethnobiology*, 28(2): 179–99.

Forster, Edward S. 1936. "Trees and Plants in Homer." *Classical Review*, 50(3): 97–104.

Foster, K. P. 1999. "The Earliest Zoos and Gardens." *Scientific American*, 281: 64–71.

Fowler, Brenda. 2001. *Iceman: Uncovering the Life and Times of a Prehistoric Man Found in an Alpine Glacier.* Chicago: University of Chicago Press.

Foxhall, Lin. 1998. "Cargoes of the Heart's Desire: The Character of Trade in the Archaic Mediterranean World." In Nick Fisher and Hand Van Wees (eds.), *Archaic Greece: New Approaches and New Evidence*, 295–309. London: Duckworth.

Foxhall, Lin. 2005. "Village to City: Staples and Luxuries? Exchange Networks and Urbanization." In Robin Osborne and Barry Cunliffe (eds.), *Mediterranean Urbanization 800–600 BC*, 233–248. Oxford: Oxford University Press.

Foxhall, Lin. 2007. *Olive Cultivation in Ancient Greece: Seeking the Ancient Economy*. Oxford: Oxford University Press.

Frischer, Bernard, et al. 2017. "New Light on the Relationship between the Montecitorio Obelisk and the Ara Pacis of Augustus." *Studies in Digital Heritage*, 1(1) Article 2: 1–121.

Frost, Honor, A. E. Werner, and W. A. Oddy. 1973. "First Season of Excavation on the Punic Wreck in Sicily." *International Journal of Nautical Archaeology*, 2(1): 33–49

Frost, Honor. 1982. "La réconstruction de la navire punique de Marsala." *Archeologia*, 170: 42–50.

Fullbright, Lori. 2012. "Woman Sues City of Tulsa for Cutting Down her Edible Garden." *News on 6,* June 15. Available online: http://www.newson6.com/story/18802728/woman-sues-city-of-tulsa-for-cutting-down-her-edible-garden (accessed March 28, 2021).

Fuller, Dorian Q. 2002. "Fifty Years of Archaeobotanical Studies in India: Laying a Solid Foundation." In S. Settar and R. Korisettar (eds.), *Indian Archaeology in Retrospect*, vol. 3, 247–363. New Delhi: Manohar.

Fuller, Dorian Q. 2003. "African Crops in Prehistoric South Asia: A Critical Review." In K. Neumann et al. (eds.), *Food, Fuel and Fields: Progress in African Archaeobotany* (Africa Praehistorica 15), 239–71. Cologne: Heinrich-Barth Institut.

Fuller, Dorian Q. 2004. "Early Kushite Agriculture: Archaeobotanical Evidence from Kawa." *Sudan and Nubia*, Bulletin No. 8: 71–4.

Fuller, Dorian Q. 2011. "Finding Plant Domestication in the Indian Subcontinent." *Current Anthropology*, 52: S347–S362.

Fuller, Dorian Q., and E. Hildebrand. 2013. "Domesticating Plants in Africa." In P. Mitchell and P. Lane (eds.), *The Oxford Handbook of African Archaeology,* 507–26. Oxford: Oxford University Press.

Fuller, Dorian Q., et al. 2004. "Early Plant Domestications in Southern India: Some Preliminary Archaeobotanical Results." *Vegetation History and Archaeobotany*, 13: 115–29.

Fuller, Dorian Q., et al. 2009. "The Domestication Process and Domestication Rate in Rice: Spikelet Bases from the Lower Yangtze." *Science*, 323: 1607–10.

Fuller, Dorian Q., et al. 2017. "The Archaeobiology of Indian Ocean Translocations." In Sila Tripati (ed.), *Maritime Contacts of the Past: Deciphering Connection Amongst Communities*, 1–23. Delhi: Delta Book World.

Gabriel, Mabel M. 1955. *Livia's Garden Room at Prima Porta*. New York: New York University Press.

Galán, José. 2018. "An Ancient Egyptian Garden for Eternal Life." Charles K. Wilkinson Lecture Series: Gardens: From Paradise to Parterre. Metropolitan Museum of Arts, New York, NY. Presented March 19.

Galen. *Claudii Galeni opera Omnia*, vol. 20. Edition by K. G. Kühn. 1821–33. Berlin and Leipzig: Libraria Car. Cnoblochii.

Galen. *Galien. Œuvres. Tome IV. Ne pas se chagriner*. Edition and translation by V. Boudon-Millot and J. Jouanna with A. Pietrobelli. 2010. Paris: Les Belles Lettres.

Gansell, A. 2014a. "Images and Conceptions of Ideal Feminine Beauty in Neo-Assyrian Royal Contexts, c. 883–627 BCE." In Brian Brown and Marian Feldman (eds.), *Critical Approaches to Ancient Near Eastern Art*, 391–420. Berlin: Walter de Gruyter.

Gansell, A. 2014b. "The Iconography of Ideal Feminine Beauty Represented in the Hebrew Bible and Levantine Ivory Sculpture." In I. J. DeHulster and J. M. LeMon (eds.), *Image, Text, Exegesis: Iconography and Interpretation in the Hebrew Bible*, 46–70. London: Bloomsbury.

Gansell, A. 2016. "Imperial Fashion Networks: Royal Assyrian, Near Eastern, Intercultural and Composite Style Adornment from the Neo-Assyrian Royal Women's Tombs at Nimrud." In J. Aruz and M. Seymour (eds.), *Assyria to Iberia: Arts and Culture in the Iron Age*, 54–64. New York, NY: Metropolitan Museum of Art.

Garfinkel, Y., M. Kislev, and D. Zohary. 1988. "Lentil in the Pre-Pottery Neolithic B Yiftah'el: Additional Evidence of its Early Domestication." *Israel Journal of Botany*, 37: 49–51.

Géczi, János. 2008. "The Roman Rose. An Anthropological Approach." *Iskolakutúra Online*, 2: 1–66.

Gellius, Aulus. *Attic Nights, Vol. 1: Books 1–5*. Edited and translated by J. C. Rolfe. 1927. Loeb Classical Library 195. Cambridge: Harvard University Press.

Genesis. *The New Oxford Annotated Bible: New Revised Standard*. Edited by Michael D. Coogan. 2018. 5th edn. Oxford: Oxford University Press.

George, Alison. 2016. "The World's Oldest Paycheck Cashed in Beer". *New Scientist*. Available online: https://www.newscientist.com/article/2094658-the-worlds-oldest-paycheck-was-cashed-in-beer/ (accessed March 25, 2021).

Georgi, L. 1982. "Pollination Ecology of the Date Palm and Fig Tree: Herodotus 1.193.4-5." *Classical Philology*, 77(3): 224–8.

Georgiou, H. 1974. "Aromatics in Antiquity and Minoan Crete: A Review and Reassessment." *Kretika Chronika*, 25: 441–56.

Germer, Renate. 1985. *Flora des pharaonischen Ägypten*. Sonderschrift Deutsches Archäologisches Institut, Abteilung Kairo 14. Mainz: Philipp von Zabern.

Giblin, J., and Dorian Q. Fuller. 2011. "First and Second Millennium AD Agriculture in Rwanda: Archaeobotanical Finds and Radiocarbon Dates from Seven Sites." *Vegetation History and Archaeobotany*, 20: 253–65.

Giesecke, Annette L. 2007. *The Epic City: Urbanism, Utopia, and the Garden in Ancient Greece and Rome*. Hellenic Studies Series 21. Washington, DC, and Cambridge, MA: Center for Hellenic Studies, Harvard University Press.

Giesecke, Annette. 2012. "Outside In and Inside Out: Paradise in the Ancient Roman House." In Annette Giesecke and Naomi Jacobs (eds.), *Earth Perfect? Nature, Utopia, and the Garden*, 118–35. London: Black Dog Publishing.

Giesecke, Annette L. 2014. *The Mythology of Plants: Botanical Lore in Ancient Greece and Rome*. Los Angeles: Getty Publications.

Giesecke, Annette. 2015. "The Good Gardener and Ideal Gardens of State." In Annette Giesecke and Naomi Jacobs (eds.), *The Good Gardener? Nature, Humanity, and the Garden*, 78–95. London: Artifice Books on Architecture.

Giesecke, Annette. Forthcoming. "Lyric Space: Sappho and Aphrodite's Sanctuary." In Laura Swift (ed.), *A Companion to Greek Lyric Poetry*, Blackwell Companions to the Ancient World. New York: Wiley-Blackwell.

Gillis, Carole, and Marie-Louise Nosch. 2007. *Ancient Textiles: Production, Crafts and Society*. Oxford: Oxbow Books.

Giovino, Mariana. 2007. *The Assyrian Sacred Tree: A History of Interpretations. Orbus Biblicus et Orientalis* 230. Fribourg and Göttingen: Academic Press and Vandenhoek & Rüprecht.

Gleason, Kathryn L. 1994. "*Porticus Pompeiana*: A New Perspective on the First Public Park of Ancient Rome." *Journal of Garden History*, 14: 13–27.

Gleason, Kathryn L. 2014. "Wilhemina Jashemski and Garden Archaeology at Oplontis." In John R. Clarke and Nayla K. Muntasser (eds.), *Oplontis: Villa A ("of Poppaea") at Torre Annunciata, Italy*. New York: American Council of Learned Societies.

Gleason, Kathryn L. 2016. "Documentation of the Garden Beds." In Thomas Noble Howe et al. (eds.), *Quaderni di Studi Pompeiani* VII: *Excavation and Study of the Garden of the Great Peristyle of the Villa Arianna, Stabiae, 2007–2012*, 67–81. Castellamare di Stabia: Nicola Longobardi Editore.

Gleason, Kathryn L. 2019. "The Lost Dimension: Pruned Plants in Roman Gardens." *Vegetation History and Archaeobotany*, 28: 311–25.

Gleba, Margarita. 2008. *Textile Production in Pre-Roman Italy*. Ancient Textiles Series 4. Oxford: Oxbow Books.

Gleba, Margarita. 2012. "Cotton." In Roger Bagnall et al. (eds.), *Encyclopedia of Ancient History*, 1813–14. Malden, MA and Oxford: Wiley-Blackwell.

Good, I.L., et al. 2009. "New Evidence for Silk in the Indus Civilization." *Archaeometry*, 51(3): 457–66.

Goodison, L., and C. Morris. 1998. "Beyond the 'Great Mother': Sacred World of the Minoans." In L. Goodison and C. Morris (eds.), *Ancient Goddesses*, 113–32. London: British Museum Press.

Goren-Inbar, Naama, et al. 2014. "Beneath Still Waters: Multistage Aquatic Exploitation of *Euryale ferox* (Salisb.) during the Acheulian." *Internet Archaeology*. Available online: http://intarch.ac.uk/journal/issue37/1/toc.html (accessed March 28, 2021).

Gott, Beth. 2005. "Aboriginal Fire Management in South-eastern Australia." *Journal of Biogeography*, 32: 1203–8.

Graeve, M. C. de. 1981. *The Ships of the Ancient Near East* (c. 2000–500 B.C.). Orientalia Lovanensis Analecta 7. Leuven: Peeters.

Granziera, Patrizia. 2001. "Concept of the Garden in Pre-Hispanic Mexico." *Garden History*, 29(2): 185–213.

Grayson, A. K., and J. Novotny. 2012. *Royal Inscriptions of Neo-Assyria Publications 3/1: Sennacherib Part I*. Helsinki: Neo-Assyrian Text Corpus.

Green, Douglas J. 2010. *I Undertook Great Works: The Ideology of Domestic Achievements in Wes Semitic Royal Inscriptions*. Tübingen: Mohr Siebeck.

Greene, Kevin. 2008. "Learning to Consume: Consumption and Consumerism in the Roman Empire." *Journal of Roman Archaeology*, 21: 64–82.

(Greg. M.) Gregory the Great. *Homilae in Hiezechihelem prophetam*. Edited by Marcus Adriaen. 1971. Corpus christianorum, series latina 142. Turnhout: Brepols.

Gregory of Nazianzus. *Discours 4–5: contre Julien*. Introduction and translation by Jean Bernardi. 1983. Sources Chrétiennes 309. Paris: Cerf.

Gregory of Nazianzus. *Julian the Emperor: Containing Gregory of Nazianzen's Two Invectives and Libaniaus' Monody, With Julian's Extant Theosophical Works*. Translated by Charles William King. 1888. London: G. Bell.

Gregory of Nyssa. *Gregorii Nysseni Epistulae*, vol. 8/2. Edited by Giorgio Pasquali. 1959. In Werner Jaeger (ed.), *Gregorii Nysseni Opera*. 2nd edn. Leiden: Brill.

Gregory of Nyssa. *Gregorii Nysseni Opera*. Edited by Werner Jaeger et al. 1959. 2nd edn. Leiden: Brill.

Gregory of Nyssa. *Gregory of Nyssa: The Letters*. Introduced, translated, and commentary by Anna M. Silvas. 2006. Supplements to Virgiliae Christianae. Leiden: Brill.

Grivet, L., et al. 2004. "A Review of Recent Molecular Genetics Evidence for Sugarcane Evolution and Domestication." *Ethnobotany Research and Applications*, 2: 9–17.

Grocock, Christopher, and Sally Grainger (eds.). 2006. Apicius. Totnes: Prospect Books.

Groenewegen-Frankfort, H. 1951. *Arrest and Movement: An Essay on Space and Time in the Representational Art of the Ancient Near East*. Chicago, IL: University of Chicago Press.

Guangping, Feng, et al. 2012. *Qin Han Shanglinyuan Zhi Wu Tu Kao (Illustrated Plants Identification of Shang lin yuan in Qin and Han Dynasty*). Beijing: SciencePress.

Guest, Clare Lapraik. 2016. *The Understanding of Ornament in the Italian Renaissance*. Leiden: Brill.

Gunn, Bee F., et al. 2011. "Independent Origins of Cultivated Coconut (Cocos nucifera L.) in the Old World Tropics." *PLOS One*. Available online: http://journals.plos.org/plosone/article?id=10.1371/journal.pone.0021143 (accessed March 28, 2021).

Guo, Xingmei, and Yiping Qiu. 2013. "Discussion on Development of Silk Weaving Trademark Process in Shang Dynasty of China." *Research of Material Science*, 2(3): 41–9.

Hald, Margrethe. 1942. *The Nettle as a Culture Plant*, Tom VI. Stockholm: Folk-Liv.

Hale, John. 2009. *Lords of the Sea: The Epic Story of the Athenian Navy and the Birth of Democracy*. New York: Viking.

Hamilakis, Y. 2013. *Archaeology and the Senses: Human Experience, Memory and Affect*. Cambridge: Cambridge University Press.

Hamlin, A. D. F. 1919. "Plant Forms in Decorative Art." *The Metropolitan Museum of Art Bulletin*,14(3): 49–50.

Hanfmann, George M. A. 1976. "On Lydian and Eastern Greek Anthemion Stelai." *Revue Archéologique*, NS 1: 35–44.

Hardy, Gavin, and Laurence Totelin. 2016. *Ancient Botany*. London and New York: Routledge.

Harlan, J. R., and D. Zohary. 1966. "Distribution of Wild Wheats and Barley." *Science*, 153: 1074–80.

Harlan. Jack R. 1992. *Crops and Man*. Madison, WI: American Society of Agronomy, Crop Science Society of America.

Harper, Donald. 1998. *Early Chinese Medical Literature: The Mawangdui Medical Manuscripts*. London: Kegan Paul.

Harris, Susanna M., and Margarita Gleba. 2015. "Bronze Age Moss Fibre Garments from Scotland—the Jury's Out." *Archaeological Textiles Review*, 57: 3–11.

Hatcher, Patrick. 2002. "Wood Associated with the A.D. 79 Eruption." In Wilhelmina Jashemski and F. G. Meyer (eds.), *The Natural History of Pompeii*, 217–24. Cambridge: Cambridge University Press.

Hawkes, David (trans.). 1985. *The Songs of the South, by Qu Yuan and Other Poets*. Harmondsworth: Penguin.

Helbaek, Hans. 1952. "Early Crops in Southern England." *Proceedings of the Prehistoric Society*, 12: 194–233.

Hemelrijk, Emily, and Greg Woolf. 2013. *Women and the Roman City in the Latin West*. Leiden: Brill.

Hepper, F. Nigel. 2009. *Pharaoh's Flowers: The Botanical Treasures of Tutankhamun*. 2nd edn. Chicago and London: KWS Publishers.

Hermetica. *Hermetica: The Greek* Corpus Hermeticum *and the Latin* Asclepius *in a New English Translation, with Notes and Introduction*. Translated by Brian P. Copenhaver. 1992. Cambridge: Cambridge University Press.

Herodotus. *The Persian Wars, Vol. 2: Books 3–4*. Edited and translated by A. D. Godley. 1921. Loeb Classical Library 118. Cambridge: Harvard University Press.

Hersch, Karen K. 2010. *The Roman Wedding: Ritual and Meaning in Antiquity*. Cambridge: Cambridge University Press.

Herva, V-P. 2006. "Flower Lovers, After All? Rethinking Religion and Human-Environment Relations in Minoan Crete." *World Archaeology*, 38(4): 586–98.

Herzhoff, Bernhard. 1990. "Phegos. Zur Identifikation eines umstrittenen Baumnamens." *Hermes*, 118: 257–72.

Hesiod. *Theogony. Works and Days. Testimonia*. Edited and translated by Glenn W. Most. 2007. Loeb Classical Library 57. Cambridge: Harvard University Press.

(Hieron.) *See* Jerome.

Higgins, Reynold. [1961] 1980. *Greek and Roman Jewellery*. 2nd edn. Berkeley and Los Angeles: University of California Press.

Hildebrand, E. 2003. "Comparison of Domestic vs. Forest-growing *Ensete ventricosum* (Welw.) *Cheesman, Musaceae* in Ethiopia: Implications for Detecting Enset Archaeologically, and Modeling its Domestication." In K. Neumann et al. (eds.), *Food, Fuel, and Fields: Progress in African Archaeobotany*. Africa Praehistorica 15, 49–70. Köln: Heinrich Barth Institut.

Hill, J. 2006. "The Historical Linguistics of Maize Cultivation in Mesoamerica and North America." In John E. Staller et al. (eds.), *Histories of Maize: Multidisciplinary Approaches to the Prehistory, Linguistics, Biogeography, Domestication, and Evolution of Maize*, 631–643. Amsterdam: Academic Press.

Hillman, G. 1975. "The Plant Remains from Tell Abu Hureyra: A Preliminary Report." *Proceedings of the Prehistoric Society*, 41: 70–3.

Hillman, G. 2000. "Abu Hureyra 1: The Epipalaeolithic." In A. M. T. Moore et al. (eds.), *Village on the Euphrates: From Foraging to Farming at Abu* Hureyra, 327–99. Oxford: Oxford University Press.

Hillman, G., et al. 1989. "Plant Food Economy During the Epipalaeolithic Period at Tell Abu Hureyra, Syria: Dietary Diversity, Seasonality, and Modes of Exploitation." In D. R. Harris and Gordon Hillman (eds.), *Foraging and Farming: The Evolution of Plant Exploitation*, 240–68. London: Unwin Hyman.

Hippocrates. *Generation. Nature of the Child. Diseases 4. Nature of Women and Barrenness*. Translated and edited by Paul Potter. 2012. Loeb Classical Library 520. Cambridge: Harvard University Press.

Hippocrates. *Hippokrates, Über die Natur des Kindes* (De genitura *und* De natura pueri), *ins Deutsche und Italienische übersetzt und textkritisch kommentiert*. Edited by Franco Giorgianni. 2006. Serta Graeca, Band 23. Wiesbaden: Ludwig Reichert Verlag.

Hippolytus. *Hippolytus Werke, Vol. 1: Hippolyts Kommentar zum Buche Daniel und die Fragmente des Kommentars zum Hohliede*, Griechischen christlichen Schriftsteller der ersten drei Jahrhunderte. Edited by Gottlieb Nathanael Bonwetsch and Hans Achelis. 1897. Leipzig: J. C. Hinrichs.

Historia Augusta. Vol. II. Caracalla. Geta. Opelllius Macrinus. Diadumenianus. Elagabalus. Severus Alexander. The Two Maximini. The Three Gordians. Maximus and Balbinus. Edited and translated by David Magie. 1924. Cambridge, MA and London: Harvard University Press.

Holleran, Claire. 2012. *Shopping in Ancient Rome: The Retail Trade in the Late Republic and the Principate*. Oxford: Oxford University Press.

Holt, Steve. 2012. "City to Tomato-growing Couple: 'You're breaking the law!'" *Take Part*, May 28. Available online: http://dev.takepart.com/article/2012/05/28/city-tomato-growing-couple-youre-breaking-law/index.html (accessed May 7, 2018).

Homer. *Homerus Odyssea*. Edited by Martin L. West. 2016. Bibliotheca scriptorum Graecorum et Romanorum Teubneriana. Berlin, Boston: De Gruyter.

Homer. *Odyssey, Vol. 1: Books 1–12*. Edited and translated by A. T. Murray and revised by George E. Dimock. 1919. Loeb Classical Library 104. Cambridge: Harvard University Press.

Homer. *Odyssey, Vol. 2: Books 13–24*. Edited and translated by A. T. Murray and revised by George E. Dimock. 1919. Loeb Classical Library 105. Cambridge: Harvard University Press.

Homer. *The Iliad, Vol. 1: Books 1–12*. Edited and translated by A. T. Murray. 1924. Loeb Classical Library 170. Cambridge: Harvard University Press.

Homeric Hymns. *Homeric Hymns. Homeric Apocrypha. Lives of Homer*. Edited and translated by Martin L. West. 2003. Loeb Classical Library 496. Cambridge: Harvard University Press.

Hood, S. 1994. *The Arts in Prehistoric Greece*. 2nd edn. New Haven: Yale University Press.

Horace. *Odes and Epodes*. Edited and translated by Niall Rudd. 2004. Loeb Classical Library 33. Cambridge: Harvard University Press.

Howes, D. 1987. "Olfaction and Transition: An Essay on the Ritual Uses of Smell". *Canadian Review of Sociology and Anthropology*, 24: 398–416.

Hu, Yunhua. 1987. *Chinese Penjing: Miniature Trees and Landscapes*. Portland: Timber Press.

Huang, Negfu (ed.). 1990. *The Great Treasury of Chinese Fine Arts: Arts and Crafts. 6., Printing, Dyeing, Weaving and Embroidery*. Beijing: Beijing Cultural Relics Publishing House.

Huckell, L. 2006. "Ancient Maize in the American Southwest: What Does It Look Like and What Can It Tell Us?" In John E. Staller et al. (eds.), *Histories of Maize: Multidisciplinary Approaches to the Prehistory, Linguistics, Biogeography, Domestication, and Evolution of Maize*, 97–106. Amsterdam: Academic Press.

Huffington Post. 2012/17. "Illegal Front Yard Garden: Canadian Couple's Kitchen Garden Targeted by Authorities [updated]." July 19, updated December 6, 2017. Available online: https://www.huffingtonpost.com/2012/07/20/illegal-kitchen-garden_n_1687558.html (accessed March 28, 2021).

Hughes, J. D., and J. V. Thirgood. 1982. "Deforestation, Erosion, and Forest Management in Ancient Greece and Rome." *Journal of Forest History*, 26(2): 60–75.

Hunt, John Dixon. 2012. *A World of Gardens*. London: Reaktion Books.

Hunt, Patrick. 1988. "Review of the International Summer School in Papyrology (Institute of Classical Studies, London)." *Bulletin of the Institute of Classical Studies*, 35: 159–61.

Hunt, Patrick. 1996. "Sensory Images in the Song of Songs." *Collected Papers of the IOSOT Congress, Paris, Sorbonne-College de France*. Beitrage zur Erforhschung des Alten Testaments und des Antiken Judentums, 28: 188–94.

Hunt, Patrick. 2004. "563 CE: Silk Worms Are Smuggled to the Byzantine Empire." In *Great Events in History: The Middle Ages 477–1453*, vol. 1. New York: Salem Press.

Hunt, Patrick. 2007. *Alpine Archaeology*, 31–32. New York: Ariel Books.

Hunt, Patrick. 2008. *Poetry in the Song of Songs: A Literary Analysis. Studies in Biblical Literature*, 96. Frankfurt: Peter Lang Verlag.

Hunt, Patrick. 2016. *When Empires Clash: Twelve Great Battles in Antiquity*. Newport, RI: Stone Tower Books.

Hunter, Michael. 2017. *Confucius Beyond the Analects*. Leiden: Brill.

Hunter, Richard L. (ed. and commentary). 1983. *Eubulus. The Fragments*. Cambridge: Cambridge University Press.

Hurcombe, Linda. 2014. *Perishable Material Culture in Prehistory: Investigating the Missing Majority*. Oxford: Routledge.

Hurvitz, Avi. 1967. "The Usage of Shesh and Bus in the Bible and Its Implication for the Date of P." *Harvard Theological Review*, 60(1): 117–21.

Hurwit, Jeffrey M. 1991. "The Representation of Nature in Early Greek Art." *Studies in the History of Art*, 32: 32–62.

Huxley, George. 1966. "Numbers in the Homeric Catalog of Ships." *Greek, Roman and Byzantine Studies*, 7(4): 313–18.

Hyginus. *The Myths of Hyginus*. Translated and edited by Mary Grant. 1960. Lawrence: University of Kansas Publications.

(IG) *Inscriptiones Graecae*, Berlin-Brandenburgische Akademie der Wissenschaften Digitale Edition. Available online: http://telota.bbaw.de/ig/digitale-edition/inschrift/IG%20 I%C2%B3%201261 (accessed March 25, 2021).

Iltis, H. 2006. "Origin of Polystichy in Maize." In John E. Staller et al. (eds.), *Histories of Maize: Multidisciplinary Approaches to the Prehistory, Linguistics, Biogeography, Domestication, and Evolution of Maize*, 22–50. Amsterdam: Academic Press.

Immerwahr, S. 1990. *Aegean Painting in the Bronze Age*. University Park: The Pennsylvania State University Press.

Isidore of Seville. *Isidori Hispalensis Episcopi Etymologiarum sive Originum libri xx*. Critical edition by Wallace M. Lindsay. 1911. Oxford: Oxford University Press.

Istros. "Istros (334)." Edited by Monica Berti and Steven Jackson. In Ian Worthington (ed.), *Brill's New Jacoby*. 2015. Available online: http://dx.doi.org/10.1163/1873-5363_bnj_a334 (accessed January 31, 2019).

Jacobsthal, Paul. 1925. "The Ornamentation of Greek Vases." *The Burlington Magazine for Connoisseurs*, 47(269): 64–75.

Jacomet, Stefanie, et al. 2002. "*Punica granatum* L. (pomegranates) from Early Roman Contexts in Vindonissa (Switzerland)." *Vegetation History and Archaeobotany*, 11(1): 79–92.

Jacomet, Stephanie. 2009. "Plant Economy and Village Life in Neolithic Lake Dwellings at the Time of the Alpine Iceman." *Vegetation History and Archaeobotany*, 18: 47–59.

Janick, Jules. 2005. "The Origin of Fruits, Fruit Growing, and Fruit Breeding." *Plant Breeding Reviews*, 25: 225–320.

Jashemski, Wilhelmina F. 1963. "The Flower Industry at Pompeii." *Archaeology*, 16(2): 112–21.

Jashemski, Wilhelmina F. 1979. *The Gardens of Pompeii Herculaneum and the Villas Destroyed by Vesuvius*. New Rochelle, New York: Caratzas Brothers.

(Jdc.) Judges. *The New Oxford Annotated Bible: New Revised Standard*. Edited by Michael D. Coogan. 2018. 5th edn. Oxford: Oxford University Press.

Jeremiah. *The New Oxford Annotated Bible: New Revised Standard*. Edited by Michael D. Coogan. 2018. 5th edn. Oxford: Oxford University Press.

Jerome. "Commentarius in Ecclesiasten" In *S. Hieronymi Presbyteri opera*, 249–361. Corpus christianorum, series latina 72. Edited by Marcus Adriaen. 1959. Turnhout: Brepols.

Jerome. *Commentarii in prophetas minores*. Edited by Marcus Adriaen. 1969–70. Corpus christianorum, series latina 76 and 76A. Turnhout: Brepols.

Jerome. *Epistulae 1–70*. Edited by I. Hilberg and Margit Kamptner (ed. al. supp.). 1910/18 Corpus scriptorum ecclesiasticorum latinorum 54. Editio altera supplementis aucta 1996. Vindobonae: Verlag der Österreichischen Akademie der Wissenshaften.

(Joh. Lyd. Men.) John the Lydian. *De Mensibus*. Edited by Richardus Wuensch. 1898. Leipzig: Teubner.

Jones, A.M.P., et al. 2013. "Morphological Diversity in Breadfruit (Artocarpus, Moraceae)." *Genetic Resources and Crop Evolution*, 60: 175–92.

Jones, B. 2007. "A Reconsideration of the Kneeling-Figure Fresco from Hagia Triada." In P. Betancourt et al. (eds.), *Krinoi kai Limenes. Studies in Honor of Joseph and Maria Shaw*, 151–158. Philadelphia: INSTAP Academic Press.

Josephus. *Antiquities, Vol. 1: Books 1–3*. Edited and translated by H. St. J. Thackeray. 1930. Loeb Classical Library 242. Cambridge: Harvard University Press.

Josephus. *The Jewish War, Vol. 3: Books 5–7*. Edited and translated by H. St. J. Thackeray. 1928. Loeb Classical Library 210. Cambridge: Harvard University Press.

Juniper, Barrie. 2001. "The Birth of the Modern Apple." *Petits Propos Culinaires*, 66: 49–54.

Juvenal. *Juvenal and Persius*. Edited and translated by Susanna Morton Braund. 2004. Loeb Classical Library 91. Cambridge: Harvard University Press.

Kakavoyiannis, Evangelos. 2007. "Memories of Phrasikleia." In George F. Bass et al. (eds.), *Great Moments in Greek Archaeology*, 332–7. Los Angeles: The J. Paul Getty Museum.

Kambylis, Athanasios. 1965. *Die Dichterweihe unh ihre Symbolik: Untersuchungen zu Hesiodos, Kallimachos, Propez und Ennius*. Heidelberg: Carl Winter Universitätsverlag.

Kandeler, Riklef. 2003. *Symbolik der Pflanzen und Farben: Botanische Kunst- und Kulturgeschichte in Beispielen*. Abhandlungen der Zoologisch-Botanischen Gesellschaft 33. Österreich: Zoologisch-Botanischen Gesellschaft.

Karttunen, Klaus. 1997. *India and the Hellenistic World*. Studia Orientalia, 83. Helsinki: Finnish Oriental Society.

Keay, Simon. 2001. "Gateway to Rome." *British Archaeology*, 57: 20–3.

Kees, Hermann. 1956. *Totenglauben und Jenseitsvorstellungen der alten Ägypter: Grundlagen und Entwicklung bis zum ende des Mittleren Reiches*. Berlin: Akademie-Verlag.

Keith, Arthur L. 1933. "Nature-Imagery in Vergil's 'Aeneid'." *The Classical Journal*, 28(8): 591–610.

Keller, F. 1866. *The Lake Dwellings of Switzerland and Other Parts of Europe*. Translated by John Edward Lee. London: Longmans, Green and Co.

Kellum, Barbara A. 1994. "The Construction of Landscape in Augustan Rome: The Garden Room at the Villa ad Galinas." *The Art Bulletin*, 76(2): 211–24.

Kenawi, Mohamed, Elizabeth Macaulay-Lewis, and Judith S. McKenzie. 2012. "A Commercial Nursery Near Abu Hummus (Egypt) and Re-use of Amphoras for the Trade in Plants." *Journal of Roman Archaeology*, 25: 195–225.

Kings, 1. *The New Oxford Annotated Bible: New Revised Standard*. Edited by Michael D. Coogan. 2018. 5th edn. Oxford: Oxford University Press.

Kirpalani, Resham. 2011. "Woman Faces Jail Time for Growing Vegetable Garden in Her Own Front Lawn." *ABC News*, July 12. Available online: http://abcnews.go.com/US/vegetable-garden-brings-criminal-charges-oak-park-michigan/story?id=14047214 (accessed March 27, 2021).

Kislev, Mordechai E. 1997. "Early Agriculture and Pleoecology of Nativ Hagdud." In Ofer Bar-Yosef and A. Gopher (eds.), *An Early Neolithic Village in the Jordan Valley. Part 1: The Archaeology of Nativ Hagdud*, 201–36. Cambridge, MA: Peabody Museum of Archaeology and Ethnology, Harvard University.

Kislev, Mordechai E., et al. 1992. "Epipalaeolithic (19,000 BP) Cereal and Fruit Diet at Ohalo II, Sea of Galilee, Israel." *Review of Palaeobotany and Palynology*, 73(1992): 161–6.

Kislev, Mordechai E., et al. 2006. "Early Domesticated Fig in the Jordan Valley." *Science*, 312 (5778): 1372–4.

Kistler, L., et al. 2014. "Transoceanic Drift and the Domestication of African Bottle Gourds in the Americas." *Proceedings of the National Academy of Sciences*, 111(8): 2937–41.

Kistler, L., et al. 2015. "Gourds and Squashes (*Cucurbita* spp.) Adapted to Megafaunal Extinction and Ecological Anachronism Through Domestication." *Proceedings of the National Academy of Sciences*, 112 (49): 15107–12.

Klee, M., et al. 2000. "Four Thousand Years of Plant Exploitation in the Chad Basin of Northeast Nigeria I: The Achaeobotany of Kursakata." *Vegetation History and Archaeobotany*, 9(4): 223–37.

Knechtges, David R. (ed.). 1982–7. *Wen Xuan*. Vols 1–2. Princeton: Princeton University Press.

Koerper, H, and A. L. Kolls. 1999. "The Silphium Motif Adorning Ancient Libyan Coinage: Marketing a Medicinal Plant." *Economic Botany*, 53(2): 133–43.

Kolata, A. 1986. "The Agricultural Foundations of the Tiwanaku State: A View from the Heartland." *American Antiquity*, 51(4): 748–62.

König, Viola. 2005. "The Representation of Landscape, Gardens, and Other Cultivated Spaces in the Codices and Lienzo (Maps) from Native Mexizo." *Anales de Anttopología*, 39(1): 79–97.

Kraft, Kraig H., et al. 2014. "Multiple Lines of Evidence for the Origin of Domesticated Chili Pepper, *Capsicum annuum*, in Mexico." *Proceedings of the National Academy of Sciences*, 111: 6165–70.

Kunze-Götte, Erika. 2006. *Myrte als Attribut und Ornament auf Attischen Vasen*. Kilchberg: Akanthus.

Kuttner, Ann L. 1999a. "Culture and History at Pompey's Museum," *TAPhA*, 129: 343–73.

Kuttner, Ann L. 1999b. "Looking Outside Inside: Ancient Roman Garden Rooms." *Studies in the History of Gardens and Designed Landscapes*, 19(1): 7–35.

Kvavadze, E., et al. 2009. "30,000 Year Old Wild Flax Fibers." *Science AAAS*, 325: 1359.

Lackenbacher, S. 1990. *Le palais sans rival: Le récit de construction en Assyrie*. Paris: Découverte.

Landgren, Lena. 2004. *Lauro Myrto et Buxo Frequentata: A Study of the Roman Garden Through its Plants*. PhD thesis, Department of Archaeology and Ancient History, Lund University, Sweden.

Landgren, Lena. 2013. "Plantings." In Kathryn L. Gleason (ed.), *A Cultural History of Gardens in Antiquity*, 75-98. London: Bloomsbury.

Lannoy, S., et al. 2002. "Etude de «pains/galettes» archéologiques français." *Civilisations*, 49: 119–60.

Lawler, Andrew. 2002. "Report of Oldest Boat Hints at Early Trade Routes." *Science AAAS*, 296: 1791–2.

Lawrence, Arnold Walter, and Richard Allan Tomlinson. 1983. *Greek Architecture*, 4th edn. New Haven: Yale University Press.

Layard, A. H. 1853. *Nineveh and Its Remains*. London: Murray and Co.

Lazenby, John F. 1993. *The Defence of Greece, 490–479 B.C.* Warminster, England: Aris & Phillips.

Lebot, V. 1999. "Biomolecular Evidence for Plant Domestication in Sahul." *Genetic Resources and Crop Evolution*, 46: 619–28.

Ledderose, Lothar. 1983. "The Earthly Paradise: Religious Elements in Chinese Landscape Art." In Susan Bush and Christian Murck (eds.), *Theories of the Arts in China*, 165–183. Princeton: Princeton University Press.

Lee, Moonsil. 2001. *Dietary Conditions and Differential Access to Food Resources Among the Various Classes During the Han Period*. MA thesis. Available online: http://d-scholarship.pitt.edu/9982/1/Moonsilthesisfinal-cracked.pdf (accessed March 28, 2021).

Lehmann, Phyllis Williams. 1953. *Roman Wall Paintings from Boscoreale in the Metropolitan Museum of Art* (Monographs of Archaeology and Fine Arts, V). Cambridge, MA: Archaeological Institute of America.

Lembach, Kurt. 1970. *Die Pflanzen bei Theokrit*. Heidelberg: Carl Winter Universitätsverlag.

Leroi-Gourhan, Arlette. 1998. "Shanidar et ses fleurs." *Paléorient*, 24: 79–88.

Lev, E., et al. 2005. "Mousterian Vegetal Food in Kebara Cave, Mt. Carmel." *Journal of Archaeological Science*, 32: 475–84.

Lewis, Charlton, and Charles Short. 1980. *A Latin Dictionary*. Oxford: Clarendon Press.

Li, Hui-Lin. 1974. "An Archeological and Historical Account of Cannabis in China." *Economic Botany*, 28(4): 437–48.

Li, Hui-lin (ed.). 1979. *Nan-fang ts'ao-mu chuang: A Fourth-Century Flora of Southeast Asia*. Hong Kong: Chinese University Press.

Li, Xi-wen, et al. 2008. "Cinnamomum." *Flora of China*, 7: 166–87.

Linder, E. 1981. "A Canaanite Thalassocracy." In Gordon D. Young (ed.), *Ugarit in Retrospect: 50 Years of Ugarit and Ugaritic*, 31–42. Winona Lake: Eisenbrauns.

Liphschitz, N., and D. Nadal. 1997. "Charred Wood Remains from Ohalo II (19000 BP), Sea of Galilee, Israel." *Journal of Israel Prehistoric Society*, 27: 5–18.

Livy. *History of Rome, Vol. 3: Books 5–7*. Edited and translated by B. O. Foster. 1924. Loeb Classical Library 172. Cambridge: Harvard University Press.

Logan, A. L. 2012. *A History of Food without History: Food, Trade, and Environment in West-central Ghana in the Second Millennium AD*. Unpublished doctoral dissertation, Department of Anthropology, University of Michigan.

Lonie, Iain M. 1981. *The Hippocratic Treatises "On Generation", "On the Nature of the Child"; "Diseases IV". A Commentary by I.M. Lonie*. Berlin and New York: Walter de Gruyter.

Lopez, L. M., et al. 2011. "Traditional Post-harvest Processing to Make Quinoa Grains (Chenopodium quinoa var. quinoa) Apt for Consumption in Northern Lipez (Potosí, Bolivia): Ethnoarchaeological and Archaeobotanical Analyses." *Archaeological and Anthropological Sciences*, 3: 49.

Lubec, G., et al. 1993. "Use of Silk in Ancient Egypt." *Nature*, 362(6415): 25.

Lucan. *The Civil War (Pharsalia)*. Edited and translated by J. D. Duff. 1928. Loeb Classical Library 220. Cambridge: Harvard University Press.

Lucian. *Anacharsis or Athletics. Menippus or The Descent into Hades. On Funerals. A Professor of Public Speaking. Alexander the False Prophet. Essays in Portraiture. Essays in Portraiture Defended. The Goddesse of Surrye*. Edited and translated by A. M. Harmon. 1925. Loeb Classical Library 162. Cambridge: Harvard University Press.

Lucian. *Dialogues of the Dead. Dialogues of the Courtesans*. Edited and translated by M. D. MacLeod. 1961. Loeb Classical Library 431. Cambridge: Harvard University Press.

Luciani, Marta. 2010. "More than just Landscapes of Pleasure: The Garden Frame in the 'Investiture' Panel at Mari." *Wiener Zeitschrift für die Kunde des Morgenlandes*, 100: 99–118.

Lucretius. *On the Nature of Things*. Edited and translated by W. H. D. Rouse and revised by Martin F. Smith. 1924. Loeb Classical Library 181. Cambridge: Harvard University Press.

Luke, Gospel of. *The New Oxford Annotated Bible: New Revised Standard*. Edited by Michael D. Coogan. 2018. 5th edn. Oxford: Oxford University Press.

Lycophron. *Callimachus: Hymns and Epigrams. Lycophron: Alexandra. Aratus: Phaenomena.* Edited and translated by A. W. Mair and G. R. Mair. 1921. Loeb Classical Library 129. Cambridge: Harvard University Press.

Mabberley, David J. 2004. "Citrus (Rutaceae): A Review of Recent Advances in Etymology, Systematics and Medical Applications." *Blumea*, 49: 486–98.

Mabberley, David J. 2017. *Mabberley's Plant Book: A Portable Dictionary of Plants, Their Classification, and Uses.* 4th edn. Cambridge: Cambridge University Press.

Macaulay-Lewis, Elizabeth R. 2006a. "Planting Pots at Petra: A Preliminary Study of *ollae perforatae* at the Petra Garden Pool Complex and at the 'Great Temple'." *Levant*, 38: 159–70.

Macaulay-Lewis, Elizabeth R. 2006b. "The Role of *ollae perforatae* in Understanding Horticulture, Planting Techniques, Garden Design, and Plant Trade in the Roman World." In J. P. Morel, J. T. Juan, and J. C. Matamala (eds.), *The Archaeology of Crop Fields and Gardens: Proceedings from 1st Conference on Crop Fields and Gardens Archaeology. University of Barcelona, Barcelona, Spain, June 1–3rd 2006*, 207–219. Barcelona: EdiPuglia.

Macaulay-Lewis, Elizabeth R. 2008. "The Fruits of Victory: Generals, Plants, and Power in the Roman World." In E. Bragg, L. Hau, and Elizabeth R. Macauley-Lewis (eds.), *Beyond the Battlefields: New Perspectives on Warfare and Society in the Graeco-Roman World*, 205–225. Cambridge: Cambridge Scholars Publishers.

Macaulay-Lewis, Elizabeth R. 2010. "Imported Exotica: Approaches to the Study of the Ancient Plant Trade." *Bolletino di archeologia on line*, 1: 16–26.

MacDonald, J. A. 2002. "Botanical Determination of the Middle Eastern Tree of Life." *Economic Botany*, 56: 113–29.

Macrobius. *Saturnalia, Vol. 1: Books 1–2.* Edited and translated by Robert A. Kaster. 2011. Loeb Classical Library 510. Cambridge: Harvard University Press.

Macrobius. *Saturnalia, Vol. 2: Books 3–5.* Edited and translated by Robert A. Kaster. 2011. Loeb Classical Library 511. Cambridge: Harvard University Press.

Malakowska, E., et al. 2015. "Comparison of Papermaking Potential of Wood and Hemp Cellulose Pulps." *Annals of Warsaw University of Life Sciences—SGGW Forestry and Wood Technology*, 91: 134–7.

Malek, Amina-Aïcha. 2018. "Mosiacs and Nature in the Roman *Domus*." In Wilhelmina F. Jashemski et al. (eds.), *Gardens of the Roman Empire*, 317–40. Cambridge: Cambridge University Press.

Mangelsdorf, P. C. 1954. "New Evidence on the Origin and Ancestry of Maize." *American Antiquity*, 19(4): 409–10.

Manning, K. R., et al. 2011. "4500-year-old Domesticated Pearl Millet (*Pennisetum glaucum*) from the Tilemsi Valley, Mali: New Insights into an Alternative Dereal Domestication Pathway." *Journal of Archaeological Science*, 38(2): 312–22.

Manning, W. H. 1975. "Roman Military Timber Granaries." *Saalburg Jahrbuch*, 32: 105–29.

Manning, W. H., and I. R. Scott. 1979. "Roman Timber Military Gateways in Britain and on the German Frontier." *Britannia*, 10: 19–62.

Marinatos, N. 1984. *Art and Religion in Thera: Reconstructing a Bronze Age Society.* Athens: D. & I. Mathioulakis.

Marinval, Philippe, et al. 2002. "Arbres fruitiers et cultures jardinées gallo-romains à Longueil-Sainte-Marie (Oise)." *Gallia*, 59: 253–71.

Markle, Minor M. 1977. "The Macedonian Sarissa, Spear and Related Armor." *American Journal of Archaeology*, 81(3): 323–39.

Marsden, Peter. 1994. *The Ships of the Port of London: First to Eleventh Centuries AD.* English Heritage Archaeological Reports, 3. London: English Heritage.

Marshall, F., and E. Hildebrand. 2002. "Cattle before Crops: The Beginnings of Food Production in Africa." *Journal of World Prehistory*, 16(2): 99–143.

Martial. *Epigrams, Vol. I*. Edited and translated by D. R. Shackleton Bailey. 1993. Cambridge: Harvard University Press.

Martial. *Epigrams, Vol. II*. Edited and translated by D. R. Shackleton Bailey. 1993. Cambridge: Harvard University Press.

Matsui, Akira, and Masaaki Kanehara. 2006. "The Question of Prehistoric Plant Husbandry during the Jomon Period in Japan." *World Archaeology*, 38: 259–73.

Matsuoka, Y., et al. 2002. "A Single Domestication of Maize Shown by Multilocus Microsatellite Genotyping." *Proceedings of the National Academy of Sciences*, 99: 8060–4.

Matthew, Gospel of. *The New Oxford Annotated Bible: New Revised Standard*. Edited by Michael D. Coogan. 2018. 5th edn. Oxford: Oxford University Press.

Matthew, K. S. (ed.), 2017. *Imperial Rome, Indian Ocean Regions and Muziris. New Perspectives on Maritime Trade*. London and New York: Routledge.

Matthews, R., et al. 2013. "Investigating the Neolithisation of Society in the Central Zagros of Western Iran." In R. Matthews and H. F. Nashli (eds.), *The Neolithisation of Iran: Formation of New Societies*, 14–34. Oxford: Oxbow Books.

Maxwell, G. S. 1976. "A Roman Timber Tower at Beattock Summit." *Britannia*, 7: 33–8.

Mayeske, B. J. B. 1972. *Bakeries, Bakers, and Bread at Pompeii: A Study in Social and Economic History*. Ann Arbor, MI: Ann Arbor: University Microfilms.

Mayor, H. Hyatt. 1943. "Carpentry and Candlelight in the Theater." *Metropolitan Museum of Art Bulletin NS*, 1(6): 198–203.

Mbida Mindzie, C., et al. 2001. "First Archaeological Evidence of Banana Cultivation in Central Africa During the Third Millennium Before Present." *Vegetation History and Archaeobotany*, 10: 1–6.

McCamant, J. F. 1992. "Quinoa's Roundabout Journey to World Use." In Nelson Foster and Linda S. Cordell (eds.), *Chilies to Chocolate: Food the Americas Gave the World*, 123–141. Tucson: University of Arizona Press.

McGee, H. 2004. *On Food and Cooking: The Science and Lore of the Kitchen*. New York: Scribner.

McGovern, Patrick E. 2003. *Ancient Wine: The Search for the Origins of Viticulture*. Princeton, NJ: Princeton University Press.

McGovern, Patrick E., et al. 2004. "Fermented Beverages of Pre- and Proto-historic China." *Proceedings of the National Academy of Sciences*, 101: 17593–8.

McGregor, James H. 2015. *Back to the Garden: Nature and the Mediterranean World from Prehistory to the Present*. New Haven: Yale University Press.

McIntosh, S. K. 1995. "Paleobotanical and Human Osteological Remains." In S. K. McIntosh (ed.), *Excavations at Jenné-jeno, Hambarketolo, and Kaniana (Inland Niger Delta, Mali), the 1981 Season*, 348–59. Berkeley: University of California Press.

McIntosh, S. K., and R. J. McIntosh. 1979. "Initial Perspectives on Prehistoric Subsistence in the Inland Niger Delta (Mali)." *World Archaeology*, 11: 227–43.

McLaughlin, Raoul. 2010. *Rome and the Distant East: Trade Routes to the Ancient Lands of Arabia*. London: Continuum Books.

Meiggs, Russell. 1982. *Trees and Timber in the Ancient Mediterranean World*. Oxford: Clarendon Press.

Melamed, Y., et al. 2011. "Extinction of Water Plants in the Hula Valley: Evidence for Climate Change." *Journal of Human Evolution*, 60(4): 320–7.

Mell, Ulrich (ed.). 2006. *Pflanzen und Pflanzensprache der Bibel: Erträge des Hohenheimer Symposions vom 26 Mai 2004*. Frankfurt/Main: Peter Lang.

Menander. *Heros. Theophoroumene. Karchedonios. Kitharistes. Kolax. Koneiazomenai. Leukadia. Misoumenos. Perikeiromene. Perinthia*. Edited and translated by W. G. Arnott. 1997. Loeb Classical Library 459. Cambridge: Harvard University Press.

Meneghini, Roberto. 2012. "The Forum Pacis (or Templum Pacis) and the Torre dei Conti." *Ricerche di Storia dell'Arte*, 108: 32–47.

Milburn, Olivia. 2016. "Rhapsodies on an Exotic Plant from Rome." *Early Medieval China* 22: 26–44.

Miller, Allison, and Barbara Schaal. 2005. "Domestication of a Mesoamerican Cultivated Fruit Tree, Spondias purpurea." *Proceedings of the National Academy of Sciences*, 102: 12801–6.

Miller, Helena. 1979. *Iconography of the Palm in Greek Art: Significance and Symbolism*. PhD thesis. University of California, Berkeley.

Miller, James I. 1969 *The Spice Trade of the Roman Empire, 29 B.C. to A.D. 641*. Oxford: Clarendon Press.

Miller, N. 2008. "Sweeter than Wine? The Use of the Grape in Early Western Asia." *Antiquity*, 82: 937–46.

Miller, N., et al. 2016. "Sign and Image: Representations of Plants on the Warka Vase of Early Mesopotamia." *University of Pennsylvania Museum Archaeology and Anthropology Papers*. Available online: http://repository.upenn.edu/penn_museum_papers/2/ (accessed March 28, 2021).

Milne, Gustav. 1982. "Recording Timberwork on the Roman Waterfront." In S. MacGrail (ed.), *Woodworking Techniques Before A.D. 1500*. Oxford: British Archaeological Reports.

Milne, Gustav. 1985. *The Port of Roman London*, 7–23. London: Batsford.

Möbius, M. 1933. "Pflanzenbilder der Minoischen Kunst in Botanischer Betrachtung." *Jahrbuch des Deutschen Archäologischen Instituts*, 48: 1–39.

Moldes, Ana B., et al. 2007. "Evaluation of Mesophilic Biodegraded Grape Must as Fertilizer." *Applied Biochemistry and Biotechnology*, 141(1): 27–36.

Moody, J. Forthcoming. "Veteran and Sacred Trees in Modern and Minoan Crete." In Joanne M. A. Murphy and Jerolyn E. Morrison (eds.), *Kleronomia: Legacy and Inheritance: Studies on the Aegean Bronze Age in Honor of Jeffrey S. Soles*. Philadelphia: INSTAP Academic Press.

Morgan, L. 1988. *The Miniature Wall Paintings of Thera: A Study in Aegean Culture and Iconography*. Cambridge: Cambridge University Press.

Moschus. *Theocritus. Moschus. Bion*. Edited and translated by Neil Hopkinson. 2015. Loeb Classical Library 28. Cambridge: Harvard University Press.

Mt. Pleasant, Jane. 2006. "The Science behind the Three Sisters Mound System." In John E. Staller et al. (eds.), *Histories of Maize: Multidisciplinary Approaches to the Prehistory, Linguistics, Biogeography, Domestication, and Evolution of Maize*, 87–98. Amsterdam: Academic Press.

Muir, John. 2009. *Life and Letters in the Ancient Greek World*. London and New York: Routledge.

Mukherjee, S. K., and R. E. Litz. 2009. "Introduction: Botany and Importance." In Richard E. Litz (ed.), *The Mango*. 2nd edn., 1–18. London: CAB International.

Mulidzi, A. R. 2007. "Winery Wastewater Treatment by Contaminated Wetlands and the Use of Treated Wastewater for Cash Crop Production." *Water Science & Technology*, 56(2): 103–9.

Munro, J. M. 1995. *Spikenard and Saffron: A Study in the Poetic Language of the Song of Songs*, JSOT Supplement, 203. Sheffield: Sheffield Academic Press.

Murr, Josef F. 1890: *Die Pflanzenwelt in der Griechischen Mythologie*. Innsbruck: Verlage der Wagner'schen Universitäts Buchhandlung.

Murr, Josef F. 1890 [969]. *Die Pflanzenwelt in der griechischen Mythologie*. Reprint, Groningen: Verlag Bouma's Boekhuis.

Murray, S. S. 2004. "Searching for the Origins of African Rice Domestication." *Antiquity*, 78 (300): 5.

Murray, S. S. 2007. "Medieval Cotton and Wheat Finds in the Middle Niger Delta (Mali)." In R. J. Cappers (ed.), *Fields of Change: Progress in African Archaeobotany*, 43–51. Groningen: Barkhuis.

Myers, K. Sara. 2005. "*Docta Otia*: Garden Ownership and Configurations of Leisure in Statius and Pliny the Younger." *Arethusa*, 38(1): 103–29.

National Research Council. 1996. *Lost Crops of Africa Vol I: Grains*. Washington: National Academy Press.

National Research Council. 2006. *Lost Crops of Africa Vol II: Vegetables*. Washington: National Academy Press.

Neumann, K. 2003. "The Late Emergence of Agriculture in Sub-Saharan Africa: Archaeobotanical Evidence and Ecological Considerations." In K. Neumann et al. (eds.), *Food, Fuel and Fields. Progress in African Archaeobotany*, 71–92. Köln: Heinrich-Barth-Institute.

Newsom, L. A., and M. C. Mihlbachler. 2006. "Mastodons (*Mammut americanum*) Diet Foraging Patterns Based on Analysis of Dung Deposits.' In S. D. Webb (ed.), *First Floridians and Last Mastodons: The Page-Ladson Site in the Aucilla River*, 263–331. Dordrecht, The Netherlands: Springer.

Nicander of Colophon. *Nicander: The Poems and Poetical Fragments*. Edited and translated by A. S. F. Gow and A. F. Scholfield. 1953. Cambridge: Cambridge University Press.

Nicol, Donald M. 1958. "The Oracle of Dodona." *Greece and Rome*, 5: 128–43.

Nielsen, Karen-Hanne Stoermose. 2005. "A Preliminary Classification of Shapes of Loomweights." In Frances Pritchard and John P. Wild (eds.), *Northern Archaeological Textiles NESAT VII: Textile Symposium in Edinburgh, 5th–7th May, 1999*, 129–35. Oxford: Oxbow Books.

Niemeier, W-D., and B. Niemeier. 1998. "Minoan Frescoes in the Eastern Mediterranean." In E. Cline and D. Harris-Cline (eds.), *The Aegean and the Orient in the Second Millennium BC*, 69–97. Aegaeum 18. Liège: Université de Liège, Histoire de l'art et archéologie de la Grèce antique; University of Texas at Austin, Program in Aegean Scripts and Prehistory.

Novák, Mirko. 2001. "The Artificial Paradise: Programme and Ideology of Royal Gardens." In Simo Parpola and Robert Whiting (eds.), *Sex and Gender in the Ancient Near East: Proceedings of the 47th Rencontre Assyriologique Internationale, Helsinki, July 2–6, 2001*, 443–459. Helsinki:

Nugent, M. 2012. "Natural Adornment by Design: Beauty and/or Function? Botanic Motifs of the Bronze Age Cycladic Islands." In M-L. Nosch and R. Laffineur (eds.), *Kosmos: Jewellery, Adornment and Textiles in the Aegean Bronze Age*. Aegaeum 33, Leuven and Liege: Peeters.

Numbers. *The New Oxford Annotated Bible: New Revised Standard*. Edited by Michael D. Coogan. 2018. 5th edn, 589–596. Oxford: Oxford University Press.

Nutton, Vivian 1985. "The Drug Trade in Antiquity." *Journal of the Royal Society of Medicine*, 78: 138–45.

Oakley, John Howard, and Rebecca H. Sinos. 1993. *The Wedding in Ancient Athens*. Madison: University of Wisconsin Press.

Oates, J., and D. Oates. 2001. *Nimrud: An Assyrian Imperial City Revealed*. London: British Institute for the Study of Iraq.

Olympiodorus. *Hesychii Hierosolymitani Presbyteri, Olympiodori Alexandrini, Leontii Neapoleos in Cypro Episcopi, opera omnia*. Edited by Jacques-Paul Migne. 1865. Patrologia Graeca 93. Paris: Migne.

Oppenheim, A. Leo. 1965. "On Royal Gardens in Mesopotamia." *Journal of Near Eastern Studies*, 24(4): 331–3.

Origen. *Origène: Homélies sur le Cantique des cantiques*. Edited and translated by Olivier Rousseau. 1854. Sources Chrétiennes 37. Paris: Cerf.

Origen. *Origenis: Opera Omnia*. Edited by Jacques-Paul Migne. 1857. Patrologia Graeca 17. Paris: Migne.

Osmun, George P. 1975. "Roses of Antiquity." *The Classical Outlook*, 52(10): 110–12.

Östenberg, Ida. 2009. *Staging the World: Spoils, Captives, and Representations in the Roman Triumphal Procession*. Oxford: Oxford University Press.

Ovalle, David. 2016. "Court Upholds Miami Shores Ban on Veggie Gardens." *Miami Herald*. August 25.

Ovid. *Art of Love. Cosmetics. Remedies for Love. Ibis. Walnut-tree. Sea Fishing. Consolation*. Edited and translated by J. H. Mozley and revised by G. P. Goold. 1929. Loeb Classical Library 232. Cambridge: Harvard University Press.

Ovid. *Fasti*. Edited and translated by James G. Frazer and revised by G. P. Goold. 1931. Loeb Classical Library 253. Cambridge: Harvard University Press.

Ovid. *Heroides. Amores*. Edited and translated by Grant Showerman and revised by G. P. Goold. 1914. Loeb Classical Library 41. Cambridge: Harvard University Press.

Ovid. *Metamorphoses, Vol. 1: Books 1–8*. Edited and translated by Frank Justus Miller and revised by G. P. Goold. 1916. Loeb Classical Library 42. Cambridge: Harvard University Press.

Ovid. *P. Ovidi Nasonis Metamorphoses. Recognovit brevique adnotatione critica instruxit R. J. Tarrant*. Edited by R. J. Tarrant. 2004. Oxford Classical Texts. Oxford: Oxford University Press.

Ovid. *Publius Ovidius Naso, Die Fasten*. Band I: Einleitung, Text und Übersetzung; Band II: Kommentar. Edited by Franz Bömer. 1957–8. Heidelberg: Carl Winter.

Ovid. *Tristia. Ex Ponto*. Edited and translated by A. L. Wheeler and revised by G. P. Goold. 1924. Loeb Classical Library 151. Cambridge: Harvard University Press.

Owyoung, Steven D. 2013. "Tianluoshan: Tea in the Neolithic Era." *Tsiosophy*. Available online: http://www.tsiosophy.com/2013/06/tianluoshan-tea-in-the-neolithic-era-3/ (accessed March 28, 2021).

Pagán, Victoria E. 2006. *Rome and the Literature of Gardens*. London: Duckworth.

Palladius. *Palladii Rutilii Tauri Aemiliani Opus agriculturae; De veterinaria medicina; De insitione*. Edited by Robert H. Rodgers. 1975. Leipzig: Teubner.

Panagiotakopulu, E., et al. 1997. "A Lepidopterous Cocoon from Thera and Evidence for Silk in the Aegean Bronze Age." *Antiquity*, 71(272): 420–9.

Papyri Graecae Magicae (PGM). *Papyri Graecae Magicae / Die grieschischen Zauberpapyri*, vol. 1. Edited by Karl Preisendanz and Albert Henrichs. 1973–4. 2nd edn. Stuttgart: Teubner.

Papyri Graecae Magicae (PGM). *The Greek Magic Papyri in Translation: Including the Demonic Spells, Vol. 1: Texts*. Edited by Hans Dieter Betz. 1966. 2nd edn. Chicago: University of Chicago Press.

Parejko, K. 2003. "Pliny the Elder's Silphium: First Recorded Species Extinction." *Conservation Biology*, 17(3): 925–7.

Parker, Grant 2008. *The Making of Roman India. Greek Culture in the Roman World.* Cambridge: Cambridge University Press.

Parpola, S. 1987. *Correspondence of Sargon II, Part I: Letters from Assyria and the West. State Archives of Assyria* 1. Helsinki: Neo-Assyrian Text Corpus.

Parrot, A. 1937. *Les Fouilles de Mari.* Paris: Geuthner.

(Ps.) Paulinus of Nola. *Paulinus Nolanus, Carmina; Paulinus Pellaeus, Oratio.* Edited by Wilhelm Hartel and Margit Kamptner (ed. al. supp.). 1999. Corpus Scriptorum Ecclesiasticorum Latinorum 30. Vindobonae: Verlag der Österreichischen Akademie der Wissenshaften.

Pausanias. *Description of Greece, Vol. 1: Books 1–2 (Attica and Corinth).* Edited and translated by W. H. S. Jones. 1918. Loeb Classical Library 93. Cambridge: Harvard University Press.

Pausanias. *Description of Greece, Vol. 2: Books 3–5 (Laconia, Messenia, Elis 1).* Edited and translated by W. H. S. Jones and H. A. Omerod. 1926. Loeb Classical Library 188. Cambridge: Harvard University Press.

Pausanias. *Description of Greece, Vol. 3: Books 6–8.21 (Elis 2, Achaia, Arcadia).* Edited and translated by W. H. S. Jones. 1933. Loeb Classical Library 272. Cambridge: Harvard University Press.

Pausanias. *Pausaniae Graeciae descriptio.* 3 vols. Edited by Maria Helena Rocha-Pereira. 1973–81. Leipzig: Teubner.

Payne, Humphry. 1971. *Necrocorinthia: A Study of Corinthian Art in the Archaic Period.* College Park, MD: McGrath Publishing.

Pherekrates. *Fragments of Old Comedy, Vol. II: Diopeithes to Pherecrates.* Edited and translated by Ian C. Storey. 2011. Loeb Classical Library 514. Cambridge: Harvard University Press.

Philippson, G., and S. Bahuchet. 1994/5. "Cultivated Crops and Bantu Migrations in Central and Eastern Africa: A Linguistic Approach." *Azania* 29/30, special edition, J. E. G. Sutton (ed.), *The Growth of Farming Communities in Africa from the Equator Southwards*, 103–20. Nairobi: British Institute in Eastern Africa.

Philostratus (the Athenian). *Apollonius of Tyana, Vol. 2: Life of Apollonius of Tyana, Books 5–8.* Edited and translated by Christopher P. Jones. 2005. Loeb Classical Library 17. Cambridge: Harvard University Press.

Philostratus the Elder. *Philostratus the Elder, Imagines. Philostratus the Younger, Imagines. Callistratus, Descriptions.* Translated by Arthur Fairbanks. 1931. Loeb Classical Library 256. Cambridge: Harvard University Press.

Philoxenus of Leucas. *Greek Lyric, Vol. 5: The New School of Poetry and Anonymous Songs and Hymns.* Edited and translated by David A. Campbell. 1993. Loeb Classical Library 144. Cambridge: Harvard University Press.

Physiologus. *Physiologus: Griechisch / Deutsch.* Edited by Otto Schönberger. 2001. Reclam Universal-Bibliothek. Stuttgart: Reclamverlag.

Physiologus. *Physiologus.* Edited by Francesco Sbordone. 1936. Milan: Società Anonima Editrice "Dante Alighieri."

Physiologus. *Physiologus.* Translated by Michael J. Curley. 1979. Chicago and London: University of Chicago Press.

Pindar. *Olympian Odes. Pythian Odes.* Edited and translated by William H. Race. 1997. Loeb Classical Library 56. Cambridge: Harvard University Press.

Piperno, Dolores. 2009. "Identifying Crop Plants with Phytoliths (and Starch Grains) in Central and South America: A Review and an Update of the Evidence." *Quaternary International*, 193: 146–59.

Piperno, Dolores R., and K. V. Flannery. 2001. "The Earliest Archaeological Maize (*Zea mays* L.) from Highland Mexico: New Accelerator Mass Spectrometry Dates and Their Implications." *Proceedings of the National Academy of Sciences*, 98: 2101–3.

Piperno, Dolores R., and Karen E. Stothert. 2003. "Phytolith Evidence for Early Holocene Cucurbita Domestication in Southwest Ecuador." *Science* 299(5609): 1054–7.

Piperno, Dolores R., et al. 2009. "Starch Grain and Phytolith Evidence for Early Ninth Millennium B.P. Maize from the Central Balsas River Valley, Mexico." *Proceedings of the National Academy of Sciences*, 106: 5019–24.

Piperno, Dolores R. 2011. "The Origins of Plant Cultivation and Domestication in the New World Tropics: Patterns, Process, and New Developments." *Current Anthropology*, 52: S453–S470.

Pitra, Jean-Baptiste (ed.). 1852–8. *Spicilegium Solesmense: Complectens Sanctorum Patrum scriptorumque ecclesiasticorum anecdota hactenus opera, selectae Graecis Orientalibusque et Latinis codicibus*, 4 vols. Paris: Didot.

Plato. *Euthyphro. Apology. Crito. Phaedo. Phaedrus.* Edited and translated by Christopher Emlyn-Jones and William Preddy. 2017. Loeb Classical Library 36. Cambridge: Harvard University Press.

Plato. *Laws, Vol. 2: Books 7–12*. Edited and translated by R. G. Bury. 1926. Loeb Classical Library 192. Cambridge: Harvard University Press.

Plato. *Platon: Phaidros*. Platon: Werke, vol. 3. Translated and commentary by Ernst Heitsch. 1997. 2nd edn. Göttingen: Vandenhoeck & Ruprecht.

Plato. *Platonis Opera, vols 1–4*. Edited by John Burnet. 1989. Oxford: Clarendon Press.

Plato. *Timaeus. Critias. Cleitophon. Menexenus. Epistles.* Edited and translated by R. G. Bury. 1929. Loeb Classical Library 234. Cambridge: Harvard University Press.

Platz-Horster, Gertrud, and Michael Vickers. 2002. *Ancient Gold Jewellery*. Mainz am Rhein: Von Zabern.

Pliny the Elder. *Natural History*. Edited and translated by H. Rackham. 1940–63 (1952–67 printing). Cambridge, MA: Harvard University Press.

Pliny the Elder. *Natural History, Vol. IV: Books 12–16*. Edited and translated by H. Rackham. 1945. Loeb Classical Library 370. Cambridge: Harvard University Press.

Pliny the Elder. *Natural History, Vol. V: Books 17–19*. Edited and translated by H. Rackham. 1950. Loeb Classical Library 371. Cambridge: Harvard University Press.

Pliny the Elder. *Natural History, Vol. VI: Books 20–23*. Edited and translated by W. H. S. Jones. 1951. Loeb Classical Library 392. Cambridge: Harvard University Press.

Pliny the Elder. *Natural History, Vol. VII: Books 24–27*. Edited and translated by W. H. S. Jones and A.C. Andrews. 1956. Loeb Classical Library 393. Cambridge: Harvard University Press.

Pliny the Younger. *Letters, Vol. I: Books 1–7*. Edited and translated by Betty Radice. 1969. Loeb Classical Library 55. Cambridge: Harvard University Press.

Plowman, Timothy. 1984. "The Origin, Evolution, and Diffusion of Coca, Erythroxylum spp., in South and Central America." In D. Stone (ed.), *Pre-Columbian Plant Migration* (*Papers of the Peabody Museum of Archaeology and Ethnology*, 76), 125–63. Cambridge: Harvard University Press.

Plutarch. *Lives, Vol. I: Theseus and Romulus. Lycurgus and Numa. Solon and Publicola*. Edited and translated by Bernadotte Perrin. 1914. Loeb Classical Library 46. Cambridge: Harvard University Press.

Plutarch. *Lives, Vol. V: Agesilaus and Pompey. Pelopidas and Marcellus*. Edited and translated by Bernadotte Perrin. 1917. Loeb Classical Library 87. Cambridge: Harvard University Press.

Plutarch. *Lives, Vol. VI: Dion and Brutus. Timoleon and Aemilius Paulus*. Edited and translated by Bernadotte Perrin. 1918. Loeb Classical Library 98. Cambridge: Harvard University Press.

Plutarch. *Lives, Vol. VII: Demosthenes and Cicero. Alexander and Caesar*. Edited and translated by Bernadotte Perrin. 1919. Loeb Classical Library 99. Cambridge: Harvard University Press.

Plutarch. *Moralia, Vol. II: How to Profit by One's Enemies. On Having Many Friends. Chance. Virtue and Vice. Letter of Condolence to Apollonius. Advice About Keeping Well. Advice to Bride and Groom. The Dinner of the Seven Wise Men. Superstition*. Edited and translated by Frank Cole Babbit. 1918. Loeb Classical Library 222. Cambridge: Harvard University Press.

Plutarch. *Moralia, Vol. III: Sayings of Kings and Commanders. Sayings of Romans. Sayings of Spartans. The Ancient Customs of the Spartans. Sayings of Spartan Women. Bravery of Women*. Edited and translated by Frank Cole Babbit. 1931. Loeb Classical Library 245. Cambridge: Harvard University Press.

Plutarch. *Moralia, Vol. IV: Roman Questions, Greek Questions. Greek and Roman Parallel Stories. On the Fortune of the Romans. On the Fortune or the Virtue of Alexander. Were the Athenians More Famous in War or in Wisdom?* Edited and translated by Frank Cole Babbit. 1936. Loeb Classical Library 305. Cambridge: Harvard University Press.

Plutarch. *Moralia, Vol. IX: Table Talk, Books 7–9. Dialogue on Love*. Edited and translated by Edwin L. Minar et al. 1961. Loeb Classical Library 425. Cambridge: Harvard University Press.

Plutarch. *Moralia, Vol. XII. Concerning the Face which appears in the Orb of the Moon. On the Principle of Cold. Whether Fire or Water is more Useful. Whether Land or Sea Animals are Cleverer. Beasts are Rational. On the Eating of Flesh*. Edited and translated by H. Cherniss and W. C. Helmbold. 1957. Cambridge, MA and London: Harvard University Press.

Plutarch. *Plutarchi vitae parallelae*, Band 3, Fasc. 2 (enthält: Lykurg, Numa, Lysandros, Sulla, Agesilaos, Pompeius, Galba, Otho). Edited by Konrat Ziegler and Hans Gärtner, Hans Auflage. 1973. Leipzig: Teubner.

Polinger-Foster, Karen. 2004. "The Hanging Gardens of Nineveh." *Iraq*, 66: 207–20.

Pollard, E.A. 2009. "Pliny's *Natural History* and the Flavian *Templum Pacis*: Botanical Imperialism in First-Century CE Rome." *Journal of World History*, 20: 309–38.

Pollux, (Julius). *Pollucis onomasticon*, vol. 1. Edited by Erich Bethe. 1900. Leipzig: B. G. Teubner.

Pokorny, Julius. 1959. *Indogermanisches Etymologisches Wörterbuch*. Bern: Franke.

Pongratz-Leisten, B. 1999. "Neujahr(sfest)." *Reallexikon des Assyriologie*, Bd. 9: 294–8.

Porter, B. N. 1993. *Images, Power and Politics: Figurative Aspects of Esarhaddon's Babylonian Policy*. Philadelphia, PA: American Philosophical Society.

Porter, R. 2000. "The Flora of the Theran Wall Paintings: Living Plants and Motifs—Sea Lily, Crocus, Iris and Ivy." In S. Sherratt (ed.), *The Wall-paintings of Thera. Proceedings of the 1st International Symposium, Petros Nomikos Conference Centre, Thera, Hellas; 30/08–4/09 1997*. Piraeus: Petros M. Nomikos and The Thera Foundation.

Portères, R. 1970. "Primary Cradles of Agriculture in the African Continent." In J. D. Fage and R. A. Oliver (eds.), *Papers in African Prehistory*, 3–19. Cambridge: Cambridge University Press.

Portères, R. 1976. "African Cereals: Eleusine, Fonio, Black Fonio, Teff, Brachiaria, paspalum, Pennisetum, and African Rice." In J. R. Harlan et al. (eds.), *Origins of African Plant Domestication*, 409–452. The Hague: Mouton.

Postgate, J. 1987. "Notes on Fruits in the Cuneiform Sources." *Bulletin of Sumerian Agriculture*, 3: 115–44.

Potter, Paul (ed. and trans.) 2012. *Hippocrates. Volume X*. Cambridge, MA: Harvard University Press.

Preller, Ludwig. 1860. *Griechische Mythologie*, vol. 1. Berlin: Weidmannsche Buchhandlung.

Preston, L. 2008. "Late Minoan II to IIIB Crete." In C. Shelmerdine (ed.), *The Cambridge Companion to the Aegean Bronze Age*. Cambridge: Cambridge University Press.

Price, Martin F., et al. 2013. *Mountain Geography: Physical and Human Dimensions*, 310–326. Berkeley: University of California Press.

Prokopius of Gaza. *Procopius Gazaeus: Epistolae et declamations*. Edited by Antonio Garzya and Raymond J. Loenertz. 1963. Studia patristica et byzantia 9. Ettal: Buch-Kunstverlag Ettal.

Purcell, Nicholas. 1996. "The Roman Garden as Domestic Building." In Ian M. Barton (ed.), *Roman Domestic Buildings*, 121–151. Exeter: University of Exeter Press.

Rackham, O. 1978. "The Flora and Vegetation of Thera and Crete before and after the Great Eruption." In C. Doumas (ed.), *Thera and the Aegean World I. Papers Presented at the Second International Scientific Congress, Santorini, Greece. August 1978*. London: Thera and the Aegean World.

Ramón-Laca, L. 2003. "The Introduction of Cultivated Citrus to Europe Via Northern Africa and the Iberian Peninsula." *Economic Botany*, 57: 502–14.

Ramos-Madrigal, J., et al. 2016. "Genome Sequence of a 5,310-Year-Old Maize Cob Provides Insights into the Early Stages of Maize Domestication." *Current Biology*, 26(23): 3195–210.

Raschke, M. C. 1978. "New Studies in Roman Commerce with the East." In H. Temporini (ed.), *Aufstieg une Niedergang der Römischen Welt. II Principat. Band 9.2*. Berlin and New York: De Gruyter.

Rasmussen, Tom, and Nigel Spivey. 2005. *Looking at Greek Vases*, 10th edn. Cambridge: Cambridge University Press.

Rathje, A. 1994. "Banquet and Ideology: Some New Considerations about Banqueting at Poggio Civitate." In R. De Puma and J. P. Small (eds.), *Murlo and the Etruscans: Art and Society in Ancient Etruria*, 25-61. Madison, WI: University of Wisconsin Press.

Rawson, Jessica. 1984. *Chinese Ornament: The Lotus and the Dragon*. London: British Museum Publications.

Reade, J. 2000. "Alexander the Great and the Hanging Gardens of Babylon." *Iraq*, 62: 195–217.

Reece, Steve. 2007. "Homer's Asphodel Meadow." *Greek, Roman, and Byzantine Studies*, 47: 389–400.

Reeder, Jane Clark. 2001. *The Villa of Livia Ad Gallinas Albas: A Study in the Augustan Villa and Garden*. Providence: Center for Old World Archaeology and Art.

(Reg.) *See* Kings, 1.

Rehak, Paul. 2002. "Imag(in)ing a Women's World in Bronze Age Greece." In N. Rabinowitz and L. Auanger (eds.), *Among Women: From the Homosocial to the Homoerotic in the Ancient World*, 34–59. Austin: University of Texas Press.

Reitzammer, Laurialan. 2016. *The Athenian Adonia in Context: The Adonis Festival as Cultural Practice*. Madison: The University of Wisconsin Press.

Renan, Ernest. 1864. *Mission de Phénicie*. Paris: Imprimerie impériale.

Retief, F. P., and L. Cilliers. 2006. "Burial Customs, the Afterlife and the Pollution of Death in Ancient Greece." *Acta Theologica*, 26(2): 44–61.

Retsö, Jan. 2003. *The Arabs in Antiquity: Their History from the Assyrians to the Umayyads*. London: Routledge.

Revelation. *The New Oxford Annotated Bible: New Revised Standard*. Edited by Michael D. Coogan. 2018. 5th edn. Oxford: Oxford University Press.

Reynolds, Philip K. 1940. "The Banana in Chinese Literature." *Harvard Journal of Asiatic Studies*, 5: 165–81.

Rezaeian, Farzin. 2012. *Recreating Pasargadae: Cyrus the Great's Paradise*. Toronto, Canada: Sunrise Visual Innovations Ltd.

Rice, E. E. 1983. *The Grand Procession of Ptolemy Philadelphus*. Oxford: Oxford University Press.

Richardson, S. 1999–2001. "An Assyrian Garden of Ancestors." *State Archives of Assyria Bulletin*, XIII: 145–216.

Richter, Giesela. 1936. *Red Figured Athenian Vases in the Metropolitan Museum of Art*, vol. 1. New Haven: Yale University Press.

Riddle, John M. 1991. "Oral Contraceptives and Early-Term Abortifacients during Classical Antiquity and the Middle Ages." *Past and Present*, 132: 3–32.

Riddle, John M. 1992. *Contraception and Abortion from the Ancient World to the Renaissance*. Cambridge, MA, and London: Harvard University Press.

Riddle, John M. 1997. *Eve's Herbs: A History of Contraception and Abortion in the West*. Cambridge, MA, and London: Harvard University Press.

Riegl, Alois. [1893] 1992. *Problems of Style: Foundations for a History of Ornament*. Translated by Evelyn Kain. Princeton: Princeton University Press.

Ritti, Tullia, et al. 2007. "A Relief of a Water-Powered Stone Saw Mill on a Sarcophagus at Hierapolis and Its Implications." *Journal of Roman Archaeology*, 20: 138–63.

Robkin, A. L. H. 1979. "The Agricultural Year, the Commodity SA and the Linen Industry of Mycenaean Pylos." *American Journal of Archaeology*, 83(4): 469–74.

Rohde, Erwin. 1925. *Psyche: The Cult of Souls and the Belief in Immortality among the Greeks*. Abingdon: Routledge and Kegan Paul Ltd.

Roques, D. 1984. "Synésios de Cyrène et le silphion de Cyrénaïque." *Revue des Études Grecques*, 97: 218–31.

Ross, J. 2014. "Art's Role in the Origins of Writing: The Seal-Carver, the Scribe and the Earliest Lexical Texts." In Brian Brown and Marian Feldman (eds.), *Critical Approaches to Ancient Near Eastern Art*, 295–318. Berlin: Walter de Gruyter.

Rossiter, Jeremy. 2013. "Olives and Olive Oil." In Roger S. Bagnall et. al. (eds.), *The Encyclopedia of Ancient History*, 1st edn., 4890–4. Chichester, UK: Blackwell Publishing Ltd.

Russell, J. M. 1987. "Bulls for the Palace and Order in the Empire: The Sculptural Program of Sennacherib's Court VI at Nineveh." *Art Bulletin*, 69: 520–39.

Russell, J. M. 1991. *Sennacherib's Palace Without Rival at Nineveh*. Chicago, IL: University of Chicago Press.

Rykwert, Joseph. 1994. "On the Palmette." *Anthropology and Aesthetics*, 26: 10–21.

Samuel, D. 2002. "Bread in Archaeology." *Civilisations*, 49: 27–36.

Sanjur, O. I., et al. 2002. "Phylogenetic Relationships Among Domesticated and Wild Species of *Cucurbita* (Cucurbitaceae) Inferred From a Mitochondrial Gene: Implications for Crop Plant Evolution and Areas of Origin." *Proceedings of the National Academy of Sciences*, 99(1): 535–40.

Sappho. *Greek Lyric, Vol. 1: Sappho and Alcaeus*. Edited and translated by David A. Campbell. 1982. Loeb Classical Library 142. Cambridge: Harvard University Press.

Sappho. *Poetarum Lesbiorum Fragmenta*. Edited by Edgar Lobel and Denys L. Page. 1955. Oxford: Clarendon Press.

Sarton, George. 1965. *Hellenistic Science and Culture in the Last Three Centuries B.C.* New York: Wiley.

Saturno, W. A., et al. 2005. "The Murals of San Bartolo, El Peten, Guatemala, Part 1: The North Wall." *Ancient America*, Number 7. Center for Ancient American Studies, Barnardsville, NC.

Scahill, David. 2009. "The Origins of the Corinthian Column." In Peter Schulz and Ralf Von Den Hoff (eds.), *Structure, Image, Ornament: Architectural Sculpture in the Greek* World, 40–53. Oxford: Oxbow Books.

Scarry, M., and V. Steponaitis. 1997. "Between Farmstead and Center: The Natural and Social Landscape of Moundville." In Kristen Gremillion (ed.), *People, Plants, and Landscapes: Studies in Palaoethnobotany*, 132–138. Tuscaloosa: University of Alabama Press.

Schafer, Edward H. 1968. "Hunting Parks and Animal Enclosures in Ancient China." *Journal of the Economic and Social History of the Orient*, 11: 318–33.

Schlinghoff, D. 1974. "Cotton-Manufacture in Ancient India." *Journal of the Economic and Social History of the Orient*, 17(1): 81.

Scholia in Aristotle. *Scholia in Thesmophoriazusas; Ranas; Ecclesiazusas et Plutum*. Edited by R. F. Regtuit. 2007. Scholia in Aristophanem, III 2/3. Groningen: Egbert Forsten.

Scholia in Nicander of Colophon. *Scholia in Nicandri Alexipharmaca, cum glossis*. Edited by Mario Geymonat. 1974. Testi e documenti per lo studio dell'antichità 48. Milano: Cisalpino-Goliardica.

Scholia in Pindar. *Scholia vetera in Pindari carmina, Vols. 1–3*. Edited by A. B. Drachmann. 1903–27. Leipzig: B. G. Teubner.

Scholia in Sophocles. *Scholia in Sophoclis Oedipum Coloneum*. Edited by Vittorio De Marco. 1952. Rome: Bretschneider.

Schulz, P., and J. Johnson. 1980. "An Early Acorn Cache from Central California." *Journal of California and Great Basin Anthropology*, 2(1): 127–8.

Semple, Ellen Churchill. 1929. "Ancient Mediterranean Pleasure Gardens." *Geographical Review*, 19(3): 420–43.

Seneca the Younger. *Moral Essays, Vol. I: De Providentia. De Constantia. De Ira. De Clementia*. Edited and translated by John W. Basore. 1928. Loeb Classical Library 214. Cambridge: Harvard University Press.

Servius. *Servii grammatici qui feruntur in Vergilii carmina commentarii, Vol. 1: Aeneidos Librorum I–V Commentarii*. Edited by Georgius Thilo and Hermannus Hagen. 1961. Hildescheim: G. Olms.

Servius. *Servii grammatici qui feruntur in Vergilii carmina commentarii, Vol. 2: Aeneidos Librorum VI–XII Commentarii*. Edited by Georgius Thilo and Hermannus Hagen. 1961. Hildescheim: G. Olms.

Shaughnessy, Edward L. 2010. "Arousing Images." In Amar Annus (ed.), *Divination and Interpretation of Signs in the Ancient World*, 61–76. Chicago: Oriental Institute.

Shaw, J. 1978. "Evidence for the Minoan Tripartite Shrine." *American Journal of Archaeology*, 82(4): 429–48.

Shaw, M. 1993. "The Aegean Garden." *American Journal of Archaeology*, 97(4): 661–85.

Sibilla, Nick. 2013. "After Public Backlash, Iowa Town Buries Plan to Ban Front Gardens." *IJ Action*, Institute for Justice, Available online: https://ij.org/action-post/after-public-backlash-iowa-town-buries-plan-to-ban-front-yard-gardens/ (accessed March 26, 2021).

Siede, Mechthild. 2010. "Lotus." In Georg Schöllgen (ed.), *Reallexikon für Antike und Christentum* vol. 23, 537–49. Stuttgart: Hiersemann.

Siede, Mechthild. 2012a. "Mohn." In Georg Schöllgen (ed.), *Reallexikon für Antike und Christentum* vol. 24, 1099–105. Stuttgart: Hiersemann.

Siede, Mechthild. 2012b. "Moly." In Georg Schöllgen (ed.), *Reallexikon für Antike und Christentum* vol. 24, 1105–112. Stuttgart: Hiersemann.

Siede, Mechthild. 2013. "Myrte." In Georg Schöllgen (ed.), *Reallexikon für Antike und Christentum* vol. 25, 370–8. Stuttgart: Hiersemann.

Singleton, V. 1994. “An Enologist’s Commentary on Ancient Wines.” In P. E. McGovern et al. (eds.), *The Origins and Ancient History of Wine*, 67–77. Philadelphia, PA: Gordon and Breach.

Sked, Jan and Chris Macdonald. n.d. “Bunya Feast.” *Australian Plants Online*. Available online: http://anpsa.org.au/APOL8/dec97-3.html (accessed March 28, 2021)).

Smartt, J. 1990. *Grain Legumes: Evolution and Genetic Resources*. Cambridge: Cambridge University Press.

Smith, Bruce D. 1995. *The Emergence of Agriculture*. New York: Scientific American Library.

Smith, Bruce. 1997. “The Initial Domestication of *Cucurbita pepo* in the Americas 10,000 Years Ago.” *Science*, 276: 932–4.

Smith, Bruce. 1998. *The Emergence of Agriculture*. New York: Scientific American Library.

Smith, Bruce. 2006a. “Eastern North America as an Independent Center of Plant Domestication.” *Proceedings of the National Academy of Science*, 103(33): 12223–8.

Smith, Bruce. 2006b. “Seed Size Increase as a Marker of Domestication in Squash (*Cucurbita pepo*).” In M. A. Zeder (ed.), *Documenting Domestication: New Genetic and Archaeological Paradigms*, 25–31. Berkeley, CA: University of California Press.

Smith, Bruce. 2006c. “Prehistoric Plant Husbandry in Eastern North America.” In C. W. Cowan et al. (eds.), *The Origins of Agriculture: An International Perspective*, 281–300. Tuscaloosa: University of Alabama Press.

Smith, Bruce D., and R. A. Yarnell. 2009. “Initial Formation of an Indigenous Crop Complex in Eastern North America.” *Proceedings of the National Academy of Sciences*, 106(16): 6561–6.

Smith, C. Earle. 1966. “Archeological Evidence for Selection in Avocado.” *Economic Botany*, 20: 169–75.

Smith, I., and D. W. Butler. 2002. “The Bunya in Queensland’s Forests.” *Queensland Review*, 9 (2): 31–8.

Smith, William, et al. 1890. “Corona.” In *A Dictionary of Greek and Roman Antiquities, Vol. 1*, 359–363. London: John Murray.

Snyder, Gary. 2016. *The Great Clod: Notes and Memoirs on Nature and History in East Asia*. Berkeley: Counterpoint.

Soles, J. 2016. “Hero, Goddess, Priestess: New Evidence for Minoan Religion and Social Organisation.” In E. Alram-Stern et al. (eds.), *METAPHYSIS: Ritual, Myth and Symbolism in the Aegean Bronze Age*, 247–254. Aegaeum 39. Liége: Peeters.

Sommer, D. J. 1999. “The Shanidar IV ‘Flower Burial’: A Re-evaluation of Neanderthal Burial Ritual.” *Cambridge Archaeological Journal*, 9: 127–9.

Song of Solomon. *The New Oxford Annotated Bible: New Revised Standard*. Edited by Michael D. Coogan. 2018. 5th edn. Oxford: Oxford University Press.

Sophocles. *Antigone. The Women of Trachis. Philoctetes. Oedipus at Colonus*. Edited and translated by Hugh Lloyd-Jones. 1994. Loeb Classical Library 21. Cambridge: Harvard University Press.

Sophocles. *Sophocles, Tragoediae II. Trachiniae, Antigone, Philoctetes, Oedipus Coloneus*. Edited by R. D. Dawe. 1996. 3rd edn. Leipzig: Teubner.

Sourvinou-Inwood, Christiane. 1979. “The Myth of the First Temples at Delphi.” *The Classical Quarterly*, 29(2): 231–51.

Staack, Thies. 2010. “Reconstructing the Kongzi shilun: From the Arrangement of the Bamboo Slips to a Tentative Translation.” *Asiatische Studien*, 64: 857–906.

Staller, J. E., and M. Carrasco. 2009. *Pre-Columbian Foodways: Interdisciplinary Approaches to Food, Culture, and Markets in Ancient Mesoamerica*. Berlin: Springer Verlag.

Stark, Barbara L. 2014. "Ancient Open Space, Gardens, and Parks: A Comparative Discussion of Mesoamerican Urbanism." In Andrew T. Creekmore III and Kevin D. Fisher (eds.), *Making Ancient Cities: Space and Place in Early Urban Societies*, 87–115. Cambridge: Cambridge University Press.

Statius. *Silvae*. Edited and Translated by D. R. Shackleton Bailey and revised by Christopher A. Parrott. 2015. Loeb Classical Library 206. Cambridge: Harvard University Press.

Steier, A. 1933. "Myrtos." In August Pauly et al. (eds.), *Paulys Real-Encyclopädie der classischen Altertumswissenschaft*, vol. 16, pt. 1: Molatzes-Myssi, 1171–83. Stuttgart: J. B. Metzler.

Stein, M. 1997. "La thériaque chez Galien: sa préparation et son usage thérapeutique." In Armelle Debru (ed.), *Galen on Pharmacology: Philosophy, History and Medicine. Proceedings of the Vth International Galen Colloquium, Lille, 16–18 March 1995*, 261–270. Leiden: Brill.

Stieber, Mary. 2004. *The Poetics of Appearance in the Attic Korai*. Austin: University of Texas Press.

Stix, Gary. 2015. "Arqueólogos resucitan bebidas alcohólicas del pueblo azteca." *Scientific American*, 9 June 2015.

Strabo. *Geography, Vol. II: Books 3–5*. Edited and translated by Horace Leonard Jones. 1923. Loeb Classical Library 50. Cambridge: Harvard University Press.

Strabo. *Geography, Vol. III*. Edited and translated by Horace Leonard Jones. 1923. Cambridge: Harvard University Press.

Strabo. *Geography, Vol. VII*. Edited and translated by Horace Leonard Jones. 1930. Cambridge: Harvard University Press.

Strabo. *Strabons Geographika*. 10 vols. Edited by Stefan Radt. 2002–11. Göttingen: Vandenhoeck & Ruprecht.

Stronach, David. 1978. *Pasargadae: A Report on the Excavations Conducted by the British Institute of Persian Studies from 1961 to 1963*. Oxford: Clarendon Press.

Stronach, David. 1989. "The Royal Garden at Pasargadae: Evolution and Legacy." In L. de Meyer and E. Haerinck (eds.), *Archaeologia Iranica et Orientalis: Miscellanea in Honorem Louis Vanden Berghe, vol.* 1, 475–502. Ghent: Peeters Presse.

Stronach, D. 1990. "The Garden as Political Statement: Some Case Studies from the Near East in the First Millennium BC." *Bulletin of the Asia Institute*, NS 4: 171–80.

Stross, Brian 2006. "Maize in Word and Image in Southeastern Mesoamerica." In J. E. Staller et al. (eds.), *Histories of Maize: Multidisciplinary Approaches to the Prehistory, Linguistics, Biogeography, Domestication, and Evolution of Maize*, 577–98. Amsterdam: Academic Press.

Struever, S. 1977. "New Developments at the Koster Site." *Central States Archaeological Journal*, 24(2): 58–64.

Su, T., et al. 2015. "Peaches Preceded Humans: Fossil Evidence from SW China." *Nature Science Reports*, 5: 16794.

Suetonius. *The Lives of the Caesars, Vol.* I. Edited and translated by J. C. Rolfe. 1914. Cambridge, MA: Harvard University Press.

Suetonius. *Lives of the Caesars. Vol. II. Claudius. Nero. Galba. Otho and Vitellius. Vespasian. Titus. Domitian. Lives of Illustrious Men: Grammarians and Rhetoricians. Poets (Terence. Virgil. Horace. Tibullus. Persius. Lucan). Lives of Pliny the Elder and Passinus Crispus.* Edited and translated by J. C. Rolfe. 1914 Cambridge, MA. and London: Harvard University Press.

Suidas. *Suidae Lexicon*. Edited by Immanuel Bekker. 1854. Berolini: G. Reimeri.

Suter, C. 2014. "Human, Divine, or Both? The Uruk Vase and the Problem of Ambiguity in Early Mesopotamian Art." In Brian Brown and Marian Feldman (eds.), *Critical Approaches to Ancient Near Eastern Art*, 545–568. Berlin: Walter de Gruyter.

Svärd, S. 2015. *Women and Power in Neo-Assyrian Palaces. State Archives of Assyria Studies*, 23. Helsinki: Neo-Assyrian Text Corpus.

Swetnam-Burland, Molly. 2010. "Aegyptus Redacta: The Egyptian Obelisk in the Augustan Campus Martius." *Art Bulletin*, 72: 135–53.

Tally-Schumacher, Kaja, and Nils Paul Niemeier. 2016. "Through the Picture Plane: Movement and Transformation in the Garden Room at the Villa ad Gallinas at Prima Porta." *Chronika*, 7: 64–76.

Taylor, Joan. 2012. *The Essenes, the Scrolls and the Dead Sea*. Oxford: Oxford University Press.

Tertullian. *De spectaculis, De idololatria, Ad nationes, De testimonio animae, Scorpiace, De oratione, De baptismo, De pudicitia, De ieiunio adversus psychicos, De anima*. Edited by Aug. Reifferscheid and Georg Wissowa. 1890. Corpus scriptorum ecclesiasticorum latinorum 20. Vienna: Tempsky.

The Greek Anthology, Vol. III. Edited and translated by W. R. Paton. 1917. Cambridge: Harvard University Press.

Themistius. *The Private Orations of Themistius*. Edited by Robert J. Penella. 2000. Transformation of the Classical Heritage 29. Berkeley: University of California Press.

Theocritus. *Theocritus. Moschus. Bion*. Edited and translated by Neil Hopkinson. 2015. Loeb Classical Library 28. Cambridge: Harvard University Press.

Theophrastus. *De causis plantarum, Vol. 1: Books 1–2*. Translated and edited by Benedict Einarson and George K. K. Link. 1976. Loeb Classical Library 471. Cambridge: Harvard University Press.

Theophrastus. *De causis plantarum, Vol. 2: Books 3–4*. Translated and edited by Benedict Einarson and George K. K. Link. 1990. Loeb Classical Library 474. Cambridge: Harvard University Press.

Theophrastus. *Enquiry in to Plants and Minor Works on Odours and Weather Signs*. Edited and translated by Arthur Hort. 1926. Cambridge, MA and London: Harvard University Press.

Theophrastus. *Enquiry into Plants, Vol. 1: Books 1–5*. Edited and translated by Arthur Hort. 1916. Loeb Classical Library 70. Cambridge: Harvard University Press.

Theophrastus. *Enquiry into Plants, Vol. 2: Books 6–9, On Odours, Weather Signs*. Edited and translated by Arthur Hort. 1916. Loeb Classical Library 79. Cambridge: Harvard University Press.

Theophrastus. *Recherches sur les plantes tome I–IV*. Edited by Suzanne Amigues. 1988–2006. Collection des universités de France. Paris: Les Belles Lettres.

Theophrastus. *Théophraste. Recherches sur les plantes. Tome V. Livre IX*. Edition and translation by Suzanne Amigues. 2006. Paris: Les Belles Lettres.

Thomason, Allison K. 2001. "Representations of North Syrian Landscapes in Neo-Assyrian Art." *Bulletin of the American Schools of Oriental Research*, 323: 63–96.

Thomason, Allison K. 2005. *Luxury and Legitimation: Royal Collecting in Ancient Mesopotamia*. London: Routledge.

Thommen, Lukas. 2012. *An Environmental History of Ancient Greece and Rome*. Translated by Philip Hill. Cambridge: Cambridge University Press.

Thompson, Dorothy Burr. 1937. "The Garden of Hephaistos." *The Journal of the American School of Classical Studies at Athens*, 6(3): 396–425.

Thompson, Dorothy Burr. 1963. *Garden Lore of Ancient Athens*. American School of Classical Studies at Athens: Excavations of the Athenian Agora, Picture book n. 8.

Thompson, Dorothy Burr. 1982. *Excavations of the Athenian Agora*. Connecticut: American School of Classical Studies at Athens.

Thorley, J. 1971. "The Silk Trade Between China and the Roma Empire at its Height, circa A.D. 90–130." *Greece & Rome*, 18(1): 71–80.

Tibullus. *Albii Tibulli aliorumque carmina*. Edited by Georg Luck. 1998. 2nd edn. Stuttgart: Teubner.

Tibullus. *Catullus. Tibullus. Pervigilium Veneris*. Edited and Translated by F. W. Cornish, J. P. Postgate, J. W. Mackail and revised by G. P. Goold. 1913. Loeb Classical Library 6. Cambridge: Harvard University Press.

Toohey, Peter G. 2010. "Death and Burial in the Ancient World." In Michael Gagarin (ed.), *The Oxford Encyclopedia of Ancient Greece and Rome*, vol. 3, 363–8. Oxford: Oxford University Press.

Totelin, Laurence M. V. 2004. "Mithradates' Antidote: A Pharmacological Ghost." *Early Science and Medicine*, 9(1): 1–19.

Totelin, Laurence M. V. 2009. *Hippocratic Recipes. Oral and Written Transmission of Pharmacological Knowledge in Fifth- and Fourth-Century Greece*. Leiden: Brill.

Totelin, Laurence M. V. 2012. "Botanizing Rulers and their Herbal Subjects: Plants and Political Power in Greek and Roman Literature." *Poenix*, 71: 122–44.

Totelin, Laurence M. V. 2015. "When Foods become Remedies in Ancient Greece: The Curious Case of Garlic and Other Substances." *Journal of Ethnopharmacology*, 167: 30–7.

Totelin, Laurence M. V. 2016. "The World in a Pill: Local Specialities and Global Remedies in the Graeco-Roman World." In Rebecca Futo-Kennedy and Molly Jones-Lewis (eds.), *The Routledge Handbook of Identity and the Environment in the Classical and Medieval World*, 151–170. London and New York: Routledge.

Totelin, Laurence M. V. 2020. "Healing Correspondence: Letters and Remedy Exchange in the Graeco-Roman World." In R. Flemming and L. Totelin (eds.), *Medicine and the Markets: Essays on Ancient Medicine in Honour of Vivian Nutton*, 17–35. Swansea, UK: The Classical Press of Wales.

Toynbee, J. M. C., and J. B. Ward Perkins. 1950. "Peopled Scrolls: A Hellenistic Motif in Imperial Art." *Papers of the British School at Rome*, 18: 1–43.

Tozzer, A. M. 1941. *Landa's relación de las cosas de Yucatán: A Translation. Papers of the Peabody Museum of American Archaeology and Ethnology, Vols X–VIII*. Cambridge: Harvard University Press.

Trendall. A. D. 1989. *Red Figure Vases of South Italy and Sicily*. London: Thames and Hudson.

Tucker, Arthur O. 1986. "Frankincense and Myrrh." *Economic Botany*, 40(4): 425–33.

Tuplin, A. 1996. *Achaemenid Studies*. Stuttgart: Franz Steiner Verlag.

Tushingham, S., and R. Bettinger. 2013. "Why Foragers Choose Acorns Before Salmon: Storage, Mobility, and Risk in Aboriginal California." *Journal of Anthropological Archaeology*, 34(4): 527–37.

Ulrich, Roger B. 2007. *Roman Woodworking*. New Haven and London: Yale University Press.

Unwin, Tim. 1991. *Wine and the Vine: An Historical Geography of Viticulture and the Wine Trade*, 1st edn. London and New York: Routledge.

Ur, J. 2005. "Sennacherib's Northern Assyrian Canals: New Insights from Satellite Imagery and Aerial Photography." *Iraq*, 67: 317–45.

Valamoti, Soultana-Maria. 2002. "Investigating the Prehistoric Bread of Northern Greece: The Archaeobotanical Evidence for the Neolithic and the Bronze Age." *Civilisations*, 49: 49–66.

Valerius Flaccus. *Argonautica*. Edited and translated by J. H. Mozley. 1934. Loeb Classical Library 286. Cambridge: Harvard University Press.

Van Der Veen, M. 2003. "When is Food a Luxury?" *World Archaeology*, 34: 405–27.

Van Der Veen, M., et al. 2008. "New Plant Foods in Roman Britain—Dispersal and Social Access." *Environmental Archaeology*, 13(1): 11–36.

Van Sickle, J. 2004. "Virgil, *Bucolics* 1.1–2 and Interpretative Tradition: A Latin (Roman) Program for a Greek Genre." *Classical Philology*, 99: 336–53.

Van Zeist, Willem, and W. A. Casparie. 1968. "Wild Einkorn Wheat and Barley From Tell Mureybit in Northern Syria." *Acta Botanica Neerlandica*, 17(1): 44–53.

Van Zeist, Willem. 1972. "Paleobotanical Results in the 1970 Season at Çayönü, Turkey." *Helinium*, 12: 3–19.

Van Zeist, Willem, and G. J. De Roller. 1991–2. "The Plant Husbandry of Aceramic Çayönü, S.E. Turkey." *Palaeohistoria*, 33/34: 65–96.

Van Zeist, Willem, and J. A. Bakker-Heeres. 1982 (1985). "Archaeological Studies in the Levant 1. Neolithic Sites in the Damascus basin: Aswad, Ghoraifé, Ramad." *Palaeohistoria*, 24: 165–256.

Van Zeist, Willem, and J. A. Bakker-Heeres. 1986. "Archaeological Studies in the Levant 2. Neolithic and Halaf Levels at Ras Shamra." *Palaeohistoria*, 26: 151–70.

VanDerwarker, A., et al. 2013, "Maize Adoption and Intensification in the Central Illinois River Valley: An Analysis of Archaeobotanical Data from the Late Woodland to Early Mississippian Periods (A.D. 600–1200)." *Southeastern Archaeology*, 32(2): 147–68.

Varro, *On Agriculture*. In *Marcus Porcius Cato, On Agriculture. Marcus Terentius Varro, On Agriculture*. Revised edition. Edited and translated by W. D. Hooper and revised by H. Boyd Ash. 1935. Cambridge, MA and London: Harvard University Press.

Velasco, Riccardo, et al. 2010. "The Genome of the Domesticated Apple (*Malus* × *domestica Borkh*)." *Nature Genetics*, 42: 833–9.

Vergil. *Aeneid: Books 7–12. Appendix Vergiliana*. Edited and translated by H. Rushton Fairclough and revised by G. P. Goold. 1918. Loeb Classical Library 64. Cambridge: Harvard University Press.

Vergil. *Appendix Vergiliana*. Edited by Wendell V. Clausen et al. 1987. Oxford Classical Texts. 5th edn. Oxford: Clarendon Press.

Vergil. *Eclogues. Georgics. Aeneid: Books 1–6*. Edited and translated by H. Rushton Fairclough and revised by G. P. Goold. 1916. Loeb Classical Library 63. Cambridge: Harvard University Press.

Vergil. *P. Vergili Maronis: Opera*. Edited by R. A. B. Mynors. 1969. Oxford: Oxford University Press.

Villiers, John. 2001. "Great Plenty of Almug Trees: The Trade in Southeast Asian Aromatic Woods in the Indian Ocean and China 500 BC–AD 1500." *The Great Circle (Australian Association for Maritime History)*, 23(2): 24–43.

Vlachopoulos, A. 2007. "*Disiecta Membra*: The Wall Paintings from the 'Porter's Lodge' at Akrotiri." In P. Betancourt et al. (eds.), *Krinoi kai Limenes: Studies in Honor of Joseph and Maria Shaw*, 131–8. Philadelphia: INSTAP Academic Press.

Vlachopoulos, A., and L. Zorzos. 2014. "*Physis* and *Techne* on Thera: Reconstructing Bronze Age Environment and Land-Use Based on New Evidence from Phytoliths and the Akrotiri Wall-Paintings." In G. Touchais et al. (eds.), *PHYSIS. L'environnement naturel et la relation homme-milieu dans le monde Égéen protohistorique*. Aegaeum 37. Liège: Peeters.

Vogelsang-Eastwood, Gillian. 1992. *The Production of Linen in Pharaonic Egypt*. Leiden: Stichtung Textile Reearch Center.

Von Rüden, C. 2014. "Beyond the East-West Dichotomy in Syrian and Levantine Wall Paintings." In Brian Brown and Marian Feldman (eds.), *Critical Approaches to Ancient Near Eastern Art, 55-78*. Berlin: Walter de Gruyter.

Von Stackelberg, Katharine T. 2009. *The Roman Garden: Space, Sense, and Society*. London and New York: Routledge.

Walberg, G. 1987. *Kamares. A Study of the Character of Palatial Middle Minoan Pottery*. 2nd rev. edn. Göteborg: Paul Åströms Förlag.

Warren, P. 1969. *Minoan Stone Vases*. Cambridge: Cambridge University Press.

Warren, P. 1976. "Did Papyrus Grow in the Aegean?" *Athens Annals of Archaeology*, 9: 89–95.

Warren, P. 1979. "The Miniature Frieze from the West House at Akrotiri, Thera and its Aegean Setting." *Journal of Hellenic Studies*, 99: 121–9.

Warren, P. 2000. "From Naturalism to Essentialism." In S. Sherratt (ed.), *The Wall-Paintings of Thera. Proceedings of the 1st International Symposium, Petros Nomikos Conference Centre, Thera, Hellas; 30/08 –4/09 1997*. Piraeus: Petros M. Nomikos and The Thera Foundation.

Watanabe, C. 2014. "Styles of Pictorial Narratives in Assurbanipal's Reliefs." In Brian Brown and Marian Feldman (eds.), *Critical Approaches to Ancient Near Eastern Art*. Berlin: Walter de Gruyter.

Watrous, L.V. 1991. "The Origin and Iconography of the Late Minoan Painted Larnax." *American Journal of Archaeology*, 60(3): 285–307.

Weisskopf, A., and Dorian Q. Fuller. 2013. "Citrus Fruits: Origins and Development." In C. Smith (ed.), *Encyclopedia of Global Archaeology*, 1479–83. New York: Springer.

Wikander, C. 1992. "Pomp and Circumstance: The Procession of Ptolemaios II." *OpAth*, 19: 143–50.

Wikander, Örjan. 2000. "Industrial Applications of Water Power." In Örjan Wikander (ed.), *Handbook of Ancient Water Technology*. Technology and Change in History, 2, 401–12. Leiden: E. J. Brill.

Wikander, Örjan. 2008. "Sources of Energy and Exploitation of Power." In J. P. Oleson (ed.), *Oxford Handbook of Engineering and Technology in the Classical World*, 136–57. Oxford: Oxford University Press.

Wild, John P. 1970. *Textile Manufacture in the Northern Roman Provinces*. Cambridge: Cambridge University Press.

Wilkinson, Alix. 1998. *The Garden in Ancient Egypt*. London: The Rubicon Press.

Willcox, G., et al. 2008. "Early Holocene Cultivation Before Domestication in Northern Syria." *Vegetation History and Archaeobotany*, 17: 313–25.

Wilson, Andrew. 2002. "Machines, Power and the Roman Economy." *The Journal of Roman Studies*, 92: 1–32.

Wilson, Gilbert Livingston. 1917. *Agriculture of the Hidatsa Indians: An Indian Interpretation. Studies in the Social Sciences*, Number 9. Bulletin of the University of Minnesota. St. Paul, MN: University of Minnesota.

Wilson, H. D., and C. B. Heiser. 1979. "The Origin and Evolutionary Relationships of 'Huauzontle' (*Chenopodium nuttalliae* Safford), Domesticated Chenopod of Mexico." *American Journal of Botany*, 66: 198–206.

Winchell, Frank, et al. 2017. "Evidence for Sorghum Domestication in Fourth Millennium BC Eastern Sudan: Spikelet Morphology from Ceramic Impressions of the Butana Group." *Current Anthropology*, 58(5): 673–83.

Winstedt, Eric Otto (ed.). 1909. *The Christian Topography of Cosmas Indicopleustes*. Cambridge: Cambridge University Press.

Winter, I. 2003. "Ornament and the Rhetoric of Abundance in Assyria." *Eretz Israel*, 27: 252–64.

Winter, I. 2006. "Representing Abundance: A Visual Dimension of the Agrarian State." In E. Stone (ed.). *Settlement and Society: Essays Dedicated to Robert McCormick Adams*, 199–225. Los Angeles/Chicago: Cotsen Institute of Archaeology.

Wiseman, D. J. 1976. "Mesopotamian Gardens." *Anatolian Studies*, 33: 137–44.
Woenig, Franz. 1886. *Die Pflanzen im Alten Ägypten: Ihre Heimat, Geschichte, Kultur und ihre mannigfache Verwendung im sozialen Leben in Kultus, Sitten, Gebräuchen, Medizin und Kunst*. Leipzig: W. Freidrich.
Wolkstein, Diane, and Samuel Noah Kramer. 1983. *Inanna, Queen of Heaven and Earth: Her Stories and Hymns from Ancient Sumer*. New York: Harper and Row.
Wu, Delin, and Kai Larsen. 2000. "Zingiberaceae." *Flora of China*, 24: 322–77.
Wujastyk, Dominik (trans.) 2003. *The Roots of Ayurveda: Selections from Sankskrit Medical Writings*. London and New York: Penguin Books.
Young, Gary K. 2001. *Rome's Eastern Trade. International Cmmerce and Imperial Policy, 31 BC–AD 305*. London and New York: Routledge.
Young, R., and G. Thompson. 1999. "Missing Plant Foods? Where is the Archaeobotanical Evidence for Sorghum and Finger Millet in East Africa?" In M. Van Der Veen (ed.), *The Exploitation of Plant Resources in Ancient Africa*, 63–72. New York: Klewer Plenum.
Younger, J. 2011. "A View from the Sea. The West House Frescoes, Akrotiri, Thera: A Description and Interpretation." In G. Vavouranakis (ed.), *The Seascape in Aegean Prehistory*. Monographs of the Danish Institute at Athens 14, 161–184. Athens: Danish Institute at Athens.
Zanker, Paul. 1990. *The Power of Images in the Age of Augustus*. Translated by Alan Shapiro. Ann Arbor, MI: University of Michigan Press.
Zarrillo S., et al. 2008. "Directly Dated Starch Residues Document Early Formative Maize (*Zea mays* L.) in Tropical Ecuador." *Proceedings of the National Academy of Sciences*, 105(13): 5006–11.
Zhan, Jianping, et al. 2016. "Phytoliths Reveal the Earliest Fine Reedy Textile in China at the Tianluoshan Site." *Scientific Reports*, 6. Available online: https://www.nature.com/articles/srep18664 (accessed March 28, 2021).
Zhang Dianxiang, and David J. Mabberley. 2008. "Citrus." *Flora of China*, 11: 90–6.
Zheng, Yunfei, et al. 2014. "Archaeological Evidence for Peach (Prunus persica) Cultivation and Domestication in China." *PLOS One*. Available online: https://www.semanticscholar.org/paper/Archaeological-Evidence-for-Peach-_Prunus-persica_-Zheng-Crawford/0da45e527882fabf7dfb8bab5fdf9de155ecaa54 (accessed March 28, 2021).
Zidar, Charles, and Wayne Elisens. 2009. "Sacred Giants: Depiction of *Bombacoideae* on Maya Ceramics in Mexico, Guatemala, and Belize." *Economic Botany*, 63(2): 119–29.
Zohary, Daniel. 1995. "Olive." In J. Smartt and N. Simmonds (eds.), *Evolution of Crop Plants*. 2nd edn., 379–82. London: Longman.
Zohary, Daniel, and Maria Hopf. 1973. "Domestication of Pulses in the Old World." *Science*, 182: 887–94.
Zohary, Daniel, and Maria Hopf. 2000. *Domestication of Plants in the Old World*. 3rd edn. Oxford: Oxford University Press.
Zohary, Daniel, and Maria Hopf (with Ehud Weiss). 2012. *Domestication of Plants in the Old World*. 4th edn. Oxford: Oxford University Press.
Zohary, Michael. 1995. *Pflanzen der Bibel*. 3rd edn. Stuttgart: Calwer.
Zumbroich, Thomas J. 2007. "The Origin and Diffusion of Betel Chewing." *Ejournal of Indian Medicine*, 1(2007/8): 87–140.

CONTRIBUTORS

Andrew Dalby, once a librarian at Cambridge University Library, lives in France, writes on food history (*Siren Feasts*; *Empire of Pleasures; Food in the Ancient World from A to Z*; *The Breakfast Book*) and translates historical sources on farming and food (*Cato on Farming*; *Tastes of Byzantium*; *Geoponika; The Treatise of Walter of Bibbesworth*). His latest book, on which he collaborated with his daughter Rachel, is *Gifts of the Gods: A History of Food in Greece* (2017).

Joanna Day is Assistant Professor in Greek Archaeology and Curator of the Classical Museum at University College Dublin, Ireland. She has published widely on Aegean Bronze Age plant iconography, in particular on crocuses and saffron. Other research interests include sensory archaeology (she co-edited the *Routledge Handbook of Sensory Archaeology*, 2019), early ceramic technology, and ancient foodways.

Annette Giesecke is a specialist in the history, meaning, and representation of ancient Greek and Roman gardens and designed landscapes in both literature and the arts. Her work extends to Near Eastern garden traditions and cultural uses of plants in antiquity. She is Professor of Classics at the University of Delaware, USA and is an Archaeological Institute of America National Lecturer. Her books include *The Mythology of Plants: Botanical Lore from Ancient Greece and Rome* (2014) and *The Good Gardener? Nature, Humanity and the Garden* (2015).

Karla Hansen-Speer is an adjunct professor at Indiana State University in Terre Haute, Indiana, USA. She earned her PhD from Washington University in St. Louis in 2006 and belongs to the Register of Professional Archaeologists (RPA). Her research involves archaeology in the American Southwest and Midwest with a specialty paleoethnobotany. She focuses on questions of anthropogenic ecology, risk management in subsistence systems, and sustainable food systems.

Patrick Hunt, PhD (Institute of Archaeology, UCL, London), has been teaching at Stanford University for twenty-eight years and has authored over twenty published books including for Penguin Group and Simon and Schuster. National Geographic has sponsored his research and he also works as a National Geographic Expeditions Expert and is an elected Fellow of both the Royal Geographical Society and the Explorers Club. He is a National Lecturer for the Archaeological Institute of America.

Jennifer Ramsay is Associate Professor at the College at Brockport, State University of New York, USA. Her area of expertise centers on archaeobotany, subsistence reconstruction, trade patterns, environmental change, and land-use patterns to gain insight into lifeways of past societies. She specializes in the Roman and late antique world, but has also analyzed and published plant material from the Neolithic through Islamic periods. She has participated in many archaeological excavation projects in Israel, Jordan, and Italy.

Mechthild Siede studied classical philology at the University of Trier, Germany, and wrote a PhD thesis on the history of phytopathology based on literary sources from Homer to Boethius. At the University of Trier, she worked with Professor Wöhrle on projects dealing with natural history and botany, among them the botanical writings of Theophrastus. Since 2011 she has also worked as a research associate at the *Reallexikon für Antike und Christentum* at the F. J. Dölger Institute, Bonn, Germany.

Kaja Tally-Schumacher is a doctoral candidate at Cornell University and is the Project Manager of the Casa della Regina Carolina garden excavation at Pompeii. Her dissertation, "Cultivating Empire: Transplanting and Translating Rome," develops a new phytocentric approach for the study of ancient Roman gardens and landscapes, thereby bringing attention to gardens as places of change, growth, and fracture.

Allison Thomason is Professor of Ancient History at Southern Illinois University Edwardsville, USA. She has written articles on Mesopotamian material culture and women at Ashur and Kanesh. Her book *Luxury and Legitimation: Royal Collecting in Ancient Mesopotamia* (2005) explores the origins of museums, botanical gardens, and zoos. Recently, she has published articles on sense-scapes in Assyria and is currently working on publications on dress, gender, and sensorial experiences in the ancient Near East.

Laurence M. V. Totelin is Senior Lecturer in Ancient History at Cardiff University, UK. Her research focuses on the history of Greek and Roman botany and pharmacology. Her key publications include *Hippocratic Recipes: Oral and Written Transmission of Pharmacological Knowledge in Fifth- and Fourth-Century Greece* (2009) and with Gavin Hardy, *Ancient Botany* (2016).

Alain Touwaide has researched the history of botany and medicinal plants in the Mediterranean world from antiquity to the Renaissance for over forty years. His many publications devoted to Greek, Latin, and Arabic manuscripts and texts on these topics include *Tractatus de herbis* (2013) and *A Census of Greek Medical Manuscripts* (2016). He is co-founder and the Scientific Director of the Institute for the Preservation of Medical Traditions.

Sarah Walshaw holds a Senior Lecturer position in the Department of History at Simon Fraser University, Canada. She holds a PhD in Anthropology (2005, Washington University in St. Louis) and has conducted archaeological and ethnobotanical fieldwork in Tanzania and Madagascar. She is interested in non-mechanized farming systems, the local logics of eastern African farmers, and the use of foodways to connect with—and contest—early globalizations in the Indian Ocean world.

INDEX

Note: Page locators in *italic* refer to figures.